SMALL ENGINE TECHNOLOGY

SMALL ENGINE TECHNOLOGY

William A. Schuster

Washtenaw Community College
Ann Arbor, Michigan

Delmar Publishers Inc.

NOTICE TO THE READER

COVER PHOTOS COURTESY OF: KOHLER COMPANY
TECUMSEH ENGINE AND TRANSMISSION GROUP

DELMAR STAFF

EXECUTIVE EDITOR: MICHAEL MCDERMOTT
SENIOR ADMINISTRATIVE EDITOR: VERNON ANTHONY
EDITORIAL ASSISTANT: PATRICIA KONCZESKI
PROJECT EDITOR: CAROL MICHELI
PRODUCTION COORDINATOR: MARY ELLEN BLACK
ART COORDINATOR: MEGAN KEANE DESANTIS
DESIGN SUPERVISOR: SUSAN MATHEWS

For information, address Delmar Publishers Inc.
3 Columbia Circle, Box 15-015
Albany, NY 12212

printed in the United States of America
published simultaneously in Canada by Nelson Canada, a division of The Thomson Corporation

1 2 3 4 5 6 7 8 9 10 XXX 99 98 97 96 95 94 93

Library of Congress Cataloging-in-Publication Data
Schuster, William A.
    Small engine technology/William A. Schuster.
      p. cm.
    Includes index.
    ISBN 0-8273-4927-0.—ISBN 0-8273-4928-9 (pbk.)
    1. Internal combustion engines, Spark ignition. I. Title.
    TJ790.S38 1993   92-35113
    621.43'4—dc20   CIP

DEDICATION

To my wife Jovanna and family Bill, Susan, and Adam.
Without their tolerance, for the many hours consumed
working at the computer, or their inspiration, to continue the
four-year effort, this writing would not have become a reality.

TABLE OF CONTENTS

SMALL ENGINE TECHNOLOGY
CHAPTER OUTLINES

PREFACE

This manuscript on small air-cooled engine technology is a result of twenty years of the author's experience as a teacher and technician. The contents are a result of experimentation in the secondary, community college, and continuing education classroom setting which has been formulated by feedback and information from the students, local dealers, other instructors, and factory schools.

The text is designed as a bridge between the intructor's lectures and the information furnished by the engine manufacturer. The service manuals, offered by the engine manufacturers, are filled with information that is very specific and indispensable when servicing engines, but the beginning technician has difficulty utilizing them.

The instructor's responsibility is to make learning as efficient as possible by providing the relevant links between the subject matter and the needs of the student. It is impossible for the instructor to be the sole source of the information, and a mechanism, such as a textbook, is needed to "fill-in" the gaps between the lecture and the engine repair manuals. The text has attempted to give the reasons "why" whenever it has demonstrated "what" to do.

The group of textbooks presently on the market lack the fine details necessary to lead a student through a desired objective, and the theory is vague in many areas. This manuscript allocates priority to indepth theoretical explanations from the belief that the excellent technician will be characterized as an efficient problem solver who can diagnose the cause of the problem and provide the most efficient repair procedure. It is impossible to understand and solve a problem without a strong basic background in the theory of operation and knowledge of the interaction of the various systems.

The small engine student population can be divided into the job entry, on-the-job training or enrichment categories. The job entry student can be further divided into the initial or retirement employment categories.

The textbook can be used in a program that is intended to be updating educational experience for current employees of servicing dealerships. It will make the mechanics more efficient in the workplace, eligible for further advancement, and more able to improve their earnings and career opportunities.

This manuscript is a collection of materials that have not been printed in a single document. The inquiries and interests of the industry and students have dictated the contents. The course will ideally meet the needs of any program offered at the secondary or post-secondary level, but it will also have information that the general public would find interesting and helpful even if they did not take a formalized course.

The most common small air-cooled engines are included in the text with emphasis on the teardown and assembly of the small Briggs & Stratton 3.5 horsepower engine. This engine is used even though it would never be economically feasible to rebuild in a repair shop. Millions of these engines were produced and are easy to obtain. The cost of the parts to rebuild them are minimal, and the transfer of knowledge to larger engines is a smooth process.

ACKNOWLEDGEMENTS

The author is very grateful for the comments and suggestions from the following individuals:

Thomas J. Bass	Bill Godwin	Bruce Mason	William (Bud) Tracy
Duane W. Bauer	Cecil Hebrew	Ross McGregor	John Vredeveld
Rob Bogue	Jake Kooperman	Louis R. Ordway	Rick Weid
Don Carter	Harold Lemon	Cornelius Reeves Jr.	William Yahr
Bill Cleary	Jim Lodwick	Dean Russell	Ed Zucal
Ralph Daily	Opie Long	George W. Schairer	
Wallace Field	John Mann	Bob Schoenhofen	

These people spent a great amount of time reading the rough draft of the manuscript, and their suggestions are integrated from the start to finish.

I want to especially thank Jack Lapides, not only for the time he devoted to reading the manuscript, but also for his never-ending assistance as my tutor in the area of technical writing.

I acknowledge the excellent instruction from Scott Moore and Bart Coffey at the Briggs & Stratton factory school; Andy Schickert, Steve Hespe, and Jeff Steffens at the Tecumseh factory school; and Dale Ten Pas, Les Heinemann, Randy Prigge, and Frank Kozlowski at the Kohler factory training school.

Recognition of invaluable comments and suggestions are bestowed on the following manuscript reviewers:

Philip Anderson
Colorado High School
Colorado City, TX

Robert Forsee
Longfellow Alternative Center
Dayton, OH

Doug Hankwitz
Kirkwood Community College
Cedar Rapids, IA

Carl Hawkins
Buchanan High School
Buchanan, MI

Leo Howard
San Joaquin Delta Community College
Stockton, CA

Ralph A. Nodwell Jr.
R & A Service, Inc.
Belleville, MI

Roy Raithel
Brevard Community College
Coca, FL

Russell Taylor
Northern Virginia Community College
Sterling, VA

Curtis Terry
Tarrant County Junior College
Fort Worth, TX

CHAPTER 1

Safety and Tools

INTRODUCTION

Proper use of tools permits the technician to efficiently and safely repair an engine. General workplace safety as well as personal tool and shop equipment safety are stressed in this chapter. The use of precision measurement tools allows the technician to gather data to make the proper and correct decision concerning the repair of an engine. Different types of measurement tools are discussed, and an explanation is presented about their use.

The tools commonly found in a technician's toolbox are listed as well as the additional tools utilized in the workplace. Specialized tools that are offered by different manufacturers to allow a more efficient and safe repair process are described.

PREVENTATIVE SAFETY

Even though a good safety program is based upon adequate instruction in the operation and maintenance of existing machinery and equipment, it is also necessary to acquire a safety mentality. Safety is a state of mind which safeguards the technician from possible danger. The technician must always be aware of the potential dangers of a machine or an operation and protect himself from the "worst case" scenarios. Safety awareness as well as acquiring safe and skillful working habits are essential components in all accident prevention programs. An "ounce of prevention" knowledge is worth a great amount more than a "pound" of first-aid training.

A positive safety attitude is essential to a safe environment, which requires a clean and orderly laboratory with regularly maintained or serviced equipment and the necessary protective devices.

GENERAL WORKPLACE SAFETY RULES

The following are some general rules for prevention of an accident or injury in the workplace:

1. *Avoid splashing gasoline solvents or other harmful liquids on yourself or on anyone else. Many of the liquids in the workplace can cause skin irritations, chemical burns, and serious eye injuries.*

2. *Clean up any spills immediately.*

3. *Keep aisles and work surfaces clear and free of parts or tools. Do not leave anything lying around that could be tripped over or knocked off a bench.*

4. *When lifting something, keep the object close to your body and lift with your legs, not with your back. Squat rather than bend to lift low-lying parts. Improper lifting can cause serious permanent back injury.*

5. *Report all accidents.*

6. *Disconnect the power from machines and equipment before performing any change of operation.*

7. *Use equipment only when the instructor is in the workplace.*

8. *Do not use tools or equipment that is defective, damaged, or broken.*

9. *Do not use any piece of equipment until you have been shown how to use it correctly and safely.*

10. *Keep tools clean and free of grease. An oily or dirty handle on a tool can cause your hand to slip and result in injury.*

11. *Use the right tool for the job. Use tools only for what they are designed to do.*

12. *Use the right size tool. Using a tool that is too small or too large for the job can damage the work or the tool and injure you.*

13. *Carry pointed or sharp tools with the point or edge held dowr.ward toward the floor. This will help avoid injuries if you bump into something or if someone bumps into you.*

14. *Do not carry tools in your pocket as they may fall out and be damaged. A tool carried in a back pocket can cause damage to the cushion if you sit in a chair.*

PERSONAL SAFETY RULES

A positive safety attitude is important to personal safety. A clean and orderly workplace, with regularly serviced equipment and the use of personal protection devices, are most important to personal safety, as well as knowing the safe and correct procedure for using the tools of the workplace. It is important to remember that the personal protective equipment does not take the place of accident prevention. Protective equipment is the back-up protection for unexpected results. It is very common for the technician to mistakenly assume that the use of personal protective devices will prevent accidents. All they will do is lessen the dangerous results of an accident.

Face and eye protection

The eye is a sensitive organ which is susceptible to injury from flying objects, molten liquids, or splashing chemicals that originate from the work done by the technician or more often from a co-worker. Most workplace environments are regulated so that eye protection is mandatory. This protection is commonly achieved by wearing safety glasses. Eyeglasses, as shown in figure 1-1, are proper protection if certain industrial standards are met, such as:

1. *Lenses at least 3-mm thick.*

2. *Lenses are capable of withstanding an impact from a 1-inch diameter steel ball dropped 50 inches.*

3. *The frames must be made of slow burning*

materials and designed to retain the lenses under impact.

4. *Sideshields are necessary.*

Hearing protection

Noise can be described as unwanted sound. Noise is just a sound wave pressure hitting the eardrum and exists at many levels that are measured in decibels. People can suffer from noise in two ways: One relates to a psychological effect that makes people short-tempered and causes a lack of concentration. Another way involves a physical effect which actually causes temporary or permanent loss of hearing. The following practices should always be observed:

1. *Keep the sound level as low as possible.*

2. *Do not operate an engine without a muffler or with a faulty exhaust system.*

3. *Use effective hearing protection in the form of ear muffs, as shown in figure 1-2, or ear plugs.*

Hand, leg, and foot protection

The hands, feet, and legs account for approximately sixty percent of all body parts injured in the

Fig. 1-1 Eye protection.

workplace. Fingers are most frequently injured for obvious reasons. The following practices should always be observed:

1. *Wear protective shoes such as hard steel-toe shoes or boots with insulating, non-skid soles and heels.*

2. *Wash hands and arms thoroughly after cleaning parts in a solvent of any kind. It may be necessary to apply skin lotion to replace the natural skin oils that are dissolved by the solvent.*

3. *If any gasoline, solvent, or other harmful liquid is splashed upon the body, wash the material off as soon as possible.*

4. *Restrict long hair of the head or face near rotating equipment. If head hair is long, wear a cap over it and keep the hair tucked neatly under the cap. A face mask can be used to prevent a long beard from becoming entangled. Long hair can be caught in a moving machine or a running engine and may result in serious injury.*

5. *Wear properly fitting workplace clothing. Loose clothes that are not tucked in or sleeves that are unbuttoned can get caught in moving machinery or a running engine.*

6. *Wear rubber or latex gloves when handling liquids.*

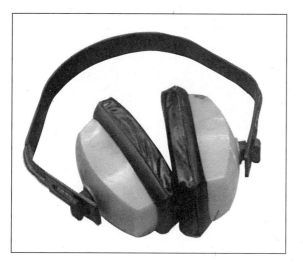

Fig. 1-2 Hearing protection.

7. *Avoid prolonged skin contact with any hazardous liquid. The skin absorbs chemicals which may be toxic (poisonous). Hazardous liquids may cause injuries or health problems and include explosive (volatile), flammable, or combustible liquids which have the potential to cause chemical burns, blindness, or respiratory problems if the vapors are inhaled. Hazardous liquids may also be fatal poisons.*

8. *Remove rings and other jewelry.*

9. *Use appropriate equipment when welding, such as proper eye protection, welding hat, protective clothing, and leather gloves.*

10. *Do not wear gloves when using a grindstone, hand grinder, or other rotating machinery.*

11. *When grasping a rotary lawnmower blade make sure the spark plug wire is removed from the spark plug and protect your hands with gloves or a shop rag.*

Respiratory protection

The workplace environment must be kept free from harmful dusts, fogs, fumes, mists, gases, smokes, sprays, or vapors. Protection can be provided by a properly maintained ventilation system or with the use of respirators, as shown in figure 1-3 adequate to remove the contaminants from the air.

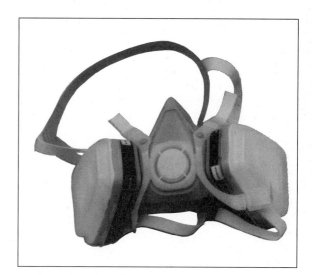

Fig. 1-3 Respiratory protection.

The following practices should always be observed:

1. Use flammable, combustible, or toxic liquids only in well-ventilated areas.

2. Exhaust gases of a running engine can be deadly. If you operate an engine in the workplace, be sure the area is well-ventilated or attach an exhaust disposal system to the exhaust pipe.

SAFETY WITH ENGINES

Power mowers, garden tractors, snowblowers, and other machines powered by air-cooled engines have become so common-place that the potential danger involved when operating and servicing them has been forgotten. The best safeguard against accidents is the use of common sense.

Operation precautions

Imprudent operation of power equipment can create hazards that can lead to personal injury and property damage. To prevent accidents, become thoroughly familiar with your machine before operating by reading the instruction manual. The following are general operating precautions:

1. Know how to stop the engine in an emergency.

2. Never allow inexperienced persons or children to operate power equipment.

3. Make sure all guards and shields are in place and secure before starting.

4. To prevent unintentional starting when working on the equipment, always first disconnect the spark plug wire.

5. Make sure hands, feet, and clothing are safely away from moveable parts when starting any equipment.

6. Never attempt to start the engine with the drive engaged. Shift the machine into neutral and set the brakes, if equipped.

7. Never tamper with the governor setting to gain more power because safe operating limits have been previously established by the manufacturer.

8. Keep people safely away from the operating area and be especially watchful for the presence of children.

9. Be on the look-out for items such as stones and metal objects that can be picked up and propelled or projected by blades.

10. Never attempt to unclog a discharge chute or free a stuck blade while the unit is operating.

11. Never let a machine idle unattended even for a moment. Stop the engine whenever you leave the operating zone of the machine.

12. Avoid steep inclines that can cause the machine to tip.

13. Never pull a rotary lawnmower up a hill when the engine is operating. If you loose your footing, your foot may slide under the deck and come in contact with the blade.

14. If you need to "prime" a carburetor, use a "pump-can" rather than pour gasoline into the carburetor. Replace the air cleaner before starting the engine to avoid a "flashback" and a possible fire.

Exhaust gas dangers

During operation, internal combustion engines discharge carbon monoxide as part of the exhaust gases. Carbon monoxide is particularly dangerous in that it is odorless, colorless, and tasteless and therefore hard to detect. Death or sickness can occur if it is inhaled for even a short period of time. Always observe the following precautions:

1. Never operate an engine inside a closed building or in any area where exhaust gases can accumulate.

2. When working in the vicinity of the engine, be careful not to breathe exhaust fumes.

3. Keep the exhaust system tight and components in good condition at all times as the engine noise can be harmful.

4. If an engine must be operated inside a workplace for test purposes, make sure the exhaust is piped safely outside.

5. The exhaust system parts get very hot, so keep hands, feet, and clothing away from them while the engine is running and, for a while after it has stopped.

6. *Never operate an engine near a building where exhaust gases can seep inside through an open window or door.*

CONSUMER PRODUCT SAFETY COMMISSION STANDARDS

All walk-behind power lawnmowers built or imported after June 30, 1982 that will be used by the consumer must comply with the Consumer Product Safety Commission (CPSC) Standards. The CPSC defines a walk-behind lawnmower as a grass cutting machine with a minimum cutting width of 12 inches. It should be noted that commercial lawnmowers are not subject to the standard. The standard states:

1. *Every rotary lawnmower must carry a certification label that states "Danger, keep hands and feet away."*

2. *The path of the blade on a rotary lawnmower must be shielded in such a manner that the unit can successfully pass specified tests, including a "foot probe" test and an "obstruction" test.*

3. *A blade control system is required that prevents the blade from operating unless the operator actuates the control.*

4. *The blade control system also requires that continuous contact with the controls are necessary to keep the blade in motion.*

5. *The blade must stop rotating within three seconds after the release of the control.*

6. *All units must be equipped with a second control, either a separate safety lever or a device incorporated into the blade control, so that two distinct actions are required to restart the blade.*

SAFETY COLORS

Certain colors are used in the workplace to denote safety locations. The common colors are red, yellow, orange, and green.

Red

Red is the basic safety identification color for:

1. *location of fire protection equipment such as an alarm, water pump, extinguishers, sprinkler*

pipes and valves, fire exits signs, and blanket boxes;

2. *portable containers that may contain flammable liquids with a flash point of 80° F or less; and*

3. *emergency stop bars, stop buttons, and emergency electrical stop switches on machinery.*

Yellow

Yellow (or yellow and black stripes) is the basic safety identification color designating:

1. *caution areas—places where injury could occur;*

2. *physical hazards such as beams, pipes, pulleys, or hand rails;*

3. *waste containers for combustible oily rags;*

4. *safe walkways or aisles; and*

5. *safety zones.*

Orange

Orange is the basic safety identification color designating:

1. *dangerous parts of a machine;*

2. *parts of equipment that may cause electrical shock; and*

3. *exposed edges of pulleys, gears, rollers, turning shafts, and cutting devices.*

Green

Green is the basic safety identification color designating:

1. *safety bulletins; and*

2. *location of first-aid kits, stretchers, gas masks, safety showers, and safety lockers.*

FIRE PROTECTION RULES

There are many combustible materials used in the workplace. The first safety concern should be to prevent a fire, while the next concern should be what to do if there is a fire. Fire extinguishers are labeled for the types of fires they should be used to fight. They should be checked to ensure that they are applicable to the most common fires found in the workplace. The following are some simple rules to apply in providing protection against fires:

1. *Store all flammable liquids in approved safety containers.*

2. *Do not operate an engine if there is fuel leaking from the carburetor, fuel pump, or fuel line.*

3. *Do not use flammable liquids, such as gasoline, to clean parts.*

4. *Use only the minimum fuel necessary in the tank when starting an engine in the workplace.*

5. *Flammable or volatile liquids such as gasoline, oil, solvents, and other cleaning solutions must be kept in labeled, capped containers.*

6. *Do not leave containers where someone could trip on or overturn them.*

7. *Never store gasoline in a breakable receptacle, such as a glass bottle or jar.*

8. *Never ignite a match or smoke in the workplace.*

9. *Flammable liquids should be stored in vented, metal cabinets when they are not in use.*

10. *Know where the fire extinguishers are located and read the instructions for each extinguisher. Know which type, or class, of fire the extinguishers are intended for and understand how to use them.*

11. *Keep the workplace clean. Entrances, emergency exits, and aisles must be kept unobstructed at all times.*

12. *When disconnecting a battery, carefully loosen and disconnect the ground cable first.*

Fire classification and extinction

Fuel, heat, and oxygen are necessary for a fire. These three components can be illustrated by the "Fire Triangle" shown in figure 1-4. The design of all fire fighting equipment is centered around the fact that once the triangle is broken, the fire will cease to exist. Removing the source of the fuel may be the simplest method, such as turning off the acetylene tank valve in extinguishing a welding torch. The oxygen can be removed, such as when a candle is extinguished by a small metal cup. Since materials need to be at a certain temperature for fire to occur, the burning object can be cooled below its combustion temperature or kindling point, such as when water is used to cool a burning object.

CLASS A FIRES Class A fires involve combustible materials like wood, paper, cloth, plastics, or other materials that may be considered flammable.

CLASS A FIRE EXTINCTION Class A fires can be put out with water or with water-based fluid or foam that will remove the heat from the triangle.

CLASS B FIRES Class B fires involve flammable liquids such as gasoline, solvents, oil, grease, oil-based paints, paint thinners, cleaning solutions, etc.

CLASS B FIRE EXTINCTION Class B fires should be put out with dry chemical extinguishers, carbon dioxide, or foam, which remove the oxygen part of the triangle. Do not use water on class B fires since most flammable liquids will float on water, and the use of water will actually help spread the fire.

CLASS C FIRES Class C fires involve electricity in objects such as motors, generators, transformers, or switch panels. Class C fires usually start as a result of an electrical short or overloaded, overheated electrical equipment.

CLASS C FIRE EXTINCTION Class C electrical fires, can be dangerous because of the risk of electrical shock. Never use water, foam, or any liquid on a class C fire unless you are certain the power source has been disconnected. A dry powder extinguisher, such as ammonium phosphate, is best to use on a class C fire to remove the oxygen from the triangle.

CLASS D FIRES Class D fires involve certain combustible metals and chemicals such as magnesium, titanium, sodium, potassium, etc.

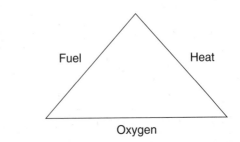

Fig. 1-4 Fire triangle.

CLASS D FIRE EXTINCTION Class D fires require a special extinguisher powder; or if the fire is small, it may be smothered with sand. Water must never be used on a class D fire as it may increase the intensity of the fire by reacting with it.

WORKPLACE CHEMICALS

Always wear eye protection when you work with hazardous liquids of any kind. The chemicals used in the workplace carry warning and caution information that should be read and understood by all users. All of the chemicals used must have a material safety data sheet (MSDS), as shown in figure 1-5, from the product manufacturer that is made available to all technicians. These sheets detail the chemical composition and precautionary information for all products that can present a health or safety hazard. The Canadian equivalent to the MSDS sheets are called workplace hazardous materials information systems (WHMIS).

All hazardous materials must be labeled indicating what health, fire, or reactivity hazard they pose and what protective equipment is necessary when handling each chemical. Waste is considered hazardous if it is on the EPA (Environmental Protection Agency) list of known harmful materials, such as asbestos, or has at least one of the following characteristics:

1. *Ignitability. If it is a liquid with a flash point below 140° F or a solid that can spontaneously ignite.*
2. *Corrosiveness. If it dissolves metals and other materials or burns the skin.*
3. *Reactivity. Any material that reacts violently with water or other materials, releases cyanide gas, hydrogen sulfide gas, or similar gases when exposed to acid solutions. This also includes materials that generate toxic mists, fumes, vapors, and flammable gases.*

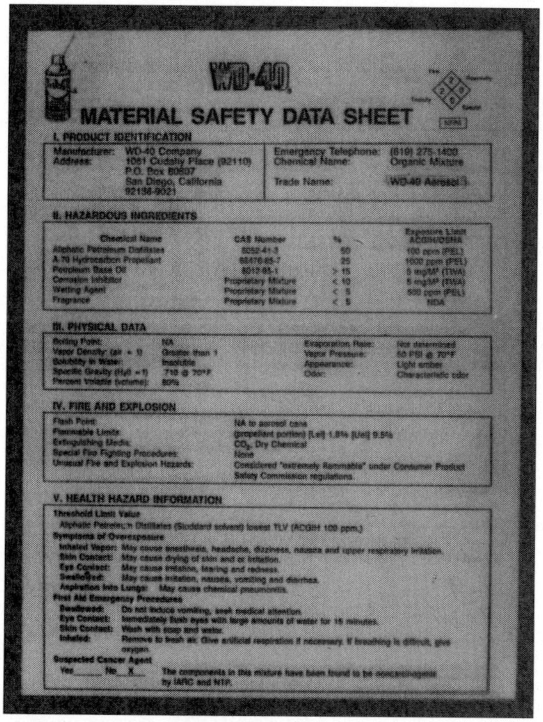

WD-40

MATERIAL SAFETY DATA SHEET

I. PRODUCT IDENTIFICATION

Manufacturer: WD-40 Company
Address: 1061 Cudahy Place (92110), P.O. Box 80607, San Diego, California 92138-9021
Emergency Telephone: (619) 275-1400
Chemical Name: Organic Mixture
Trade Name: WD-40 Aerosol

II. HAZARDOUS INGREDIENTS

Chemical Name	CAS Number	%	Exposure Limit ACGIH/OSHA
Aliphatic Petroleum Distillates	8052-41-3	50	100 ppm (PEL)
A-70 Hydrocarbon Propellant	68476-85-7	25	1000 ppm (PEL)
Petroleum Base Oil	8012-95-1	> 15	5 mg/M³ (TWA)
Corrosion Inhibitor	Proprietary Mixture	< 10	5 mg/M³ (TWA)
Wetting Agent	Proprietary Mixture	< 5	500 ppm (PEL)
Fragrance	Proprietary Mixture	< 5	NDA

III. PHYSICAL DATA

Boiling Point: NA
Vapor Density (air = 1): Greater than 1
Solubility in Water: Insoluble
Specific Gravity (H_2O = 1): .718 @ 70°F
Percent Volatile (volume): 80%
Evaporation Rate: Not determined
Vapor Pressure: 50 PSI @ 70°F
Appearance: Light amber
Odor: Characteristic odor

IV. FIRE AND EXPLOSION

Flash Point: NA to aerosol cans
Flammable Limits: (propellant portion) [Lel] 1.8% [Uel] 9.5%
Extinguishing Media: CO_2, Dry Chemical
Special Fire Fighting Procedures: None
Unusual Fire and Explosion Hazards: Considered "extremely flammable" under Consumer Product Safety Commission regulations.

V. HEALTH HAZARD INFORMATION

Threshold Limit Value
Aliphatic Petroleum Distillates (Stoddard solvent) lowest TLV (ACGIH 100 ppm.)
Symptoms of Overexposure
Inhaled Vapor: May cause anesthesia, headache, dizziness, nausea and upper respiratory irritation.
Skin Contact: May cause drying of skin and or irritation.
Eye Contact: May cause irritation, tearing and redness.
Swallowed: May cause irritation, nausea, vomiting and diarrhea.
Aspiration Into Lungs: May cause chemical pneumonitis.
First Aid Emergency Procedures
Swallowed: Do not induce vomiting, seek medical attention.
Eye Contact: Immediately flush eyes with large amounts of water for 15 minutes.
Skin Contact: Wash with soap and water.
Inhaled: Remove to fresh air. Give artificial respiration if necessary. If breathing is difficult, give oxygen.
Suspected Cancer Agent
Yes ____ No X The components in this mixture have been found to be noncarcinogenic by IARC and NTP.

Fig. 1-5 Material safety data sheet (MSDS).

BATTERY SAFETY

Caution must be used when handling starting batteries because they contain acid that can "eat" through clothing, burn skin, and cause blindness. A battery gives off a highly flammable hydrogen gas when it is charged and may cause an explosion if a spark is in the area.

Follow these precautions when working with a battery:

1. *Handle a battery with care and use a battery strap when moving it.*
2. *Always hold the battery upright and set it securely so there is no danger of spilling the acid.*
3. *When mixing battery solutions, always pour acid into water.*
4. *If acid is splashed, flush the area immediately with plenty of cold water.*
5. *Never charge a frozen battery.*

EQUIPMENT SAFETY

The electric drive motor, bench grinder, drill press, Oxygen/Acetylene torch, valve grinder, lifting devices, and air tools are common equipment that have special rules for safe operation.

A good reminder for equipment safety would be to clearly post the rules in the proximity of each device. Each of the following discussions have some safety precautions repeated so that they can be used in this manner.

ELECTRIC DRILL MOTOR

The electric drill motor is used to power the drill bit, as shown in figure 1-6. A drill is a tool for creating holes. A typical drill is the twist drill, which is a round bar with helical grooves. There are two cutting edges, and the chips pass up the helical grooves and away from the working area.

There are certain precautions necessary when using an electrical motor such as the drill motor. The following practices should always be observed:

1. *Always wear eye protection.*

2. *When the power cord is detached from the receptacle, always pull the plug and not the cord.*

3. *Stand on a dry floor whenever working with*

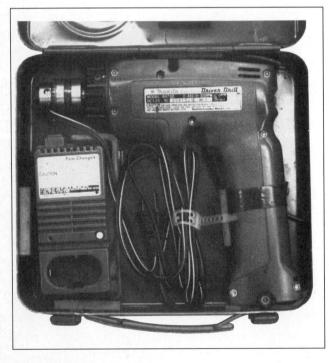

Fig. 1-6 Electric drill motor.

electricity. A wet or damp floor may result in electrocution.

4. *Keep the power cord disconnected from the power source when not in use.*

5. *Check the power cord to make sure the insulation is in good condition. Do not use the tool if the insulation is broken, cut, or damaged.*

6. *Electrical power tools must be grounded or double insulated to prevent a possible electric shock.*

7. *Do not use electrical equipment around batteries or oxyacetylene welding equipment or any flammable liquids. A spark from a switch or a motor could cause an explosion, a fire, or both.*

8. *Always disconnect the tool before any changes or adjustments are made.*

9. *Wait until the tool has come to a complete stop before setting it on the workbench or other resting place.*

DRILL PRESS

A drill press is used to drill a hole in material where positioning and depth angle are crucial, as shown in figure 1-7. The following practices should always be observed:

1. *Always wear approved safety glasses when near the drill press.*

2. *Do not operate the machine unless the belt and pulley safety guards are in place.*

3. *The drill bit must be sharp, straight, and in good condition.*

4. *Check the drill speed and use the right speed for the work, which depends on the type of material and the size of the drill.*

5. *Use the chuck key to fasten the cutting tool in the chuck.*

6. *Never leave a chuck key in the drill chuck.*

7. *Always check twice to insure that the chuck key has been removed before the machine is activated.*

8. Always use a clamp to hold the work firmly on the table.

9. Secure the table locking clamp firmly to prevent the table from shifting or sliding down the column.

10. Adjust the table so the clearance hole in the table is lined up directly under the drill bit.

11. Keep hands out of the area directly under the drill bit.

12. Use machining lubricant when necessary so that the drill bit does not overheat.

13. Do not use too much pressure to force the drill into the material.

14. Turn the drill press off before any changes or adjustments are made.

15. Do not use hands to clear away the chips or metal turnings. A brush works best.

16. Do not use compressed air to remove chips from a machine. The flying metal chips can cause a dangerous situation.

17. If the work piece is caught in the drill, stop the drill press immediately to realign the materials.

Fig. 1-7 Drill Press.

BENCH GRINDER

A bench grinder has one or more grinding wheels made of abrasive material that is/are bonded together, as shown in figure 1-8. When the wheel is rotated by an electric motor, objects held against the wheel are ground down. The grinder can be used to shape and sharpen tools. Some grinders have a combination of one grinding wheel and one wire buffing wheel. When the wire wheel revolves, objects held against it are buffed and polished. When operating a bench grinder, the following practices should always be observed:

1. Always wear approved safety glasses when near the bench grinder.

2. The transparent safety shield on the grinder should be present and in the correct position.

3. Before turning on the grinder, check the abrasive wheel to observe that the stone is in good condition and clean.

4. Stand to one side of the grinder as you switch it on and until it reaches full speed.

5. If the grinder starts to shake or vibrate, move to the side and turn it off immediately.

6. Keep your hands and fingers away from the grinding wheel while it is moving.

7. Never use a rag or wear gloves to hold an object while buffing or grinding as the glove can be pulled into the wheel along with your hand.

Fig. 1-8 Bench Grinder.

8. *Never try to grind free-handed without a tool rest. The tool rest is adjusted properly when it is no further than 1/8-inch from the face of the grinding stone.*

9. *When adjustments to the tool rest are made, make sure the grinder is turned off and the grinding stone has stopped completely.*

10. *Never use the side of the wheel for grinding; all grinding must occur on the face of the wheel.*

11. *Small objects should be held with a pliers or vise-grips instead of your fingers.*

12. *When using the wire buffing wheel, always buff on the lower part of the wheel and with the rotation. This will avoid the object being kicked back or knocked out of the hand.*

13. *Don't buff a sharp corner or a leading edge as the wire wheel can catch sharp corners and edges and knock the object out of the hand.*

14. *Use a light, steady pressure against the wheel as too much pressure can damage the grinding wheel.*

15. *Grind or buff in short intervals because prolonged, continuous grinding will heat the material enough to cause a burn.*

16. *Wash your hands to remove the grains of abrasive and metal particles; avoid touching your face or eyes until the hands have been washed.*

VALVE FACE GRINDER

The valve face grinder is used to cut the intake and exhaust valve so that it can be used again in the rebuilt engine, as shown in figure 1-9.

1. *Always wear approved safety glasses when near the valve grinder.*

2. *Check to be sure all the safety guards are in place and working properly.*

3. *Make sure the machine is turned off and completely stopped by the person who used it previously.*

4. *Work in well-lighted areas.*

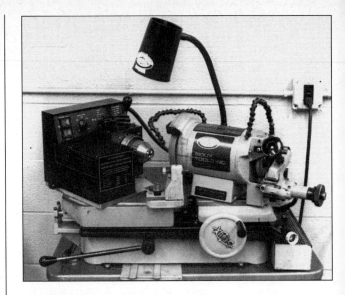

Fig. 1-9 Valve face grinder.

AIR DRIVEN POWER TOOLS AND BLOW GUN

Equipment operated by compressed air are pneumatic devices, as shown in figure 1-10. The following practices should always be observed:

1. *Always wear approved safety glasses when near the air powered tools.*

2. *Compressed air systems have an air regulator valve to control the air pressure. Check to be sure the regulator is adjusted to the correct pressure for the attachment.*

3. *Never point the compressed air blow gun at yourself or another person.*

4. *Protect you lungs by wearing an air filter or a particle mask when dust is blown into the air.*

5. *Keep air hoses out of the way so that they cannot trip anyone.*

6. *Before the pneumatic device is connected, make sure its control is in the "off" position.*

7. *If there is a torque control on the tool, adjust it correctly.*

8. *Final tightening of fasteners started by an air tool should be done with a hand wrench or a torque wrench.*

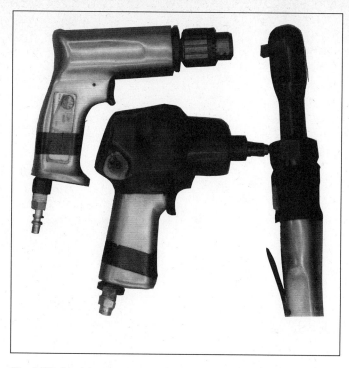

Fig. 1-10 Air driven power tools.

9. *Operate the air tool with short bursts of power.*

10. *Use only special impact sockets and adapters with the air impact wrench. A normal socket may shatter and fly off, causing potential danger to surrounding workers.*

11. *The air line pressure should be regulated so that there is no more than 30 psi (pounds per square inch) of pressure available to the blow gun in order to avoid excessive noise and the possibility of propelling dirt or metal chips at high velocities that may damage or plug a workpiece or injure a bystander.*

OXYGEN/ACETYLENE TORCH

The Oxygen/Acetylene Torch is a type of torch which produces intense heat that can be used for heating rusted parts, welding metal-to-metal, cutting through metal, and in methods for extracting bolts broken off at or below the surface level of the hole. A functional fire extinguisher must be on hand

before the torch is lighted. The following practices should always be observed:

1. *Eye protection must be worn that has a tinted face shield. The intense light that results from the torch operation must be shielded from the eyes. Never look directly at the oxyacetylene flame without eye protection.*

2. *Use welding gloves that cover the coat sleeves to protect the hands and wrists from heat and sparks.*

3. *Pants should not have any cuffs that might catch the burning metal particles.*

4. *Leather boots should be worn to protect the feet.*

5. *Cover all body surfaces with clothing.*

6. *Occasionally check all fittings of the cylinders with a soap solution to identify any leaks. Correct all leaks before proceeding.*

7. *Proper ventilation is necessary during welding to protect the respiratory system from the poisonous fumes of the fluxes and metal alloys.*

8. *Weld in an area away from others who could be injured by the intense heat or hot particles. Make sure that no one in close proximity is using a flammable material.*

9. *Contact lenses should not be worn when welding because the lenses are suspended on the fluids of the eye and any smoke or fumes may lodge between the lens and the eye. Since the lenses are plastic, an intense "flash" may melt the lens to the eye.*

10. *Check the area for rags, paper, oil, and greases or other flammable materials and remove them to a safe distance.*

11. *Never bring a lit torch near a fuel tank, even if it is empty.*

LIFTING DEVICES

Heavy parts can be lifted by chain hoists or cranes. Another term for a chain hoist is a chain fall, and cranes are often called cherry pickers. The following practices should always be observed:

1. *Locate the crane or chain hoist directly over the balance point so there is no pull in any direction except straight up.*

2. *Use only approved types of straps, cables, chains, fasteners, and brackets.*

3. *The length of the fastener should be at least 1.5 times the diameter of the fastener.*

4. *Never get underneath an engine or work on an engine that is being held by a crane.*

TOOL IDENTIFICATION & SAFETY

Many different types of handtools are used by the technician. Identification and safety precautions of these tools are important.

SCREWDRIVERS

A screwdriver is a common tool used to turn a screw with a slotted or recessed head that allows it to be tightened or loosened. Screwdrivers are classified by the length of the shank from the handle to

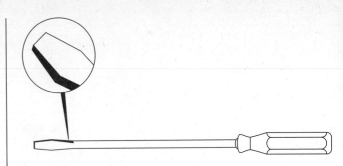

Fig. 1-11 Common screwdriver.

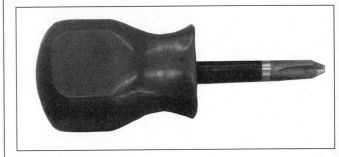

Fig. 1-12 Phillips screwdriver.

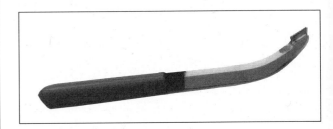

Fig. 1-13 Offset screwdriver.

the drive end of the blade and the shape of the drive blade. Some screwdriver tips are magnetized to hold the screw while starting.

1. **Common screwdriver**—*It is composed of a handle and a shank. The end of the shank is flattened to form a blade that fits squarely into the screw slot, as shown in figure 1-11.*

2. **Phillips screwdriver**—*It has two slots at right angles. The Phillips screwdriver is commonly used when it is desirable to reduce the chance of the screwdriver slipping from the head and marring the surrounding surface. This tool, as*

Fig. 1-14 TORX® screwdriver tip.

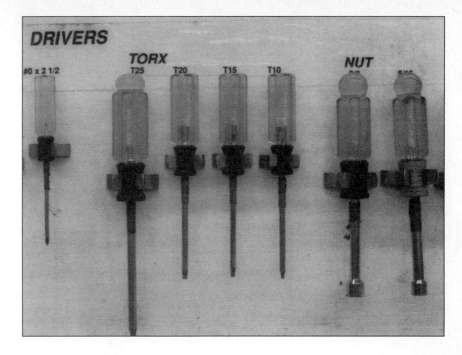

Fig. 1-15 TORX® screwdriver sizes.

shown in figure 1-12, is manufactured in four sizes: #1, #2, #3, and #4, to match various sized fasteners.

3. **Offset screwdriver**—It is used on screws where there is limited space and the screw is hard to reach, as shown in figure 1-13. Bits are normally at right angles to each other, allowing the screw to be turned a quarter-turn at a time by using opposite ends alternately.

4. **TORX®1 screwdriver**—It is a specially designed slot that permits more precise torquing of the screw with a low incidence of slipping of the workpiece, as shown in figure 1-14. Common TORX® sizes are: T10, T15, T20, T25, as shown in figure 1-15.

---

[1]TORX® is a registered trademark of Cam-Car Screw & Manufacturing Company

SAFETY TIP: SOME OF THE PRACTICAL TIPS AND SAFETY RULES THAT SHOULD BE FOLLOWED ARE:

1. THE SCREWDRIVER TIP SHOULD FIT THE SLOT PROPERLY.

2. ALWAYS HOLD THE SCREWDRIVER PERPENDICULAR TO THE HEAD AS YOU TWIST IT.

3. THE SCREWDRIVER SHOULD NOT BE USED FOR ANY JOB OTHER THAN TURNING SCREWS.

4. DO NOT HOLD THE WORK IN ONE HAND AND THE SCREWDRIVER IN THE OTHER. KEEP BOTH HANDS BEHIND THE BLADE OF THE SCREWDRIVER.

continued

HAMMERS

The hammer has many shop uses and can be identified by its weight and type. Among the different kinds of hammers, the ball-peen is used most frequently in the small engine workplace. When a hammer is used, it should be gripped near the end of the handle and swung so that the hammer face contacts the object squarely. If the materials may be dented by the hammer, a soft-faced hammer should be used. The different types of hammers that may be found in the workplace are as follows:

1. *Ball-peen hammer*—A steel hammer that has a flat face for hammering and the ball part for rounding off rivets, as shown in figure 1-16.

2. *Brass-faced hammer*—A soft-faced hammer with the weight and impact force of steel, as shown in figure 1-17.

3. *Plastic/rubber combination hammer*—A soft-faced hammer that will not damage fragile parts and will protect machined surfaces.

4. *Rubber mallet (hammer)*—A soft-faced hammer that will not damage fragile parts and will protect any machined surfaces, as shown in figure 1-18.

5. *Rawhide hammer*—A soft-faced hammer that will not damage fragile parts and will protect any machined surfaces, as shown in figure 1-19.

Fig. 1-16 Ball-peen hammer.

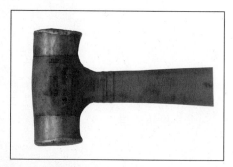

Fig. 1-17 Brass faced hammer.

PLIERS

The pliers are a gripping and cutting tool, available in different sizes and shapes. The different types of pliers used in the workplace are:

1. *Combination pliers*—Used to hold parts, rather than using your hand, to prevent injuries and to bend or twist thin materials, as shown in figure 1-20.

2. *Adjustable pliers*—Have many different channels to allow for different opening sizes, as shown in figure 1-21.

3. *Needle nose pliers*—Used for gripping small objects such as pins and clips, positioning small parts, and bending or forming wire, as shown in figure 1-22.

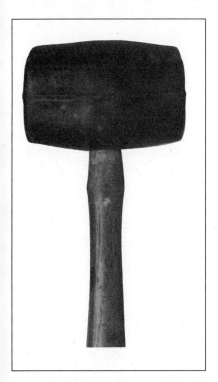

Fig. 1-18 Rubber mallet.

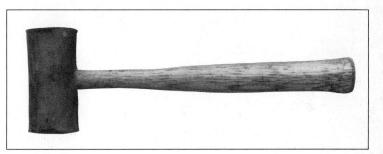

Fig. 1-19 Rawhide hammer.

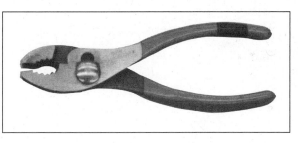

Fig. 1-20 Combination pliers.

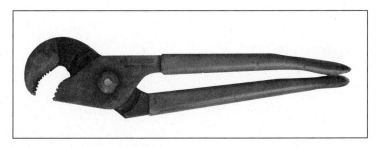

Fig. 1-21 Adjustable pliers.

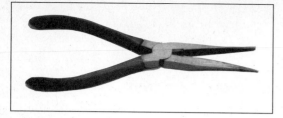

Fig. 1-22 Needle nose pliers.

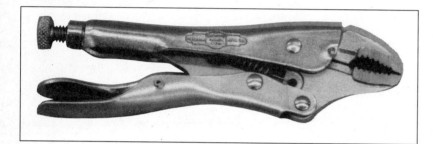

Fig. 1-23 Vise grip (locking) pliers.

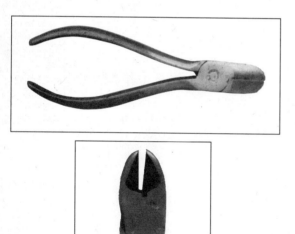

Fig. 1-24 Diagonal cutting pliers.

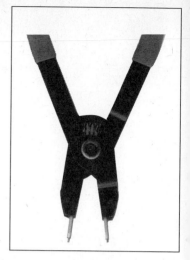

Fig. 1-25 Snap ring pliers.

4. ***Vise grip (locking) pliers***—*Used like the combination pliers for holding, but the clamping action can be locked, as shown in figure 1-23.*

5. ***Diagonal cutting pliers***—*Contain extra-hard cutting jaws for cutting wire and cotter pins 1/8-inch or smaller, as shown in figure 1-24 A & B.*

6. ***Snap ring pliers***—*Used to spread snap rings to just the right amount to remove or install them. There are two versions: one for inside retainers and one for outside retainers, as shown in figure 1-25.*

SAFETY TIPS: SOME OF THE PRACTICAL TIPS AND SAFETY RULES THAT SHOULD BE FOLLOWED ARE:

1. DO NOT USE PLIERS TO LOOSEN OR TIGHTEN GAS LINES, BOLTS, OR NUTS.

2. WHEN CUTTING A WIRE, COVER THE END TO PREVENT THE CUT PIECE FROM BEING PROPELLED ABOUT THE WORKPLACE.

Fig. 1-26 Open end wrench.

Fig. 1-27 Box end wrench.

Fig. 1-28 Combination wrench.

WRENCHES

A wrench is a tool for twisting or holding a bolt head or nut. Bolt heads and nuts generally have six flat sides around the outer surface. The "flats" permit the use of a wrench to tighten or loosen a nut or bolt.

1. *Open end wrench*—Used to fit snugly on two sides of a bolt or nut so the bolt or nut can be tightened or loosened easily, as shown in figure 1-26.

2. *Box end wrench*—The jaws of the wrench fit completely around the bolt or nut to provide better contact, as shown in figure 1-27. The tool is either a 12-point or 6-point wrench, depending on the number of notches on the gripping surface.

3. *Combination wrench*—The wrench has an open end jaw on one end and a box end on the other, as shown in figure 1-28. Both ends are the same size.

4. *Adjustable wrench*—The adjustable wrench

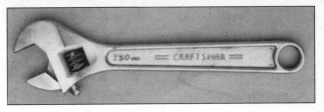

Fig. 1-29 Adjustable wrench.

Fig. 1-30 Hex (Allen) wrench.

has an adjustable jaw that allows the use of one tool to fit different sizes of bolts or nuts, as shown in figure 1-29. Even though this wrench is handy, it should be used as little as possible in lieu of the box end wrench because it will not fit as snugly around the bolt and, in addition, it very well may round the sides of the bolt.

5. *Hex (Allen) wrench*—The wrench which fits a set screw that may be found on many pulleys having hexagonal recesses, as shown in figure 1-30. The size of the wrench is the distance across the flats.

6. *Socket wrench*—An essential tool for working on an engine. The socket is firmly and snugly seated before you apply force to the handle, as shown in figure 1-31. Different types of socket wrenches are the ratchet, speed handle, and "T"-handle.

SAFETY TIPS: SOME OF THE PRACTICAL TIPS AND SAFETY RULES THAT SHOULD BE FOLLOWED ARE:

1. THE WRENCH OPENING FOR THE NUT OR BOLT HEAD SHOULD MATCH IT CLOSELY SO THAT THE
continued

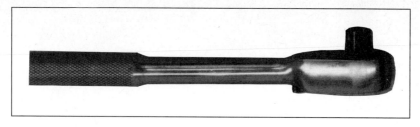

Fig. 1-31 Socket wrench.

WRENCH DOES NOT SLIP OFF THE WORKPIECE OR ROUND THE EDGES OF THE BOLT OR NUT

2. IT IS ADVISABLE TO PULL RATHER THAN PUSH ON A WRENCH TO AVOID SKINNED KNUCKLES.

3. DO NOT STRIKE THE END OF THE WRENCH WITH A HAMMER OR ANY OTHER OBJECT AS IT MAY BREAK THE TOOL OR DESTROY THE FASTENER.

4. DO NOT USE A PIPE OR ANOTHER WRENCH ON THE END OF A WRENCH FOR ADDITIONAL LEVERAGE BECAUSE THE LENGTH OF A WRENCH IS DESIGNED TO BE USED WITH NORMAL FORCES.

TORQUE WRENCHES

The torque wrench is a special wrench that indicates the amount of torque, or twisting force, applied to a nut or bolt. Many critical fasteners have the torque value calculated so that the bolts or nuts do not loosen during engine operation. If the fastener is tightened too much, the bolt may be strained and break.

The torque wrench can be calibrated to read in pound-feet or pound-inch, depending on the size of fastener. There are three main types:

1. *"I"-Beam torque wrench—When the force is applied to the handle during the tightening procedure of a bolt, a deflecting "I"-beam will move and point to a number which represents the twisting force, as shown in figure 1-32.*

2. *Dial torque wrench—The torquing forces are indicated on a dial located on the torque wrench, as shown in figure 1-33.*

3. *Clicking torque wrench—Used extensively where visual observation of the torque reading is not convenient, as shown in figure 1-34. The amount of torque is preset on a dial located on the handle, and a click is heard when the torque is reached.*

SAFETY TIPS: SOME OF THE PRACTICAL TIPS AND SAFETY RULES THAT SHOULD BE FOLLOWED ARE:

1. THE TORQUE WRENCH SHOULD BE PULLED OR PUSHED WITH A STEADY FORCE. AVOID THE USE OF SHORT, JERKY MOTIONS.

2. USE THE PROPER TORQUE WRENCH. A TORQUE WRENCH IS MOST ACCURATE WHEN USED AT MID-SCALE.

3. WHEN READING A TORQUE WRENCH, LOOK STRAIGHT DOWN AT THE SCALE.

4. PULL OR PUSH ONLY ON THE HANDLE OF THE TORQUE WRENCH AND DO NOT ALLOW IT TO TOUCH ANYTHING.

5. IF A PATTERN OF BOLTS OR NUTS ARE TO BE TORQUED IT IS BEST TO FOLLOW A FOUR-STEP PROCEDURE: STEP ONE, TORQUE ALL BOLTS TO ONE-THIRD OF THE FINAL VALUE IN A CROSSING PATTERN; STEP TWO, TORQUE ALL BOLTS TO TWO-THIRDS OF FINAL TORQUE IN THE CROSSING PATTERN; STEP THREE, TORQUE ALL BOLTS TO FINAL TORQUE IN THE CROSSING PATTERN; STEP FOUR, CHECK THE PATTERN AGAIN IN CASE ONE BOLT WAS MISSED.

6. THE TORQUE WRENCH SHOULD NEVER BE USED

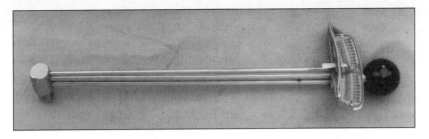

Fig. 1-32 "I"-Beam torque wrench.

Fig. 1-33 Dial torque wrench.

Fig. 1-34 Clicking torque wrench.

Fig 1-35 Flat chisel.

TO LOOSEN A BOLT. THE EXCESSIVE FORCES WILL DAMAGE THE TOOL'S CALIBRATION.

7. THE CLICKING TORQUE WRENCH SHOULD BE STORED WITH THE DIAL AT THE LOWEST READING. THE INTERNAL CALIBRATED SPRING WILL BE DAMAGED IF STORED UNDER TENSION AND INACCURATE TORQUE VALUES WILL FOLLOW.

8. IF A TORQUE WRENCH IS DROPPED, HAVE IT CHECKED FOR ACCURACY BEFORE USING IT AGAIN.

9. TORQUE READINGS ARE ONLY CORRECT IF THE SCALE IS READ WHILE THE WRENCH IS MOVING. IF THE BOLT OR NUT SEIZES WHILE TIGHTENING, BACK OFF THE NUT AND TIGHTEN IT AGAIN WITH A STEADY SWEEP OF THE HANDLE.

CHISELS

Chisels, such as a flat chisel, are used to cut metal by driving the chisel with a hammer, as shown in figure 1-35. The chisel can be used for cutting sheet metal, shearing off rivets and bolt heads, and splitting rusted nuts.

SAFETY TIPS: SOME OF THE PRACTICAL TIPS AND SAFETY RULES THAT SHOULD BE FOLLOWED ARE:

1. USE SHARP COLD CHISELS WITHOUT MUSHROOMED ENDS.

2. HOLD THE CHISEL FIRMLY AGAINST THE WORK WHEN STRIKING IT WITH A HAMMER.

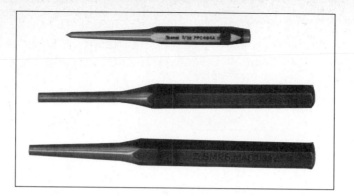

Fig. 1-36 Center punch, Pin punch, Starting punch.

Fig. 1-37 Aligning punch.

PUNCHES

Punches are used for driving out pins, aligning holes in parts, and marking the starting point for drilling a hole.

1. **Starting punch**—*Used to loosen a frozen bolt or pin, as shown in figure 1-36 top. Once the bolt or pin is loosened, a pin punch should be used to drive it out.*

2. **Pin punch**—*Used to drive out shafts, pins, and bearings without the punch jamming in the hole, as shown in figure 1-36 middle.*

3. **Center punch**—*Used to mark parts before disassembly so they can be reassembled in the same relative position, as shown in figure 1-36 bottom. It can also be used for marking hole locations for drilling and eliminate the drill from "wandering."*

4. **Aligning punch**—*Used to align holes during an assembly, as shown in figure 1-37.*

SAFETY TIPS: SOME OF THE PRACTICAL TIPS AND SAFETY RULES THAT SHOULD BE FOLLOWED ARE:

1. BEFORE USING A PUNCH, REMOVE ANY MUSHROOMED PARTS WITH A GRINDER.

2. HOLD THE PUNCH AT RIGHT ANGLES TO THE WORK TO PREVENT THE PUNCH FROM SLIPPING SIDEWAYS.

3. USE A PUNCH HOLDER WHEN THERE IS ROOM FOR IT.

4. STRIKE THE PUNCH SQUARELY WITH THE HAMMER.

5. NEVER STRIKE THE OPPOSITE END WITH A HAMMER, AS THE NORMAL STRIKING END HAS SOFTENED METAL, WHILE THE OTHER SIDED IS HARDENED.

6. AN ALIGNING PUNCH SHOULD NEVER BE USED AS A CENTER PUNCH.

7. NEVER USE A PIN PUNCH TO START A PIN SINCE THIS TOOL HAS A SLIM SHANK AND A HARD BLOW MAY CAUSE IT TO BREAK OR BEND.

FILES

A file is a hand cutting tool made of high-carbon steel. It is like many little sharp-edged chisels lined up next to each other. As the face of the file is moved across the metal piece, the cutting edges, or teeth, remove shavings. Files only cut in one direction, which is away from the tang, or the part that

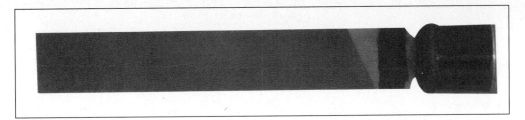

Fig. 38 Single cut file.

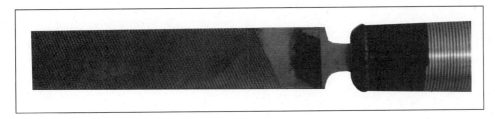

Fig. 1-39 Double cut file.

fits into the handle. <u>Pressure on the return stroke will dull the file.</u>

Files can be found in different shapes, such as flat, round, three-sided, etc., so that they can fit into a variety of areas. The primary types of files are:

1. ***Single Cut***—*Used when a smooth finish is desired or when hard materials are to be finished, as shown in figure 1-38. It has a single row of parallel teeth running diagonally across the face.*

2. ***Double Cut***—*Used for fast removal of metal and easy clearing of chips, as shown in figure 1-39. It is made of two intersecting rows of teeth.*

SAFETY TIPS: SOME OF THE PRACTICAL TIPS AND SAFETY RULES THAT SHOULD BE FOLLOWED ARE:

1. NEVER USE A FILE WITHOUT A HANDLE ON IT. THE PART OF THE FILE THAT PROJECTS OR FITS INTO THE HANDLE IS CALLED A TANG. SINCE THE FILE CUTS ON THE FORWARD STROKE ONLY, IF THERE IS NO HANDLE ON THE TANG IT COULD BE DRIVEN INTO THE PALM OF THE HAND.

2. WHEN A FILE IS USED, THERE MAY BE METAL FILINGS, METAL CHIPS, OR DUST AND

SPLINTERS. RATHER THAN USE YOUR HAND TO WIPE AWAY ANY METAL WASTE, UTILIZE A BRUSH OR RAG.

3. A FILE SHOULD NEVER BE USED AS A PRY BAR OR HAMMER. SINCE THE FILE IS HARD, IT CAN SNAP EASILY, CAUSING SMALL PIECES TO FLY.

4. KEEP THE FILE CLEANED BY USING A FILE CARD FREQUENTLY. A SHARP TOOL WILL GIVE MORE CONSISTENT RESULTS.

HACKSAW

Hacksaws are used for cutting metal. The tool consists of a blade held in an adjustable metal frame, as shown in figure 1-40. Hacksaw blades are made of high-speed molybdenum or tungsten-alloy steel that has been hardened and tempered. Only the teeth of the blade is hardened, and the remaining part of the blade is flexible. The blades are manufactured in various pitches (number of teeth per inch), such as 14, 18, 24, and 32. The 18-pitch blade is the one recommended for general use.

The blade only cuts in one direction and should not have any pressure applied when moving it in the opposite direction.

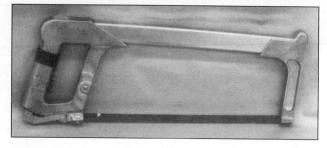

Fig. 1-40 Hacksaw.

Fig. 1-41 Bench vise.

Fig. 1-42 Drill vise.

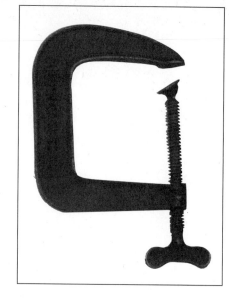

Fig. 1-43 C-clamp.

CLAMPS OR VISE

A **bench vise** is used to hold things in the work-place, and the jaws should be covered with soft metal to prevent marring the surfaces, as shown in figure 1-41. Light hammering may be done only on the stationary jaw of the vise.

A **drill vise** is used to hold round, rectangular, square, and odd shaped pieces for any operation that can be performed in a drill press, as shown in figure 1-42. It is a good workplace practice to clamp the vise to the drill table when drilling holes over 3/8-inch in diameter.

A **C-clamp** is a moveable or portable device for holding pieces of material together while working on them, as shown in figure 1-43.

> **NOTE: SOME OF THE PRACTICAL TIPS AND SAFETY RULES THAT SHOULD BE FOLLOWED ARE:**
>
> 1. NEVER USE A HAMMER TO TIGHTEN OR LOOSEN A VICE.
>
> 2. USE A VISE LARGE ENOUGH FOR THE PARTS OR TYPE OF WORK TO BE HELD.

TWIST DRILLS

Twist drills, as shown in figure 1-44, are tools for making holes and are constructed from a round bar with grooves cut in it. As the two cutting edges produce chips, they pass up the helical grooves and away from the working area. The drills may be used in a portable electric drill motor or in a drill press. A center punch should always be used to mark the location of the hole to be drilled in the metal piece.

Drill sizes are designated in four systems:

1. *Fractional size drill*—*Ranging from 1/64-inch (0.016-inch) to 3-1/4-inch (3.250-inch) varying in steps of 1/64-inch from one size to the next.*

2. *Number size drill*—*Ranging from #1 (0.228-inch) to #97 (0.006-inch).*

3. *Letter size drill*—*Ranging from letter A (0.234-inch) to letter Z (0.413-inch).*

4. *Metric size drill*—*Ranging from 0.5 mm to 20 mm.*

Fig. 1-44 Twist drills.

SAFETY TIPS: SOME OF THE PRACTICAL TIPS AND SAFETY RULES THAT SHOULD BE FOLLOWED ARE:

1. KEEP THE DRILLS COOL WHILE CUTTING BY LUBRICATING THE CUTTING SURFACES WITH OIL.

2. USE ONLY SHARP DRILL BITS. A DULL BIT MAY STIMULATE OR TEMPT THE OPERATOR TO APPLY INCREASED FORCE TO THE DRILL, WHICH COULD RESULT IN AN ACCIDENT OR INJURY.

3. WHEN DRILLING LARGE HOLES, USE THE STEP METHOD BY DRILLING A SMALL HOLE FIRST AND THEN PROGRESSING TO A LARGER OR FINAL SIZE DRILL.

TAP AND DIE SET

A tap and die set contains many different bolt or thread diameters as well as different pitches, as shown in figure 1-45.

A tap is a tool used to cut internal threads and is made from high-quality, hardened, ground, tool steel. The tap is similar to a screw, but flutes or grooves range the length of the thread that allow chips cut from the metal to escape.

A die is run over the outside of a rod to make external or outside threads. A die of the correct size is placed in the handle and is turned. Many dies are often adjustable in size so that the outside diameter can be enlarged or reduced slightly.

EXTRACTORS

Extractors are used to remove a bolt that has broken off level or below the surface of the hole, as shown in figure 1-46. One type of extractor is called an EZy-Out and consist of a coarse spiral thread. A hole is drilled into the broken bolt and the extractor is driven into the hole. As the extractor is turned, the coarse spiral thread cuts into the bolt and turns in the direction that removes it. Extractors sets contain different size extractors for different size bolts.

PULLER TOOLS

Special pullers are designed to separate pressed-together parts with an even distribution of force so that the parts are not "cocked" or damaged during separation, as shown in figure 1-47. An external puller grips the back of the object with the jaws, and

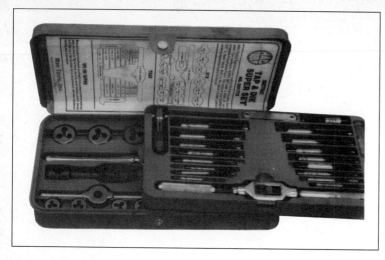

Fig. 1-45 Tap and Die set.

Fig. 1-46
Extractor.

Fig. 1-47 Puller tool.

the center screw pushes against the stationary part, such as a shaft.

MEASURING TOOLS

Many parts of the engine have exact dimensions. Small variances in these sizes must be measured so that proper decisions can be made prior to assembly.

Since accuracy and precision are necessary, tools are designed for this purpose.

The diameter of a cylinder can be accurately measured with a yardstick, but the amount of precision displayed would not be sufficient. A specification might state that an unworn cylinder bore is 2.563-inches and that any more than 0.003-inch wear is unacceptable for reassembly. These measurements are in thousandths of an inch, the tool necessary to measure these parts must be precise, and the reading of the tool by the technician must be accurate.

Some of the common precision measurement tools used in the workplace are discussed in the following sections.

FEELER GAGES

Feeler gages are strips of hardened steel that are rolled to the proper thickness with extreme accuracy, as shown in figure 1-48. A set of gages contains a variety of thicknesses in thousandths of an inch. The tool is necessary for checking the distance of a space, such as a valve clearance or point gap setting.

Wire feeler gages are similar to the flat gage except that they are made of steel wire that vary in thickness. They are useful in checking spark plug gaps.

Fig. 1-48 Feeler gages.

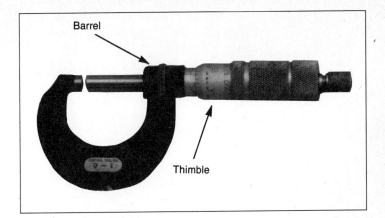

Fig. 1-49 Micrometer.

Feeler gages are precision measuring instruments and must be carefully handled to keep them accurate. The gage should never be forced into an opening that is too small. Before using a feeler gage, it should be wiped with a clean oiled cloth that will remove any dirt that may prevent an accurate reading.

MICROMETER

The micrometer is a precision tool which will accurately measure in thousandths of an inch, as shown in figure 1-49. It consists of a stationary frame and a moveable spindle. The spindle is turned by the thimble and, as the thimble moves outward, the numbers on the barrel are exposed.

English system

If all the numbers were exposed, they would be 0,1,2,3,4,5,6,7,8,9,0. The distance from the first zero to the last is exactly one inch. If the spaces between the numbers are counted, there would be ten spaces, or one-tenth of an inch (0.1-inch) as shown in figure 1-50.

Since the distance between each number on the

barrel is one-tenth of an inch, and there are four spaces between the graduations, one-tenth of an inch is divided by four to indicate 0.025-inch. Each small space represents twenty-five thousandths of an inch.

There are also numbers on the thimble of the micrometer ranging from 0 through 24. If the spaces are counted between these graduations around the thimble, 25 spaces are indicated. For every one complete turn (25 spaces) of the thimble, one more small line on the barrel is uncovered. Since each small line indicates 0.025-inch then each mark on the thimble is worth 0.001-inch.

Understanding the value of the markings makes it easier to interpret the reading on a micrometer. The reading on the micrometer in figure 1-51 is 0.111-inch because:

	Quantity		Value	=	
Micrometer range	0"-1"		0.000"	=	0.000"
Number showing on barrel	1	*	0.100"	=	0.100"
Spaces uncovered after last number	0	*	0.025"	=	0.000"
Number on the thimble	11	*	0.001"	=	0.011"
Total					0.111"

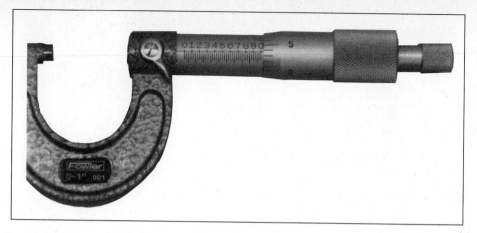

Fig. 1-50 The distance from the first zero to the last is exactly one inch.

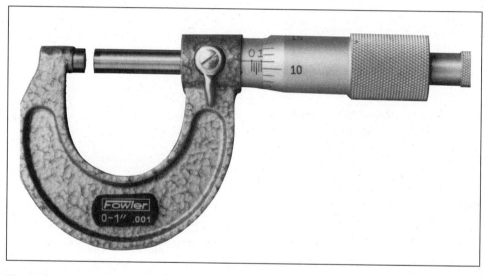

Fig. 1-51

Another measurement shown in figure 1-52, is taken with a micrometer that ranges from 2 inches–3 inches. The measurement is 2.562 inches because:

	Quantity		Value	=	
Micrometer range	2"-3"		2.000"	=	2.000
Number showing on barrel	5	*	0.500"	=	0.500"
Spaces uncovered after last number	2	*	0.025"	=	0.050"
Number on the thimble	12	*	0.001"	=	0.012"
Total					2.562"

The problem with using a micrometer is that unless it is used often, the retention of the process is low. One alternative method used to remember how the reading is derived at is to make an analogy that each number on the barrel is worth $1.00, while each space is worth 25¢, and each mark on the thimble is worth 1¢. A measurement, as shown in figure 1-53, would indicate at least three dollars, but less than four, because the number three is uncovered. After the number, there are 3 complete spaces showing, worth 25¢ each. Since the fourth space is not

Fig. 1-52 The reading on the micrometer is 0.111"

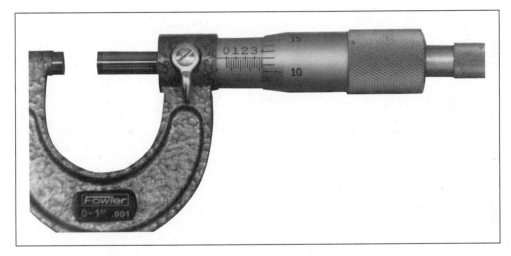

Fig. 1-53 The measurement is 2.562".

totally uncovered, the cents are added from the number on the thimble, which is 12. The total of the "cash-on-hand" is:

Dollars	$3.00
Quarters	$0.75
Pennies	$0.12
Total "cash-on-hand"	**$3.87**

The $3.87 designates the three numbers to be used as the decimal, eg., $3.87 = 0.387. All that is needed is the number added to the left of the decimal, and this is determined by the range of the micrometer used.

The digital micrometer is gaining popularity because of its easy reading of the measurement value. The actual measurement appears in the "counter" on the frame, as shown in figure 1-54, as well as on the barrel and thimble.

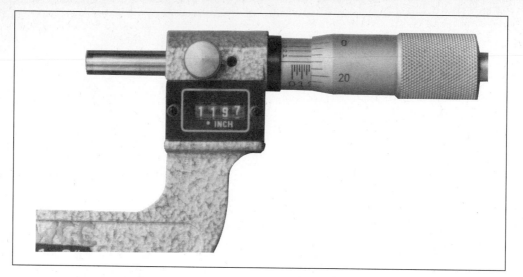

Fig. 1-54 Digital micrometer.

Metric system

The barrel and thimble of a metric micrometer are used to measure hundredths of a millimeter. The metric micrometers span the range of 25,0 mm so that one micrometer measures 0 mm-25,0 mm, while another is needed for a measurement that ranges from 25,0 mm-50,0 mm, etc.

Every revolution of the thimble opens up the tool by 0,5 mm (a comma is used in place of a decimal in metric system). The barrel is marked off in millimeters above the line and half millimeters (0,50 mm) below the line. The thimble is marked and divided into 50 equal parts each worth 0,01 mm. An example of a reading, as shown in figure 1-55 would be:

	Quantity	Value	=	
Micrometer range	0-25 mm	0,00 mm	=	0,00 mm
Top line marks showing on barrel	18	* 1,00 mm	=	18,00 mm
additional half mm (bottom line)	0	* 0,50 mm		0,00 mm
Number on the thimble	46	* 0,01 mm	=	0,46 mm
Total				18,46 mm

DIAL CALIPER

The dial caliper, as shown in figure 1-56, allows a quick method of making a direct inside or outside measurement of an object. The use of the dial caliper in small air-cooled engine repair is accept-

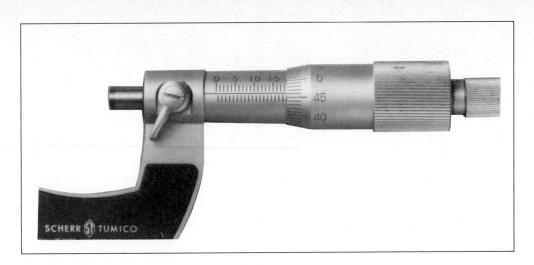

Fig. 1-55 Metric micrometer.

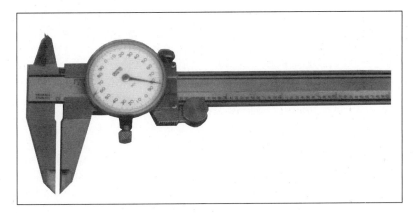

Fig. 1-56 The dial caliper.

able because one tool can take the place of four or five separate micrometers. The cost savings and the ease of reading the dial make it the tool of choice. The dial caliper is accurate to a thousandths of an inch and can range from 0–6 inches. The reading on the dial is 0.342-inch.

Another type of caliper is the digital caliper. Electronic technology provides a readout in the English or Metric system and is accurate to the half of a thousandths, as shown in figure 1-57.

SAFETY TIPS: SOME OF THE PRACTICAL TIPS AND SAFETY RULES THAT SHOULD BE FOLLOWED ARE:

1. ALWAYS LEAVE A GAP BETWEEN THE FIXED AND ADJUSTABLE JAW ON THE DIAL CALIPER WHEN STORING TO ALLOW FOR THE METAL EXPANSION AND CONTRACTIONS.

continued

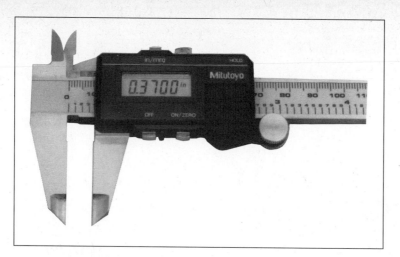

Fig. 1-57 Digital caliper.

DIAL INDICATOR

The dial indicator is a gage that is able to measure small differences in movement, as shown in figure 1-58. The dial indicator has a movable plunger that moves a needle on the face of the dial to indicate movement as minute as one-thousandth of an inch. This gage can be used in a holding tool to measure the crankshaft endplay of an engine.

PLASTIGAGE

The easiest way to check a bearing clearance is with the use of Plastigage, as shown in figure 1-59, that is available in three sizes ranging from 0.001-inch–0.003-inch, 0.002-inch–0.006-inch and 0.004-inch–0.009-inch. It is a round piece of plastic that will be flattened when squeezed between two parts. The less the clearance, the greater the squeezing effect.

For example, in a certain engine the maximum allowable clearance between the bearing attaching the connecting rod to the crankshaft is 0.003-inch. When the rod is removed from the crankshaft, a piece of Plastigage is placed along the width of the crankshaft rod bearing surface while the crankshaft is held stationary. The connecting rod is then connected to the crankshaft and the bolts are torqued to the proper specification. Do not pivot the rod on the crankshaft for it will smear the initial impression of the Plastigage on the rod cap when the bolts were torqued. The rod cap is then removed and the flattened Plastigage is measured with the graduations on the package, which relates the total bearing clearance in thousandths of an inch.

GENERAL TOOL REQUIREMENT

The basic tools necessary to work on the small air-cooled engine range from general tools to very specific tools. The purpose of the tools is to enable an efficient, safe method of operation while working

Fig. 1-58 Dial Indicator.

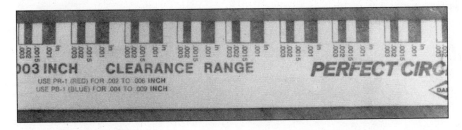

Fig. 1-59 Plastigage.

on an engine. Many tools are used for purposes other than those for which they were intended because of a lack of the correct tool, which gives rise to an increased incidence of broken tools and physical injuries. The tools listed below are the common ones, but not all that are needed to work on the well known air-cooled engines. The tools are divided into tool box, shop, and specialized tools categories.

TOOL BOX TOOLS

Combination wrenches

shown in figure 1-60
1/4", 11/32", 3/8", 7/16",
1/2", 9/16", 5/8", 11/16",
3/4"

Chapter 1, Safety and Tools **31**

3/8-inch drive tools

shown in figure 1-61
3/8" drive ratchet
3/8" drive sockets
> 3/8", 7/16", 1/2", 9/16", 5/8"
> 11/16", 3/4"

3/8" drive-3" extension
3/8" drive deep sockets
> 3/8", 7/16", 1/2", 9/16"

3/8" drive spark plug sockets
> 5/8", 3/4", 13/16"

3/8" to 1/4" reducer

1/4-inch drive tools

shown in figure 1-62
1/4" drive ratchet
1/4" drive sockets
> 3/16", 7/32", 1/4", 9/32", 5/16", 11/32",
> 3/8", 7/16", 1/2"

1/4" drive-4" extension

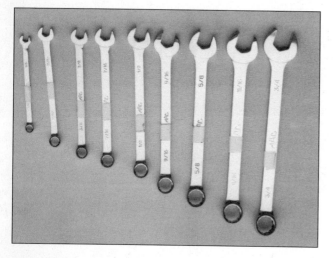

Fig. 1-60 Combination wrenches.

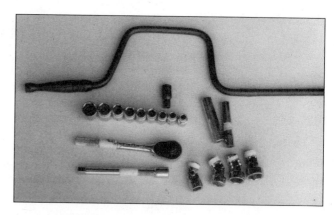

Fig. 1-62 1/4" drive tools.

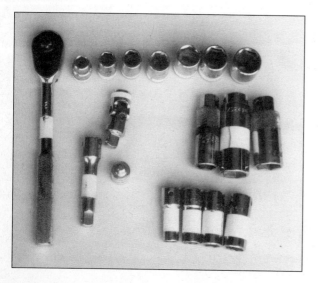

Fig. 1-61 3/8" drive tools.

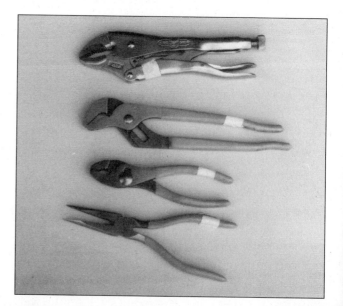

Fig. 1-63 Pliers.

1/4" drive #4 TORX® socket
1/4" drive deep sockets
      1/4", 11/32"
1/4" drive universal joint flex sockets
      3/8", 7/16", 1/2", 9/16"
1/4" drive speed wrench

Hex (Allen) wrench set

Pliers

shown in figure 1-63
Combination pliers

Needle nose pliers
Vise grip (locking) pliers
Diagonal cutting pliers
Snap ring pliers

Hammers

shown in figure 1-64
Rubber/Plastic combination hammer
Ball-peen hammer

Screwdrivers

shown in figure 1-65
common large screwdriver
common small screwdriver
Philips large screwdriver
Philips small screwdriver

Blade adaptor puller

Inspection mirror

shown in figure 1-66

Safety Glasses

shown in figure 1-66

Plastic "Ziploc®" baggies for parts

Air blow gun

shown in figure 1-66

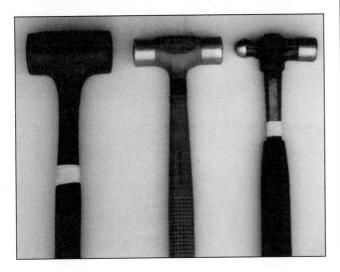

Fig. 1-64 Hammers.

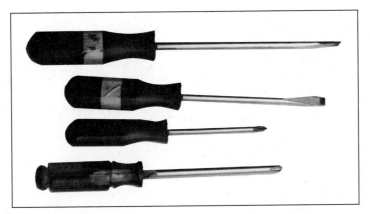

Fig. 1-65 Screwdrivers.

Matches

Knife

ADDITIONAL SHOP TOOLS

Impact screwdriver
shown in figure 1-67

Soldering equipment

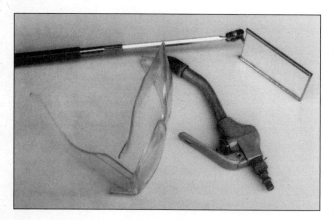

Fig. 1-66 Inspection mirror, Safety glasses, Air blow gun.

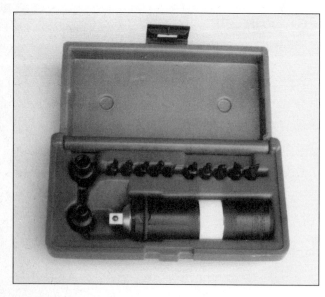

Fig. 1-67 Impact screwdriver.

Shop vise

Twist drills

Starting punch

Pin punch

Tap and Die set

Extractor set

Torque wrench

Continuity light or ohmmeter

Snap ring pliers (inner and outer)

Hex (Allen) wrench set

Hacksaw

Flat and round files

Pen light
shown in figure 1-68

METRIC TOOLS
Metric tools are necessary for the wide range of international engines found in the workplace.

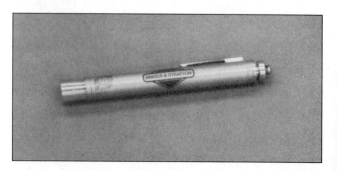

Fig. 1-68 Pen light.

Combination wrenches

6 mm, 7 mm, 8 mm, 9 mm,
10 mm, 11 mm, 12 mm, 13 mm,
14 mm, 15 mm, 16 mm, 17 mm,
18 mm, 19 mm

3/8-inch drive metric sockets

10 mm, 11 mm, 12 mm, 13 mm,
14 mm, 15 mm, 16 mm, 17 mm,
18 mm, 19 mm, 20 mm, 21 mm,

Hex (Allen) metric wrenches

2 mm, 2.5 mm, 3 mm, 4 mm,
5 mm, 6 mm, 8 mm

1/4-inch drive metric tools

1/4" drive metric sockets
5 mm, 5.5 mm, 6 mm, 7 mm,
8 mm, 9 mm, 10 mm, 11 mm,
12 mm, 13 mm, 14 mm

BRIGGS & STRATTON SPECIALIZED TOOLS

The following is a list of the more common tools and kits offered by Briggs & Stratton Corporation:

Briggs & Stratton basic tool kit (#19300)

Briggs & Stratton offers a kit, as shown in figure 1-69, which includes a group of tools recommended for repair of their engines. These are recommended in addition to your common tools. The tools in the kit are designed to make some the repairs much easier and more efficient. Many of the tools can be used on engines produced by other engine manufacturers. The tool kit part number is 19300 and contains the following tools:

Part #	Tool Name	Tool Use
19063	Valve Spring Compressor	Used to compress the valve springs for easy removal and installation of the intake and exhaust valve.
19069	Flywheel Puller	The tool used to remove the flywheel on most small B&S air-cooled engines.
19070	Piston Ring Compressor	Used to compress the piston rings so that they can be installed into the cylinder. This tool used for 5HP engines and under.
19122	Valve Guide Reject Gage	This tool is used to determine if the valve guide is worn. If the gage does not fit into the hole, the guide is not worn. This reject gage is used on all 1/4-inch diameter holes.
19151	Valve Guide Reject Gage	This tool is used similar to #19122 except that it should be used on all valve guides that are 5/16-inch diameter holes.
19165	Flywheel Puller	The tool used to remove the flywheel on most large bore B&S air-cooled engines.
19167	Flywheel Holder	This tool holds the flywheel while removing or installing the flywheel nut or starter clutch.
19203	Flywheel Puller	This tool removes the flywheel on the largest bore B&S engines.
19200	Tachometer	The tool used to correctly set the idle and top no load RPM.
19229	Tang Bender	The tool used to adjust the governor top no load RPM and governed idle on Model series 130000 through 280000.
19230	Piston Ring Compressor	Used to compress the piston rings so that they can be installed into the cylinder. This tool used for larger than 5HP engines.
19244	Starter Clutch Wrench	The tool that removes and torques the rewind starter clutch.
19256	Brake Adjustment Gage	The tool necessary to adjust the band brakes on the compliance model engines.
19258	Valve Lapper	The tool used to lap the valve face to the valve seat.

continued

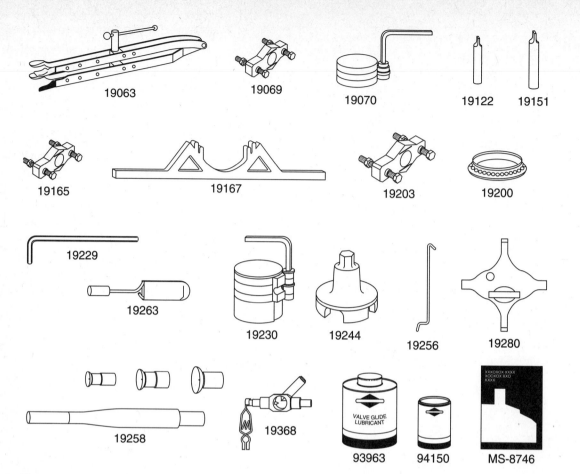

Fig. 1-69 Briggs & Stratton basic tool kit (#19300).

19263	Carburetor Screwdriver	The tool that allows easy adjustment of the Vacu-Jet and Pulsa-Jet carburetor mixture screw.
19280	Screwdriver	A tool that consists of a four-bladed screwdriver that is useful when removing and installing the nozzles on the Flo-Jet carburetor.
19368	Ignition Tester	The tool used to test the condition of the ignition system by checking for spark.
93963	Valve Lubricant	Solution to lubricate the valve stems and valve guides, spark plug threads, muffler bolts, and cylinder head bolts.

94150	Valve Lapping Compound	Double-sided canister that contains both fine and course compound to lap in the valves to the valve seats.

Briggs & Stratton valve guide repair kit (1/4-inch) (#19269)

The valve guide repair kit is available to recondition worn 1/4-inch valve guides, as shown in figure 1-70. The kit contains all the tools necessary for removing and installing a brass bushing or performing the process of oversizing an aluminum bushing and installing a new brass one. The valve guide repair kit part number is 19269 and includes the following parts:

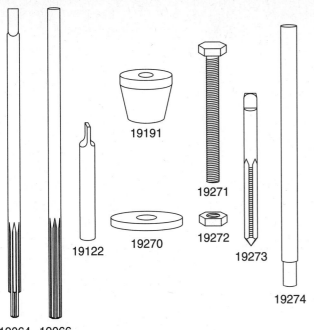

19064   19066

Fig. 1-70 Briggs & Stratton valve guide repair kit (1/4")
(#19269).

(**NOTE:** VALVE GUIDE REPAIR KIT 19232 IS USED FOR
SERVICING 5/16-INCH VALVE GUIDES USED IN
ENGINES ABOVE 5 HP.)

Part #	Tool Name	Tool Use
19064	Reamer-Counterbore	This tool is used to oversize a plain aluminum valve guide to the size necessary for the installation of the new brass bushing.
19066	Reamer-Finish	This tool is used to dimension the new bushing, after it is installed, to exactly 1/4-inch diameter.
19122	Valve Guide Reject Gage	This tool is used to determine if the valve guide is worn. If the gage does not fit into the hole, the guide is not worn. This reject gage is used on all 1/4-inch diameter holes.
19191	Guide	This tool centers the counterbore reamer (19064) in the worn valve guide for an accurate tooling.
19270	Washer	This part is used with parts 19271 and 19272 to work as a combination worn brass valve guide puller.
19271	Puller Bolt	This part is used with parts 19270 and 19272 to work as a combination worn brass valve guide puller.
19272	Puller Nut	This part is used with parts 19271 and 19270 to work as a combination worn brass valve guide puller.
19273	7 mm tap	This tool cuts a thread into the worn brass bushing so that the puller bolt (19271) can be screwed in. A 7 mm is used so that the threads are cut into the brass bushing without damaging the aluminum block.
19274	Bushing Driver	This tool is used to press the new brass bushing into the hole.

Briggs & Stratton valve seat cutter kit

A valve seat cutter kit, manufactured by Neway (part number 19237), is available for reconditioning both 30° and 45° angle valve seats. An additional valve seat cutter kit (19343) is used on the Quantum and Vanguard overhead valve engines.

Briggs & Stratton torque wrench

A torque wrench (part number 19197) is available for torque ranges of 5- to 200-inch-pounds.

Briggs & Stratton dial caliper

A dial caliper (part number 19199) is available that ranges from 0 inch to 6 inches.

Briggs & Stratton telescoping gage

A telescoping gage (part number 19198) is available that measures engine bore from 2 inches to 3.5 inches.

Briggs & Stratton flywheel strap wrench

A flywheel strap wrench (part number 19373) is available as an alternative method of securing the flywheel for removal of the flywheel nut or starter clutch.

TECUMSEH SPECIALIZED TOOLS

The following is a list of the more common tools and kits offered by Tecumseh corporation:

Tecumseh basic tool kit (#670195)

Tecumseh offers a kit which includes a group of tools recommended for repair of their engines. These are recommended in addition to your common tools. The tools in the kit are designed to make some repairs much easier and more efficient. Many of the tools can be used on engines produced by other engine manufacturers. The tool kit part number is 670195 and contains the following tools:

Part #	Tool Name	Tool Use
670305	Strap Wrench	Holds flywheel during flywheel nut installation and removal.
670103	Knockoff Tool	Used to remove the flywheel when no puller holes are available.
670266	Knockoff Tool	Used to remove the flywheel when no puller holes are available.
670306	Flywheel Puller	Used to remove the aluminum flywheel on the small and large bore engines.
670202	Oil Seal Protector	Protects oil seals from damage during installation.
670263	Oil Seal Protector	Protects oil seals from damage during installation.
670264	Oil Seal Protector	Protects oil seals from damage during installation.
670295	Oil Seal Protector	Protects oil seals from damage during installation.
670277	Oil Seal Protector	Protects oil seals from damage during installation.
670260	Oil Seal Protector	Protects oil seals from damage during installation.
670294	Oil Seal Protector	Protects oil seals from damage during installation.
670292	Oil Seal Protector	Protects oil seals from damage during installation.
670293	Oil Seal Protector	Protects oil seals from damage during installation.
670237	Valve Spring Compressor	Used in the valve in head engine.
670292	Air Gap Gage	Used to measure the 0.0125-inch gap required between the magnets of the flywheel and the external armatures.
670232	Timing Tool	Used on some 2-cycle engines.
670253	Float Tool	Measures carburetor float level.

LAWN-BOY SPECIALIZED TOOLS

The following are a list of the more common tools and kits offered for Lawn-boy engines:

Lawn-boy basic tool kit—"M" Engine (#LB684232)

Lawn-boy offers a kit which includes a group of tools recommended for repair of their engines. These are recommended in addition to your common tools. The tools in the kit are designed to make some the repairs much easier and more efficient. Many of the tools can be used on engines produced by other engine manufacturers. The tool kit part number is #LB684232 and contains the following tools:

Part #	Tool Name	Tool Use
LB613592	Piston Ring Compressor	Tapered ring compressor makes installation of piston into the bore easy.
LB613599	Governor Protector	This governor shaft seal protector is used to install the governor shaft seal without damage.
LB684233	Wrist Pin Remover	The connecting rod and crankshaft are one piece. Locating the piston in such a way as to safely drive out the wrist pin.

LB613590	Crankshaft Remover	Removes both the flywheel and blade end crankshaft seals.
LB613598	Seal Protector	Prevents nicking and premature failure of seal.
LB613595	Drive Handle	Used with LB613594 to install crankshaft seals.
LB613594	Seal Installer	Used with drive handle to install both crankshaft seals.
LB611600	Puller	Used to remove the flywheel and blade brake clutch.
LB613587	Governor Adjusting Tool	Used to bend governor linkage to adjust for proper engine RPM.
LB613596	Seal Installer	Installs the seal to the proper depth.
LB613593	Flywheel Holder	Since piston stop may not be used on the "M" engine, this tool prevents damage to the flywheel during flywheel installation.
LB613589	Mounting Plate Locator	Used to align the blade brake clutch mounting plate to the crankshaft to prevent scuffing and wear.
LB613591	Seal Remover	Used to remove the governor shaft seal.
LB604659	Air Gap Gauge	Adjust the air gap between flywheel magnets and coil laminations, CD pack, and alternator assemblies.

Lawn-boy basic tool kit—"D" and "F" Engine (#LB683625)

Lawn-boy has a tool kit that groups many common tools that make repairs to their engines more accessible. Many of the tools can be used on engines produced by other engine manufacturers. The tool kit part number is #LB683625 and contains the following tools:

Part #	Tool Name
LB605081	Bearing Installer
LB612344	Governor Adjusting Gage
LB604659	Air Gap Gauge

LB602887	Crankshaft Guide
LB612103	Piston Stop
LB426814	Spark Tester
LB605082	Bearing Remover
LB609968	Crankshaft Gage
LB610510	Piston Ring Compressor
LB609967	Piston Ring Compressor
LB612533	Snap Ring Plier
LB612087	Seal Protector
LB611600	Clutch Puller
LB611591	Mounting Plate Locator
LB612231	Carburetor Adjusting Wrench
LB608976	Seal Installer
LB611592	Clutch Removal Tool
LB611702	Ignition Shorting Gage
LB611703	Control Cable Adjusting Gage

MISC. TOOL

This tool will help service engines and can be easily made.

Crankshaft turning tool

This tool, as shown in figure 1-71, is useful when it is necessary to turn the crankshaft. It is made out of an old connecting rod with some additional washers added.

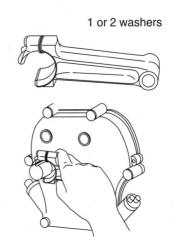

1 or 2 washers

Fig. 1-71 Crankshaft turning tool.

SUMMARY

Even though a good safety program is based upon adequate instruction in the operation and maintenance of existing machinery and equipment, it is also necessary to acquire a safety mentality. A positive safety attitude is essential to a safe environment, which requires a clean and orderly laboratory with regularly maintained or serviced equipment and the necessary protective devices. It is important to remember that the personal protective equipment does not take the place of accident prevention.

All walk-behind power lawnmowers built or imported after June 30, 1982, that will be used by the consumer must comply with the Consumer Product Safety Commission (CPSC) Standards.

There are many combustible materials used in the workplace. The first safety concern should be preventing a fire, while the next concern should be what to do if there is a fire.

The electric drive motor, bench grinder, drill press, oxygen/acetylene torch, valve grinder, lifting devices, and air tools are common equipment that have special rules for safe operation. A good reminder method for equipment safety would be to clearly post the rules in the proximity of each device.

Imprudent operation of power equipment can create hazards that can lead to personal injury and property damage. To prevent accidents, become thoroughly familiar with your machinery before operating by reading the instruction manual.

Many parts of the engine have exact dimensions. Small variances in these sizes must be measured so that proper decisions are made prior to assembly. Since accuracy and precision are necessary, tools are designed for this purpose.

The basic tools necessary to work on the small air-cooled engine range from general tools to very specific tools. The purpose of the tools is to enable an efficient safe method while working on an engine.

Questions

1. Safety can best be described as:

A. the use of safety glasses.
B. a clean, orderly workplace.
C. a positive safety attitude.
D. avoiding all potential dangerous environments.

2. A dust mask worn on the face will protect the lungs from:

A. everything.
B. most airborne particles.
C. dangerous exhaust gases.
D. any hazardous vapors.

3. Safety devices are installed on power equipment so that:

A. accidents will not happen.
B. the chance of accidents will be lower without them.
C. the price and profits will be increased.
D. the equipment will be protected.

4. Exhaust gas contains a deadly, colorless, and tasteless gas called:

A. Carbon Dioxide.
B. Carbon Tetrachloride.
C. Carbon Monoxide.
D. Oxygen Distillates.

5. One of the Consumer Product Safety Commission (CPSC) standards for post-1982 consumer mowers is:

A. that the engine will stop whenever the operator changes the grass catch bag.
B. that the blade speed will be no faster than 2000 RPM above the specified idle speed.
C. that a warning label must be attached to the mower.
D. that shoes worn by all operators of commercial mowers must be of the steel toe type.

6. **To extinguish a fire:**

   A. remove the fuel.
   B. remove the heat.
   C. remove the oxygen.
   D. all of the above.

7. **A class A fire extinguisher can be used on:**

   A. a wood fire.
   B. a gasoline fire.
   C. an electric motor fire.
   D. all of the above.

8. **A class B fire extinguisher can be used on:**

   A. a wood fire.
   B. a gasoline fire.
   C. a paper fire.
   D. all of the above.

9. **A class C fire extinguisher can be used on:**

   A. a wood fire.
   B. a gasoline fire.
   C. an electric motor fire.
   D. all of the above.

10. **If a chemical is rated as non-hazardous on the MSDS sheet, it is:**

    A. safe to drink.
    B. very corrosive.
    C. not irritating to the skin.
    D. still possible to cause injury.

11. **A lead acid battery may explode if:**

    A. water is added to the cells.
    B. a spark is near the charging battery.
    C. it is transported incorrectly.
    D. it is completely discharged.

12. **When using an electric drill motor:**

    A. wear eye protection.
    B. change the drill bit while the motor is plugged in.
    C. wet the floor that you stand on to prevent the sparks from igniting nearby materials.
    D. all of the above.

13. **When using the bench grinder, keep the safety shield attached and in the proper position so that:**

    A. safety glasses are not necessary.
    B. abrasive particles do not "fly" at your face.
    C. the machine will "power-up" properly.
    D. all of the above.

14. **An air powered impact wrench:**

    A. should be used only with a special socket.
    B. should not have more than 20 PSI line air pressure.
    C. Should be used to fully tighten a bolt or nut.
    D. all of the above.

15. **The oxygen/acetylene torch can be used for:**

    A. welding.
    B. cutting metal.
    C. heating of rusted parts.
    D. all of the above.

16. **A screwdriver should be used to:**

    A. open a paint can.
    B. turn a screw.
    C. scrape carbon from a cylinder head.
    D. open wooden crates.

17. A pliers may be used:

A. in place of a wrench.
B. for positioning small parts.
C. in place of a hammer.
D. as a torque wrench.

18. A torque wrench:

A. is used to loosen bolts.
B. is used to apply a certain amount of twist to a fastener.
C. is accurate at all readings.
D. all of the above.

19. Center, starter, and aligning are all types of:

A. wrenches.
B. pliers.
C. punches.
D. chisels.

20. Which of the following statements are true about a hacksaw?

A. It will cut metal or wood.
B. It will cut on the forward and back stroke.
C. The total blade is composed of hardened steel.
D. It can be used to turn a screw.

CHAPTER 2

Air-Cooled Engine Technology Basics

INTRODUCTION

A well-trained technician is one who knows proper operation of the engine and the relationship of each engine part to the whole. This chapter will be a study of the fundamental requirements for the air-cooled engine. Various measurements that relate to the internal combustion engine, such as bore, stroke, piston displacement, compression, torque, and horsepower will be described.

ENGINE COMPONENTS (INTERNAL)

An understanding of engine components and their function is necessary for a common terminolo-gy when presenting certain theories. The following are some of the major components in the internal combustion engine:

COMBUSTION CHAMBER

The combustion chamber is enclosed and formed by two units called the cylinder and cylinder head. The ignition, or "setting on fire," of the air-fuel mixture takes place in the ignition compartment of the combustion chamber, which is the space remaining in the cylinder head when the piston is at the top of its travel in the cylinder. After the air-fuel mixture is ignited by the spark plug, the burning of the fuel causes the piston to travel downward in the cylinder. The combustion chamber, or location where the burning of the fuel and expansion of gases takes place, includes the entire hollow of the cylinder,

Fig. 2-1 An aluminum alloy cylinder.

i.e., the ignition compartment plus the rest of the space through which the piston travels.

Cylinder

The cylinder is a long, round object with flat ends and, in the internal combustion engine, is hollow and houses the piston. Because of the high heat and stress of piston motion, the cylinder must be constructed of high strength material such as aluminum alloy, cast-iron insert in aluminum, or totally cast-iron, as shown in figures 2-1, 2-2, and 2-3.

The cylinder construction is a considerable factor in determining maximum engine life. The sand-casted aluminum cylinder has the inherent shortest wear life, while the aluminum cylinder with the cast-iron liner has an extended wear life but the remaining part of the block is constructed with lightweight aluminum. The cast-iron lined aluminum engines are constructed by pouring liquefied aluminum around cast-iron inserts, as shown in figure 2-4. The fully cast-iron cylinder provides the engine with the longest possible wear life.

Cylinder head

The cylinder head is usually a one-piece aluminum casting that is bolted to the top of the cylinder block, as shown in figure 2-5 (page 46). The cylinder together with the cylinder head form the combustion chamber in which burning of fuel takes place.

On engines that have the valves in the cylinder block, the cylinder head is a simple piece containing cooling fins and the spark plug. In valve-in-head engines, such as the overhead valve design, the cylinder incorporates the valves, valve ports, valve seats, and valve operating mechanism. A gasket is placed between the cylinder head and the cylinder block to completely seal the combustion chamber. A metal head gasket, which allows easy heat transfer to the cooler cylinder, provides a more uniform cylinder and cylinder block temperatures for a cooler-running engine.

Fig. 2-2 A cast-iron cylinder insert in an aluminum alloy cylinder.

Fig. 2-3 A cast-iron cylinder.

Fig. 2-4 The aluminum is poured around the cast-iron liners to provide longer engine life than the plain aluminum engine block.

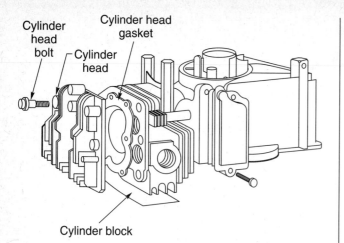

Fig. 2-5 The cylinder head is a one-piece aluminum casting that is bolted to the top of the cylinder block.

PISTON

The piston transmits the force of the burning and expanding gases in the cylinder through the connecting rod to the engine crankshaft, as shown in figure 2-6. Pistons usually are made of high-grade aluminum alloy because of its light weight and high conductivity of heat. The piston underside usually contains ribs, or other means of obtaining more surface area, for contact with the lubricating oil which is splashed on it, facilitating piston cooling. Grooves are arranged on the piston's outer surface in order to permit installation of piston rings.

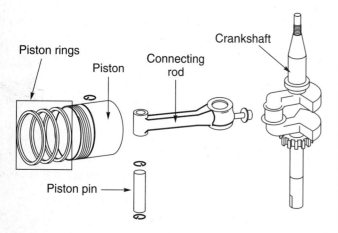

Fig. 2-6 The piston transmits the force of combustion through the connecting rod to the crankshaft.

WRIST OR PISTON PIN

The wrist or piston pin is a short steel shaft that is mounted inside the piston. It attaches the piston to the upper end of the connecting rod, as shown in figure 2-7.

PISTON RINGS

The piston rings are installed on the piston to form a seal between the piston and the cylinder wall, as shown in figure 2-8. The metal of the cylinder wall and the piston have different expanding rates, so a comparatively large clearance between them must be provided. The piston rings are usually made of high grade cast-iron that provides a seal to hold the combustion pressures in, retain as much oil as possible in the crankcase, and conduct heat from the piston head to the cylinder walls.

If the functions are not carried out, the engine's overall performance will be severely affected. If the rings do not adequately seal against the cylinder wall, the force of the combustion will be reduced, which will result in a loss of power. If the rings lose their scraping ability, oil will enter the combustion chamber. On each intake stroke, atmospheric pres-

Fig. 2-7 The wrist or piston pin.

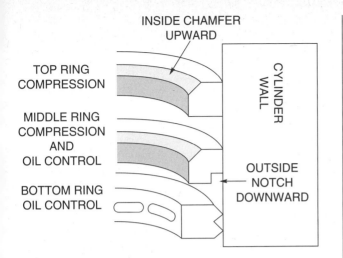

INSIDE CHAMFER UPWARD

TOP RING COMPRESSION

MIDDLE RING COMPRESSION AND OIL CONTROL

BOTTOM RING OIL CONTROL

CYLINDER WALL

OUTSIDE NOTCH DOWNWARD

Fig. 2-8 The piston rings are installed between the piston and the cylinder wall.

sure forces gases and oil up past worn rings into the combustion chamber. The oil is then burned, producing a thick, bluish-white smoke and burnt oil odor that causes excessive oil consumption. If the rings lose their ability to transfer heat, internal engine damage will result.

CONNECTING ROD

The connecting rod furnishes a means of converting the reciprocating motion (up and down) of the piston to a rotating movement of the crankshaft, as shown in figure 2-9. The small end is attached to the piston, and the larger end is usually split to provide for assembly on the crankshaft.

The larger end consists of two pieces. The upper half is the rod to the crankshaft. The lower half is called the rod cap, which is used to bolt the rod to the crankshaft. The connecting rod and its cap are manufactured as a unit and must always be kept together.

CRANKSHAFT

The crankshaft converts the up and down movement of the piston from the power stroke of the piston into rotational force called torque, which

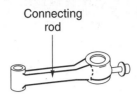

Connecting rod

Fig. 2-9 The connecting rod connects the piston to the crankshaft.

turns the pulley, blade, etc., as shown in figure 2-10. The distance from the centerline of the crankshaft to the centerline of the rod journal is the throw. Opposite the throw, a counter weight arrangement is usually found, which balances the forces of the spinning crankshaft.

Counter weights

The counter weights on the crankshaft reduce the load on the crankshaft bearing surfaces, as shown in figure 2-11. The counter weights are either forged as integral parts on the crankshaft or can be a separate mechanism that is geared to the crankshaft. The counter weight masses balance the reciprocating (up and down) forces from the movement of the piston and connecting rod.

Crankpin, or rod journal

The crankpin, or rod journal, is a round metal surface on the crankshaft to which the connecting rod is attached.

Fig. 2-10 The crankshaft converts the up and down movement of the piston to rotary motion.

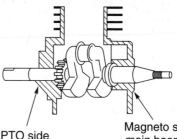

PTO side
main bearing

Magneto side
main bearing

Fig. 2-12 The main bearings support the crankshaft.

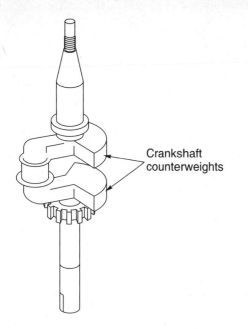

Crankshaft
counterweights

Fig. 2-11 The counter weights on the crankshaft reduce the load on the bearings by balancing the reciprocating movement of the piston.

MAIN BEARINGS

The function of the main bearings is to support the crankshaft and to transmit oil to and from the main bearings. In the single cylinder engine there are two main bearings: one on the magneto side of the crankcase and one on the PTO (power take off) side, as shown in figure 2-12.

The bearing material must allow the steel crankshaft to rotate freely with very little friction.

Friction types

Friction can be classified as solid friction and fluid friction.

SOLID FRICTION Solid friction can be divided into sliding friction and rolling friction. Sliding friction is caused by the moving of one solid body across the surface of another, like the friction produced when the piston ring moves up and down in the cylinder of the engine or the crankshaft revolving on the crankcase bearing surfaces. Rolling friction is caused when two surfaces are rolled on or

against each other, such as the type of friction produced by a roller or ball bearing.

FLUID FRICTION Fluid friction is the resistance of a liquid to flowing. An example of fluid friction is the slow flow of a thick engine oil, as compared to a light oil, due to the difference in fluid friction between the molecules. Fluid friction produces less wear and heat than sliding friction and accounts for the reason lubricating fluids are used to separate the contacting surfaces of the internal engine components.

Bearing classification

The material of the bearing surface is determined by the cost of the engine and the demand or stresses on the crankshaft. Bearings are classified in two major groups called the:

1. *friction, or plain, bearings*

2. *antifriction bearings*

FRICTION BEARINGS The friction-type of bearings are those that have sliding contact between their surfaces. A sliding friction will develop if the rubbing surfaces are not lubricated. If the bearing is supplied with enough oil, the oil will form a layer between the bearing and the rotating part so that the metal surfaces are prevented from touching. The oil molecules will slide on each other, causing a sliding friction that reduces the horsepower output. The bearing surface can be composed of aluminum alloy or bronze.

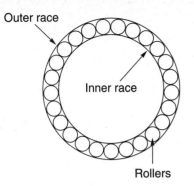

Fig. 2-13 Antifriction bearing, such as a roller bearing, uses rolling friction instead of sliding friction.

Fig. 2-14 DU bearing on left and ball bearing on right.

The most economical is the aluminum bearing surface, which is manufactured simply by drilling, reaming, and polishing a hole in the aluminum crankcase to the proper size. A bronze bearing can be used in an engine to prolong the bearing surface wear, but it will add to the total cost of the engine.

ANTIFRICTION BEARINGS Antifriction bearings, such as a ball or roller bearing, can substitute a rolling friction for a sliding friction, as shown in figure 2-13. The rolling friction reduces the amount of resistance, which increases the power output of the engine and decreases the amount of lubrication necessary. In a two-cycle engine where the lubricating oil is dissolved in the air-fuel mixture, the amount of lubricant deposited on the main bearing and the connecting rod bearing is reduced, compared to a splash or pressurized lubricating system, and the use of roller or ball bearings is crucial for extended engine life.

The load carrying aspect of the antifriction bearing is greater than the plain bearing, and it is used commonly on the PTO side of any engine where a pulley may be attached to drive machinery.

Some engines employ "long-life" DU type bearings that are self-lubricating and combine the best of two bearing materials into one, as shown in figure 2-14. Bronze and a second material, called Polytetrafluoroethylene lead (PTFE), which is much like TEFLON (lowest coefficient of friction of any known solid), are combined with a steel backing to form the DU type bearing. The DU bearing has the advantage of being self-lubricated. Such an advan-

tage is not found with a ball bearing or an aluminum plain bearing.

During initial engine operation, the crankshaft journal absorbs a microscopic film of the PTFE lead from the DU bearing. As this initial film is depleted, the rotating crankshaft journal coming in contact with the bronze interstructure creates heat. Because this heat is localized and of a non-damaging nature, the PTFE lead expands to the surface, preventing a critical breakdown of the bearing surface. These features provide an overall advantage of giving the DU bearing a greater resistance level against wear caused by grit and against seizure (scoring) caused by a low oil level.

CRANKCASE

The crankcase is the housing which encloses the various mechanisms surrounding the crankshaft, as shown in figure 2-15. The cylinder and external accessories are attached to it. This housing must be oil tight and provided with a breather system to relieve the rapidly changing pressures within, as shown in figure 2-16.

The name used to denote the other part of the crankcase, used to completely seal it, can vary depending on the application. A chamber, called a **sump**, is provided on the lower side of the **vertical** crankshaft crankcase to collect or store the oil supply. A **cover** is provided on the side of the **horizontal** crankshaft crankcase to allow access to the inner parts. A **base** is provided on a **horizontal cast-iron** engine.

Fig. 2-15 The crankcase is the housing that surrounds the crankshaft and holds the lubricating oil and other internal parts.

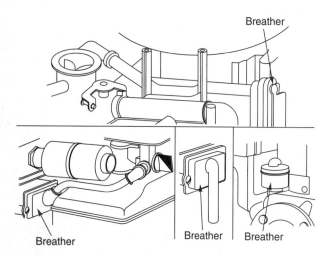

Fig. 2-16 The crankcase breather is necessary to relieve the crankcase pressures during combustion.

VALVE MECHANISMS

Valves in the four-stroke-cycle engine control the intake and exhaust of the gases into and out of the combustion chamber and also seal the chamber during the combustion stroke. The valves are opened by an upward push that is transmitted through the camshaft lobes to the valve lifters and then to the valve stem. The valves are closed by the pressure of the valve springs, as shown in figure 2-17.

Valves

The intake valves are constructed of a chromium-nickle alloy steel, while the exhaust valves, which operate in higher temperatures, are made of special heat-resisting alloy steels like Stellite, which is a metal alloy composed of chrome, cobalt, and tungsten that is used for some exhaust valves because it is hard and has a high melting point. The valve head has a face on it that is ground to either a 45° or a 30° angle, as shown in figure 2-18.

Valve seats

Valve seats are inserted into the cylinder block. The valve seats must be harder than the metal of the cylinder block because of the heavy pounding and wear incurred. The valve seats are ground to an angle that mates the valve face so that a perfect seal is created when the valve contacts the seat.

Valve cooling

The valves are cooled by the transfer of heat through the valve face to the valve seat every time they are closed. Some valves are furthered cooled by rotation. A valve rotator is added so that the valve rotates slightly every time it opens. The heat will be

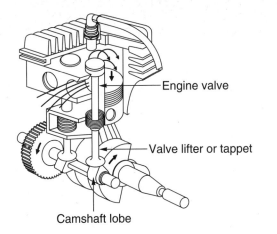

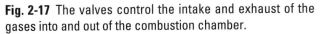

Fig. 2-17 The valves control the intake and exhaust of the gases into and out of the combustion chamber.

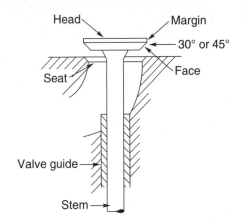

Fig. 2-18 The valve head has a face on it that is ground to either a 45° or 30° angle.

Fig. 2-19 A flywheel is a weight attached to the crankshaft to maintain the inertia of its spinning.

distributed all around the valve, causing a more even cooling of the valve face and heating of the valve seat. Besides better cooling, the rotating valve wears more evenly because carbon deposits are less likely to accumulate and hinder the valve from sealing completely.

Valve lifters

Valve lifters, sometimes know as tappets, are driven up by projections, or lobes, on the camshaft. As the camshaft rotates, a lobe contacts the lifter and moves it upward against the valve stem.

ENGINE COMPONENTS (EXTERNAL)

Many external components are added to the assembled crankcase to perform a variety of functions.

IGNITION PARTS

Ignition parts that are necessary to generate electrical energy, from the mechanical energy produced by the engine, to the spark plug at the precise moment for proper combustion. Some of the compo-

nents include the flywheel, coil or armature, and the mechanical or solid state breaker points.

Flywheel

A flywheel is a weight or mass attached to the crankshaft to maintain the inertia of its spinning during the three non-power producing strokes in a four-stroke-cycle engine, as shown in figure 2-19. Since there is only one power stroke for every two crankshaft revolutions, the flywheel inertia also will help to smooth the abrupt crankshaft movement during the power stroke.

The flywheel in a small air-cooled engine can be made of either cast-iron and attached to the magneto side of the crankshaft or can be the combination of an aluminum flywheel and the steel blade attached to the PTO side of the crankshaft.

Coil or armature

The ignition coil is an electrical device used to increase the low voltage induced in the primary circuit to a high voltage in the secondary or spark plug

Fig. 2-20 A coil or armature.

wire that is high enough to jump a spark plug and fire an air-fuel mixture. The coil consists of many windings in both the secondary and primary circuits. The wires are wound around an iron core that helps concentrate the magnetic field of the magnets as they spin past the core, or legs, as shown in figure 2-20.

Points and condenser

The points are a pair of electrical contacts used within the ignition system to interrupt the current flow in the primary ignition circuit. They consist of

a movable point arm and a stationary base. One of the contact points remains stationary, while the other moves. The opening and closing is controlled by the lobes on the crankshaft or camshaft.

The condenser is a electrical capacitor that is used to store electrical energy and is primarily used in the primary ignition circuit to provide a clean cutoff of current when the points open, as shown in figure 2-21. The condenser absorbs the surges of high voltage and minimizes arcing across the points, which greatly increases their life.

Solid state

Electronic components have replaced the action of the points and condenser in the solid state ignition systems. The solid state components are sealed to keep any dust and moisture from reducing their life, as shown in figure 2-22.

CARBURETOR PARTS

The liquid gasoline must be changed to a vapor and mixed in the proper proportions with the air in order for combustion to occur. The carburetor directs the incoming air through it and allocates the correct amount of fuel to the air flow, as shown in figure 2-23.

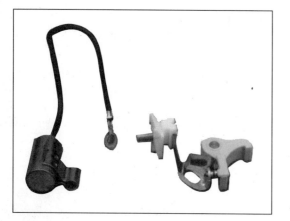

Fig. 2-21 Points and condenser.

Fig. 2-22 Solid state ignition.

STARTER PARTS

The starter is used to start the engine components in motion so that the combustion process begins. The starter can be an electrical one, as shown in figure 2-24, that operates on either a 12-volt battery source of electricity or a 120-volt household plug source. Other types of starters are the manual type, where the operator pulls on a rope to start the motion of the engine, as shown in figure 2-25.

ENGINE SHROUD AND SHEET METAL PARTS

The metal engine shroud and sheet metal are necessary parts of the air-cooling system, as shown in

Fig. 2-24 Electric starter.

Fig. 2-23 Carburetor parts.

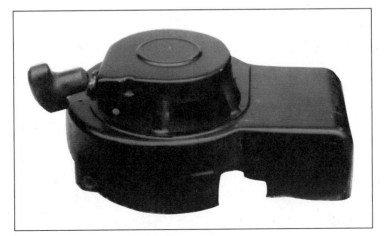

Fig. 2-25 Manual rope starter.

Fig. 2-26 Engine shroud and sheet metal.

figure 2-26. The air that is impelled by the flywheel must be directed to the warm areas of the engine for long life operation. The passageways must be free from any obstructions.

ENGINE CLASSIFICATION

Engines are classified by the method in which heat is generated from the combustion of fuel in developing mechanical power. There are two such categories:

1. *External combustion engine*
2. *Internal combustion engine*

EXTERNAL COMBUSTION ENGINE

An example of an external combustion engine is the steam engine, in which burning of the fuel, in an externally-fired boiler, generates steam, which is transmitted to the engine cylinder where it is expanded against the piston, as shown in figure 2-27. The piston then transmits power to the crankshaft. Characteristically, this type of engine is quite heavy per horsepower produced and is relatively inefficient.

INTERNAL COMBUSTION ENGINE

In an internal combustion engine, the air-fuel mixture is burned and expanded within the combustion chamber, which is that portion of the cylinder above the piston at top dead center (TDC), as shown in figure 2-28 (page 56). The combustion pressure is applied directly to the head of the piston. The power of the engine is captured by allowing the burning and expanding gases enclosed in the cylinder to function directly against a reciprocating piston.

Five events are necessary for proper combustion in an internal combustion engine, and these are:

1. *Power event*
2. *Exhaust event*
3. *Intake event*
4. *Compression event*
5. *Ignition event*

Power event

The power part of the combustion process involves the burning of the air-fuel, creating a release of energy that heats the air molecules in the combustion chamber to approximately 2000° to 3000° F. The intense heat vaporizes any moisture and rapidly expands the air to create a high pressure that drives the piston downward. The stored chemical energy in the fuel is converted to heat and mechanical energy during the power part of combustion.

Exhaust event

The exhaust part of the combustion process relates to the discharge of the products of the burned fuel from the combustion chamber along with a great deal of unused heat energy.

Intake event

The intake part of the combustion process is where a relatively cool air mass is drawn into the combustion chamber as either a combination of vaporized fuel and air, or just air.

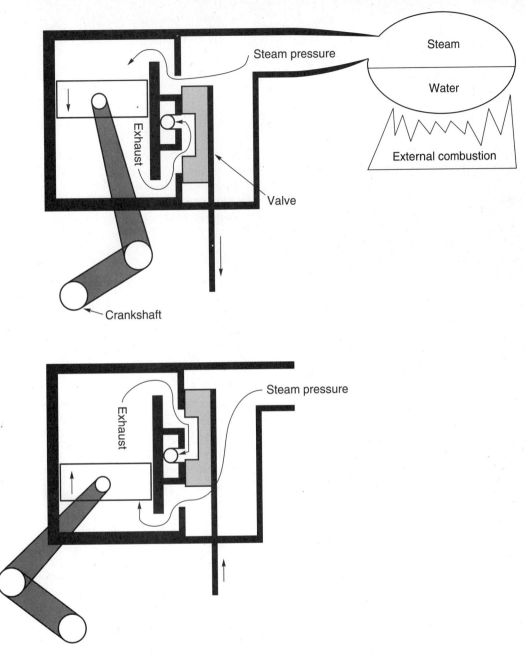

Fig. 2-27 A steam engine is an example of an external combustion engine.

Compression event

The compression part of the combustion process is used to compact the intake gases so that the molecules are closer to each other, and also increase the temperature of the charge so that the chemical reaction (combustion) will take place quicker. A chemical reaction, such as the burning of the air-fuel mixture, will be more efficient and occur quicker when compressed.

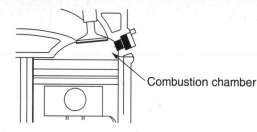

Fig. 2-28 The combustion chamber is where the air-fuel mixture is burned in the internal combustion engine.

Ignition event

The ignition part of the combustion process deals with the initiation of the chemical reaction. In the case of a spark ignition engine, the air-fuel mixture has been compressed and mixed in the proper ratio so that an intense spark from the spark plug ignites the air-fuel mixture and starts the reaction.

In a diesel engine where only air is brought in and compressed, the combustion is started by the injection of liquid fuel into highly compressed air in the combustion chamber. The heat caused by the compression vaporizes the fuel instantly and the high temperature ignites it.

INTERNAL COMBUSTION ENGINE TYPES

Reciprocating internal combustion engines are divided into two general classes according to the number of piston strokes required to produce one power stroke. The two-stroke-cycle engine has the ignition of the fuel charge on every up-stroke of the piston, whereas in the four-stroke-cycle engine, the fuel charge is ignited every other up-stroke.

FOUR-STROKE-CYCLE ENGINE

In the four-stroke-cycle engine, the five parts of combustion occur for every four strokes (two up-strokes and two down-strokes) of the piston. Two revolutions of the crankshaft are required to accomplish the five parts of combustion so that every other time the piston approaches the top of the up-stroke, combustion and power occur. The four strokes are:

Power stroke (down-stroke)

Exhaust stroke (up-stroke)

Intake stroke (down-stroke)

Compression stroke (up-stroke)

The four strokes are a continuous cycle that repeat over and over. Traditionally, the order of the strokes has been presented as intake, compression, power, and exhaust. Since the progression is circular rather than linear, it makes little difference which stroke is first, but it is easier to explain the valve coordination when the power stroke is explained initially.

Power stroke (down-stroke)

The power stroke of the engine is when the piston moves from the top to the bottom of the cylinder with both valves closed, as shown in figure 2-29.

OPERATION—POWER STROKE As the piston moves slightly past TDC, the pressure of the burning gases exerts a force of several tons on the piston head and propels it downward in the cylinder on the power stroke. The temperature of these gases is between 2000° and 3000° F. This temperature, if prolonged, would be harmful to the internal components and, therefore, it must be lowered immediately after being produced. If the gases expand normally in the cylinder on the power stroke, they will cool considerably when the hot gases are exposed to the cylinder walls as the piston descends, and the temperature will be within safe limits when they are discharged through the exhaust system.

If the combustion is normal, the force will be applied in the form of a rapid "push." The piston is driven downward in the cylinder, and the force is converted to rotary motion by the crankshaft.

The force pushing down on the piston head created during the power stroke supplies the energy to drive the piston in the exhaust, intake, and compression strokes. The power stroke rotates the crankshaft, whose momentum is maintained by the weight of the flywheel, lawnmower blade, and balancing weight on the crankshaft. This provides for a smooth, continuous turning of the crankshaft and movement of the piston between power strokes.

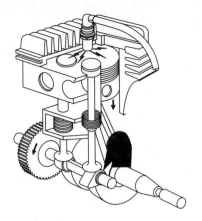

Fig. 2-29 During the <u>power stroke</u>, the burning gases exert a downward force on the piston.

VALVE TIMING—POWER STROKE The exhaust valve actually opens from **50° to 75° before the bottom** of the power stroke so that the pressures of the power stroke can be used to start the flow of gases through the exhaust system.

Exhaust stroke (up-stroke)

The exhaust stroke of the engine is when the piston moves from the bottom to the top of the cylinder with the exhaust valve open and the intake valve closed, as shown in figure 2-30.

OPERATION-EXHAUST STROKE The exhaust gases are completely removed from the combustion chamber as the piston approaches the top of its stroke while the exhaust valve is open.

VALVE TIMING—EXHAUST STROKE The intake valve is opened anywhere from **8° to 30° before the top of the exhaust stroke** to assist the intake of unburned air-fuel mixture of gases. The velocity of the escaping gases during the exhaust stroke creates a reduced pressure in the cylinder at the end of the exhaust stroke so the intake valve is opened before the exhaust valve closes to start the flow of fresh mixture into the cylinder.

Intake stroke (up-stroke)

The intake stroke of the engine is when the piston moves down from the top to the bottom of the cylinder with the intake valve open and the exhaust valve closed, as shown in figure 2-31.

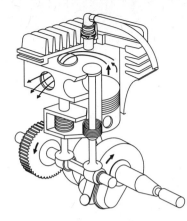

Fig. 2-30 During the <u>exhaust stroke</u>, the exhaust gases are removed from the combustion chamber.

OPERATION-INTAKE STROKE During the intake stroke, the piston is pulled downward in the cylinder by the rotation of the crankshaft and the flywheel. This establishes a lower pressure in the combustion chamber than the air surrounding the carburetor, which is at atmospheric pressure. The air is pushed through the carburetor to the reduced pressure area (commonly referred to as air being "sucked" into the engine). As the air passes through the carburetor, a suitable amount of fuel is added to the air flow. The amount of air-fuel mixture entering will depend on the throttle and choke plate position.

VALVE TIMING-INTAKE STROKE The intake valve remains open for the complete stroke. The exhaust valve is closed and must not leak for maximum intake of air through the carburetor.

Compression stroke (up-stroke)

The compression stroke of the engine is when the piston moves from the bottom to the top of the cylinder with the intake and exhaust valve closed, as shown in figure 2-32.

OPERATION-COMPRESSION STROKE The upward movement of the piston in the cylinder during the compression stroke, with both the intake and the exhaust valves closed, compresses the air-fuel charge to obtain the desired burning and expansion characteristics.

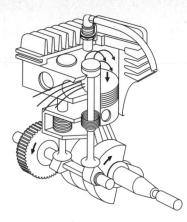

Fig. 2-31 During the intake stroke, the unburned air-fuel mixture enters the combustion chamber.

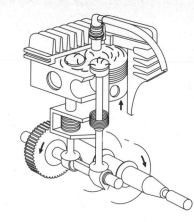

Fig. 2-32 During the compression stroke, the air-fuel mixture is compressed.

As the piston approaches the top of the compression stroke, the air-fuel charge is set afire by means of an electric ignition system in order to insure complete burning by the time the piston is slightly past TDC of the power stroke. The time of ignition occurs before the piston is at the top of the travel on the compression stroke and can vary depending on the requirements of the engine. This timing specification can be stated in degrees of crankshaft movement before the TDC of the compression stroke as in 6° BTDC (**B**efore **T**op **D**ead **C**enter) or in the measurement of piston travel as in 0.122-inch BTDC, as shown in figure 2-33. Since the charge in the cylinder is compressed, the resistance between the spark plug's electrodes is increased, and the voltage provided must be sufficient to jump the high resistance gap.

VALVE TIMING-COMPRESSION STROKE The exhaust valve remains closed while the camshaft is designed to close the intake valve during the compression stroke. The intake valve closes as the piston is moving upward on the compression stroke to allow as much fresh fuel charge into the cylinder as possible.

COMPRESSION RELEASE MECHANISM Many air-cooled four-stroke-cycle engines are equipped with a compression release mechanism to reduce cylinder pressure during the starting process. This is accomplished by slightly lifting the exhaust or intake valve during the compression stroke to release some of the pressure. The decompression

mechanism can be a design that will allow full compression at faster speeds, as shown in figure 2-34, or can be a permanent part of the camshaft lobe. Engines that are equipped with a permanent compression release lobe will lift the exhaust or intake valve during the faster speeds, but the time interval the valve is opened is so short that it has little effect on the total compression.

FOUR-STROKE-CYCLE ENGINE VARIATIONS

For many years the "L" head engine design, with its valves in the block, has been the choice of most of the lawn and garden engine manufacturers. However, the overhead valve (OHV) engine is gaining in popularity because of its compact, high-efficiency properties that offer longer life, its reduced oil consumption, and less frequent maintenance needs.

"L" head valve arrangement

In the "L" head engine, both the intake and exhaust valve are located on the same side of the cylinder, as shown in figure 2-35. The name is derived from the shape of the combustion area, which is shaped like an inverted letter "L" when the piston is at bottom dead center. The valve operating mechanism is located in the crankcase under the valves and a single camshaft operates all the valves. This arrangement makes the valve train lubrication relatively simple.

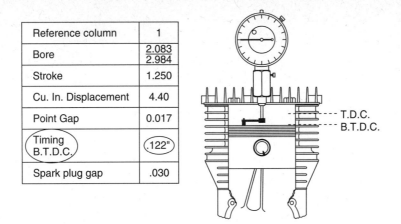

Reference column	1
Bore	2.083 / 2.984
Stroke	1.250
Cu. In. Displacement	4.40
Point Gap	0.017
Timing B.T.D.C.	.122"
Spark plug gap	.030

Fig. 2-33 The spark should occur when the piston is 0.122 inches before top dead center on this engine.

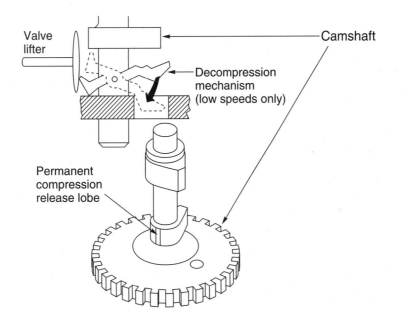

Fig. 2-34 The decompression mechanism can be a permanent part of the camshaft lobe or designed so decompression is bypassed at higher speeds

In this engine, the spark plug is located above the valves in a pocket into which the valves move. The pocket serves as the top of the combustion chamber, where the air-fuel mixture is compressed as the piston reaches the end of the compression stroke.

Overhead valve arrangement

In the overhead valve engine (OHV), both the intake and exhaust valves are mounted in an inverted position in the cylinder head directly above the cylinder, as shown in figure 2-36. This arrangement yields a greater volumetric efficiency and a higher compression ratio.

The overhead valve design creates a high efficiency compression ratio of 8.5:1 compared to a 6.0:1 in the "L" head engine. This increase in compression over the side-valve, "L" head engine is achieved by positioning the valves above, rather than to one side, of the cylinder bore and requires a slightly higher minimum octane fuel (85) compared to a minimum of 77 for the "L" head engine. Both of these fuel requirements are low enough so that unleaded regular gasoline (usually rated with an octane number of 87) can be used. The higher compression ratio permits a more complete burning of the fuel with less deposits remaining. It provides longer valve life and improved cooling for the valve train.

The more complete and even-burning fuel combustion in the overhead valve engine contributes to better fuel economy and "cleaner" exhaust emissions.

The position of the valves allows uniform thickness of material around the cylinder bore, improving cooling. The increased cooling eliminates the heat-induced-bore distortion, which is the major cause of oil consumption, and provides a greater tolerance for multiviscosity oils.

The valve mechanism requires continuous lubrication, and an oil pump and oil filter are incorporated in the lubrication system which allows longer time between oil changes.

Air-cooled vs liquid-cooled engine

Liquid-cooled engines better regulate the cylinder heat, which converts to potential longer engine life. The combustion chamber will operate efficiently at a certain range in temperature. If the temperature is excessive, the cylinder components may warp and the combustion pressures are lost. If the temperature is too low, the gaseous fuel charge may condense on the cylinder walls and wash the lubrication away, causing premature wear. Liquid-cooled engines surround the combustion area with a "blanket" of cooling fluid which absorb the combustion noises and quiet the engine operating noise.

Even though air-cooling is not as efficient in removing the heat as the liquid-cooling, the air-cooled engine is primarily used on many small applications because the disadvantages of the cooling system are balanced by other factors. The air-cooled engine is more economical to produce and has a lighter weight-to-power ratio. A liquid-cooling system would add many components to the engine that would reduce the accessibility of many parts and increase maintenance costs.

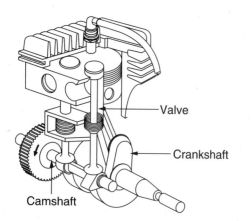

Fig. 2-35 In "L"-head engine, the intake and exhaust valve are located on the same side next to the cylinder block.

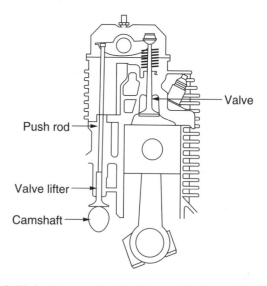

Fig. 2-36 In the overhead valve engine (OHV), both the valves are mounted in the cylinder head.

Engine noise standards

Engine operating noise is a concern of many engine manufacturers, and various designs are used to reduce it. The features are commonly found on the premium commercial engines, but shortly they the will be incorporated in the smaller models. Larger mufflers are used that quiet the engine exhaust noise, but they also make other engine noises more prevalent. Valve lash gaps are reduced, and helical-cut camshaft gears are used to reduce noise levels.

Laminated engine housings and plastic components are used to absorb the engine noises. Cam-ground, lubricant-impregnated pistons with very low clearances between the cylinder and piston are used to reduce any piston slap.

Hydraulic valve lifters are used to automatically eliminate all lash in the valve train, thus eliminating the noise. The lifter stays in constant contact with the valve stem or push rod regardless of the thermal expansion or component wear.

TWO-STROKE-CYCLE ENGINE

In the two-stroke-cycle engine, the five parts of combustion occur for every two strokes (one up-stroke and one down-stroke) of the piston. One revolution of the crankshaft is required to accomplish the five parts of combustion so that every time the piston approaches the top of the up-stroke, combustion and power occur.

Two-stroke vs four-stroke engines

The two-stroke-cycle engine is best suited for chain saws and other products for a number of important reasons:

	4-stroke-cycle-engine	2-stroke-cycle-engine
1. Number of major moving parts.	Nine major parts to wear, maintain, and replace.	Three major parts to wear, maintain, and replace.
2. Lubrication design.	Splash or pump. The same oil is used over with- out filtering.	Spray. The oil is used one time only. New oil is induced for each revolution of the crankshaft.
3. Number of complete revolutions necessary to produce one power stroke.	Two revolutions or four strokes of the piston.	One revolution or two strokes of the piston, which pro- vides a faster acceleration.
4. Versatility of operation.	Limited slope operation because the bearings receive the less lubrication when engine is tilted at an angle.	Lubrication is not affected by any operating angle.
5. Weight and size of engine for equal dis- placement and power.	Heavy and bulkier in relation to power output.	Relatively light and small in relation to power output.
6. Ease of starting.	Two complete revolutions of the crankshaft are required to pro- duce one ignition event.	Less effort because one complete revolution on the crankshaft produces one ignition event.
7. Engine life.	Since there is a power stroke every other up- stroke of the piston, the com- bustion heat is dissipated better, which permits a longer life.	Since there is a power stroke every upstroke of the piston, the engine operates at a higher tempera- ture, which will reduce the longevity of the engine.

The two-stroke-cycle engine does not produce twice as much power as the four-stroke-cycle engine of the same size even though the two-stroke-cycle engine has twice as many power strokes with both engines running at the same speed. In the two-stroke-cycle engine, when the transfer and exhaust ports are opened by the piston, there is always some mixing of the fresh air-fuel mixture and the burned exhaust gases. This reduces the amount of fresh air-fuel mixture that can enter. Also, only part of the piston stroke is devoted to getting air-fuel mixture into the cylinder. This further reduces the amount of air-fuel mixture that can enter and the amount of power developed.

In the four-stroke-cycle engine, nearly all of the burned gases leave the cylinder during the exhaust stroke. More air-fuel mixture can enter because a complete piston stroke is devoted to getting air-fuel mixture into the cylinder. The result is a more powerful piston stroke.

Power/exhaust stroke (down-stroke)

The power/exhaust stroke of the two-stroke-cycle engine uses both the combustion chamber on top of the piston as well as the crankcase located below the piston, as shown in figure 2-37.

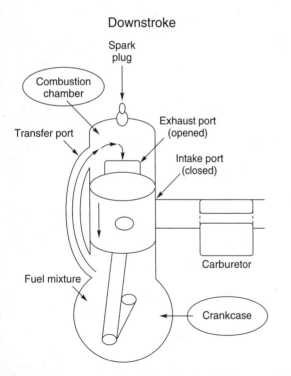

Downstroke

Fig. 2-37 The power/exhaust <u>downstroke</u> of the two-stroke-cycle engine uses both the combustion chamber and the crankcase.

BELOW PISTON (DOWN-STROKE) The crankcase volume is reduced as the piston descends in the cylinder, and the gasoline-air-oil mixture is compressed. The mixture is prevented from exiting through the carburetor by a valve that is closed, and it cannot enter the combustion chamber until the intake port is uncovered by the piston.

ABOVE PISTON (DOWN-STROKE) The ignited air-fuel mixture creates a pressure on the piston that moves it downward in the cylinder. When the piston moves downward, the exhaust port is uncovered and the pressure of the combustion chamber escapes, relieving the pressure on the piston. The piston moves down further and the intake ports are uncovered, allowing the pressurized fresh air-fuel charge to enter the cylinder. The cylinder is designed so that the incoming fresh charge pressure pushes all of the exhaust gases from the combustion chamber and introduces a new air-fuel mixture.

Compression/intake stroke (up-stroke)

The compression/intake stroke of the two-stroke-cycle engine uses both the combustion chamber on top of the piston as well as the crankcase located below the piston, as shown in figure 2-38.

BELOW PISTON (UP-STROKE) As the piston rises in the cylinder, the crankcase volume increases, which causes a low pressure area (compared to the atmosphere and combustion chamber) that allows air to be pushed into the carburetor through the air cleaner. The fuel is mixed with the air in the carburetor, and oil is injected into the air stream or is already mixed with the fuel. The intake mixture of air-fuel-oil is then pushed into the crankcase until a valve closes.

ABOVE PISTON (UP-STROKE) As the piston rises in the cylinder, the volume of the combustion chamber is reduced so that the pressure of the air-fuel gaseous mixture is increased. The pressure is contained in the combustion chamber by the sealing of the piston rings, the intake, and exhaust ports. Just before the piston reaches the top of its travel, the air-fuel mixture is ignited by a spark so that the

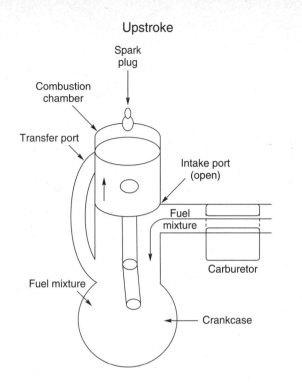

Upstroke

Spark plug

Combustion chamber

Transfer port

Intake port (open)

Fuel mixture

Carburetor

Fuel mixture

Crankcase

Fig. 2-38 The compression/intake <u>upstroke</u> of the two-stroke-cycle engine.

maximum pressure (which takes time to develop) will happen on the down-stroke of the piston.

TWO-STROKE-CYCLE ENGINE VARIATIONS

The design variations do not change the basic theory, but only how the theory is applied. The different two-stroke-cycle designs used in the small air-cooled engine are:

1. *Cross scavenged with reed valve type*

2. *Loop scavenged with reed valve type*

3. *Loop scavenged with third port*

Cross scavenged with reed valve type

The cross scavenged two-stroke-cycle with reed valve is a design used in many outboard engine applications. The design is characterized by a **piston**

head that has a **deflector** on it and a **one-way reed valve** that allows the intake of air-fuel mixture from the carburetor into the crankcase but prevents the vapor from escaping back through the valve, as shown in figure 2-39.

OPERATION As the piston descends in the cylinder on the power stroke, the crankcase mixture of air and fuel is compressed until the piston uncovers the intake port. When the intake port is uncovered, the crankcase charge rushes into the combustion chamber and pushes out the exhaust gases. The piston is designed with a deflector to swirl the intake mixture upward to better scavenge or remove the exhaust gases. The swirling also prevents the fresh intake charge from directly exiting the exhaust port before it is burned.

The reed valve controls the intake and retention of air-fuel in the crankcase. As the piston moves upward in the cylinder, the reduction in pressure in the crankcase allows air and fuel to enter from the carburetor by opening the reed valve. As the piston descends in the cylinder and the crankcase volume decreases, the reed valve closes so the crankcase pressures cannot exit through the carburetor.

Loop scavenged with reed valve type

The difference between the loop scavenged two-stroke-cycle engine and the cross scavenged is the design of the piston head. The loop scavenged **piston is flat** because the **intake ports are located directly across from each** other and 90° from the exhaust port, as shown in figure 2-40. The intake charge enters the combustion chamber from each side and the air masses collide and loop upward. This resulting turbulence cleans the combustion chamber of all exhaust gases. This type of design allows better intake efficiency because of two intake ports rather than one, which results in more horsepower for the size.

Loop scavenged with third port

The loop scavenged two-stroke-cycle with **a piston controlled third port** can eliminate the need for a reed valve and allows a more compact design that

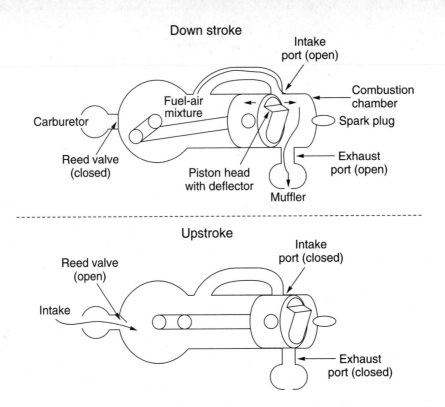

Down stroke

Intake
port (open)

Combustion
chamber

Fuel-air
mixture

Carburetor

Spark plug

Reed valve
(closed)

Exhaust
port (open)

Piston head
with deflector

Muffler

Upstroke

Intake
port (closed)

Reed valve
(open)

Intake

Exhaust
port (closed)

Fig. 2-39 The cross-scavenged two-stroke-cycle engine has a piston head with a deflector on it.

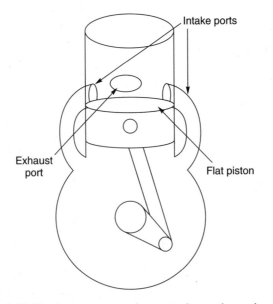

Intake ports

Exhaust
port

Flat piston

Fig. 2-40 The loop-scavenged two-stroke-cycle engine has a flat piston head and two intake ports directly across from each other.

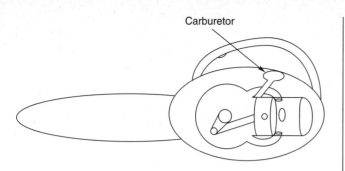

Fig. 2-41 The loop scavenged two-stroke-cycle engine with a piston controlled third port used on chainsaw.

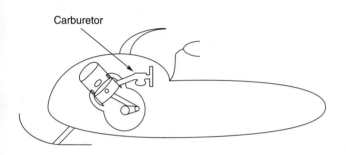

Fig. 2-42 The loop scavenged two-stroke-cycle engine with a piston controlled third port used on snowmobile.

is used to power many chainsaw, snowblower, and snowmobile applications. The piston controls the intake of fuel and air into the crankcase and permits the carburetor to be mounted on the side of the crankcase, as shown in figure 2-41 and 2-42.

CLEAN BURNING TWO-STROKE-CYCLE ENGINE

The Orbital Engine Company of Perth, Australia has licensed some companies in the United States to experiment and produce a clean-burning, fuel efficient two-stroke engine. On a conventional two-stroke-cycle engine, some fuel will escape through the exhaust, creating pollution and reducing fuel efficiency.

The orbital engine's innovation is the addition of a compact pneumatic fuel injector that shoots a supply of compressed air and fuel into the combustion chamber at the top of the compression stroke. Since the intake stroke does not contain any gasoline, the exhaust gases are removed with very little loss of

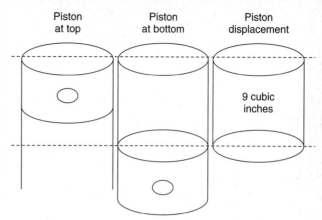

Fig. 2-43 Piston displacement is the volume of space through which the piston travels.

unburned hydrocarbons. The application of this type of engine can be beneficial on outboard engines and lawn and garden equipment when the federal emission regulation become more strict.

ENGINE POWER FACTORS

The power an engine develops is affected by factors such as:

1. *piston displacement*
2. *compression ratio*
3. *amount of air-fuel intake*
4. *valve timing*
5. *degree of compression*
6. *ignition timing*
7. *type of fuel*
8. *engine friction*
9. *temperature and humidity*

PISTON DISPLACEMENT

The volume of space through which the piston travels from the very top of its stroke to the bottom limit is called the piston displacement, as shown in figure 2-43, which can be determined by multiplying the stroke by the area of the circle made by the cylinder wall.

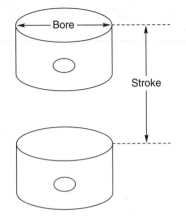

Fig. 2-44 The stroke is the distance the piston moves from top to the bottom of the cylinder. The bore is diameter of the cylinder circle.

The stroke is the distance the piston is propelled from the top to the bottom of its movement in the cylinder and can be measured in inches or millimeters, as shown in figure 2-44.

The **area of a circle** = $\frac{1}{4}\pi D^2$; where "D" represents the bore or diameter of the cylinder and π = **3.142**. Thus the piston displacement can be calculated as follows:

Piston displacement = stroke x $\frac{1}{4}\pi$ bore² x number of cylinders.

For a typical single cylinder 3.5 HP Briggs & Stratton engine with a bore of 2.562 inches and a stroke of 1.750 inches, the numbers can be inserted into the formula to calculate the piston displacement:

Piston displacement = 1.750"
$$* \frac{(3.142 * 2.562"^2)}{4} * 1 \text{ cyl} = \underline{9.025 \text{ cubic inches}}$$

For a dual cylinder 18 HP Briggs & Stratton engine with a bore of each cylinder 3.438 inches and a stroke of 2.281 inches, the piston displacement is 42.35 cubic inches:

Piston displacement = 2.281"
$$* \frac{(3.142 * 3.438"^2)}{4} * 2 \text{ cyl} = \underline{42.35 \text{ cubic inches}}$$

COMPRESSION RATIO

The compression ratio is a comparison of two quantities and is stated as a numerical relationship.

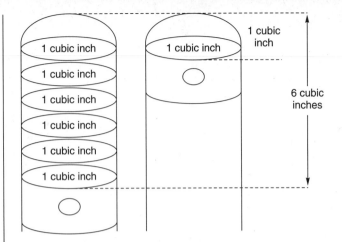

Fig. 2-45 An engine with a compression ratio of 6:1.

A compression ratio is based on the engine combustion chamber volume while the piston is at the bottom of its travel compared with the volume when the piston is at the top of its travel. If the total displacement of the combustion chamber, while the piston is at the lowest point in the cylinder is six cubic inches and after the piston moves to the top in the cylinder, the total volume of the combustion chamber is now one cubic inch, the compression ratio would be described as six to one (6:1), as shown in figure 2-45.

The higher the compression ratio, the faster the rate of combustion because the gaseous molecules are packed more closely together, causing the chain reaction to occur more swiftly.

AMOUNT OF AIR-FUEL INTAKE

A number of factors determine the amount of air-fuel mixture that enters the cylinder. These include:

1. *atmospheric pressure*

2. *throttle plate opening*

3. *carburetor venturi size*

4. *engine speed*

5. *exhaust gas flow*

The theoretical total volume of the intake charge can be calculated from the piston displacement. If time were not a factor, the volume of air entering the cylinder would exactly match the total volume. Since this perfect condition can never be achieved, the term volumetric efficiency is used to define the ratio by comparing the theoretical amount of fuel charge to the actual fuel charge received in the combustion chamber. This means that a Briggs & Stratton 3.5 horsepower with a cylinder displacement of about 9 cubic inches should expect the delivery of that same volume of air-fuel mixture to the combustion chamber during the intake stroke. In actuality, this can never be achieved. The air-fuel intake is always less than its capacity.

Atmospheric pressure

Air is brought into the cylinder only when the atmospheric pressure is greater than the pressure inside the cylinder. At sea level, more air can be pushed into the cylinder than at a higher altitude, where the pressure is less.

Throttle plate opening

The throttle plate in the carburetor controls the amount of fuel mixture entering the cylinder. When the engine is idling with no load, the throttle plate is practically closed. This allows only a small amount of air-fuel mixture to enter the cylinder. Opening the throttle plate allows a greater amount of air to enter.

Carburetor venturi size

The carburetor venturi is necessary to create a low pressure area so that fuel is pushed from the fuel reservoir to the rapid flow of air in the air horn of the carburetor. The venturi size can vary depending on the performance requirements of the engine. Some chokeless engines have a small venturi, necessary to create the strong suction required to raise the priming fuel charge for easy starting. This small venturi utilizes the suction at slow speeds, but sacrifices the volume of the intake charge (and horsepower) at higher speeds.

Engine speed

The faster the engine runs, the more difficult it is to maintain a proper air-fuel balance. At a slow engine speed, there is enough time during piston strokes for a full charge of fuel mixture to enter the cylinder. With high speeds, the piston moves up and down the cylinder so rapidly that even with the throttle plate wide open, it is impossible to charge the cylinder fully.

Exhaust flow

Anything that interferes with the flow of exhaust gases out of the engine causes a back pressure in the combustion chamber. This creates a resistance to the piston when it returns upward in the exhaust stroke and reduces the engine power and speed. Some of the exhaust gases mix with the incoming fuel mixture and further reduce the engine's power.

If carbon has been deposited around the exhaust valve, it will interfere with the flow of exhaust gases out of the cylinder. In addition, the carbon deposits may prevent the exhaust valve from closing completely which, in turn, will alter the vacuum created by the intake stroke of the piston and decrease the amount of fuel sucked into the cylinder.

VALVE TIMING

The intake valve must open and close at very precise moments to insure the best possible fuel charge. If the valve timing is off by an incorrect valve lash adjustment, the engine will be deprived of a full air-fuel charge.

There is a period of time when the intake valve is open at the same time the exhaust valve is still open, as shown in figure 2-46. This period is called **valve opening lap** and its purpose is to:

1. *increase the amount of mixture induced into the cylinder by taking advantage of the suction caused by the momentum of the exhaust gases;*

2. *facilitates the more complete discharge of waste gases, forcing them out by the cooler incoming mixture; and*

3. *aid in cooling the cylinder internally by circulating the incoming mixture, which has a comparatively low temperature.*

The piston descending in the cylinder with the exhaust valve closed will create a suction and air will flow through the intake valve port. The restriction, due to the comparatively small intake valve port opening relative to the cylinder volume, will provide considerable velocity to the intake mixture

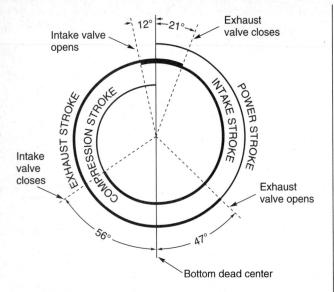

Fig. 2-46 The period of time when the intake and exhaust valves are both open is called the valve opening lap.

at this point. The intake valve will close from 50° to 75° after the bottom center of the compression stroke. A more complete filling of the cylinder and an increased power output is accomplished by taking advantage of the momentum of the intake mixture due to its velocity.

DEGREE OF COMPRESSION

The degree of compression for the intake gases is affected by more than the calculated compression ratio. Other factors are the amount of air-fuel mixture that enters the cylinder during the intake stroke and the loss of pressure from leaks in the valves or between the piston and cylinder.

When the throttle plate is only partially opened, the quantity of fuel and air flow are decreased. This reduction affects the power directly by reducing the fuel charge that can be ignited and, also indirectly, by reducing the degree of compression that can take place with the reduced volume of gases.

Some of the air-fuel mixture can be lost by slipping past the piston and cylinder wall, through a leaking valve or a leaking cylinder head gasket. This leakage lowers the degree of compression which affects the power output of the engine.

IGNITION TIMING

The ignition of the fuel charge must occur at the proper time for full power to be achieved from the engine. The ideal situation is to completely burn the fuel so that the full force of the expanding gases would be forced on the piston when it starts the downward power stroke.

A period of time is necessary for the fuel to begin burning so the ignition occurs before the piston is on the downward power stroke; most commonly, it begins while the piston is still moving upward on the compression stroke.

When an engine operates at a fast speed, the spark must occur earlier than at slower speeds. To maximize the power capabilities of the engine, it is advisable to have the ignition system adjust the timing at different speeds. Many small engines have only one ignition setting that is set (according to the service manual) to a midway position that allows easy starting and also efficiently harnesses the engine's power at higher speeds.

TYPE OF FUEL

The flame should travel across the combustion chamber at a uniform rate, taking about 1/250 of a second to complete the combustion. Fuels are blended with additives to control this burn rate so that "explosions" or quick burns do not put undue stress on the piston and connecting rod and/or create intense cylinder temperatures that can destroy components.

ENGINE FRICTION

The friction between many of the engine's moving parts consume part of the power developed by the engine. Friction is defined as the force which acts between two bodies at their surfaces of contact that resist their sliding upon each other. All engine parts, no matter how polished their surfaces, disclose many uneven, rough surfaces when examined under a microscope. The uneven surfaces are responsible for the resistance to motion, which is called friction. Friction produces heat, causes excessive wear, and reduces the efficiency of the engine.

TEMPERATURE AND HUMIDITY

With higher temperatures, air molecules are farther away from each other so that the intake of oxygen molecules necessary for maximum combustion is reduced. The available engine power will be decreased at *higher temperatures.* Engine power decreases by 1 percent for every 10° F deviation.

The amount of water vapor present in the air (humidity) will also affect the maximum power output. An engine may lose as much as 5 percent of its power under conditions of high humidity. In the air-fuel mixture, a certain quantity of oxygen molecules are present that affect the combustion reaction. If too much water vapor (high humidity) is contained in the air-fuel mixture, the amount of oxygen available is reduced, which may affect the total combustion power output.

ENGINE OUTPUT MEASUREMENTS

Engines are useful in our lives because they do work for us. They provide the forces to move things by the power they produce. The words work, force, and power can be described in a practical sense or in a detailed physical science definition. The common use of the words are primarily emphasized.

WORK, FORCE, TORQUE, AND POWER CONCEPTS

Engines are used to accomplish work. Engine forces that cause the crankshaft to twist are called **torque** and the resulting movement is called **work.** The work was accomplished in a definite period of time, depending on the **power** of the engine. The terms work, force, torque, and power are all terms that interrelate with each other and should be understood.

Force

Force can be described as a push or pull on an object, and if the force is great enough to move the object a certain distance, then we can say that work is done. Force can be measured in Pounds in the English system or Newtons (named after the scientist) in the Metric system.

Sir Isaac Newton (1642–1727) formulated laws which explain the way objects move. One of his laws states: *An object which is moving or at rest does not change its state of motion unless a force acts on it.* It takes a force to start and stop motion. Objects in motion tend to remain in motion, and it takes a force to change the speed or direction of an object in motion.

Work

Work is when a force is applied to an object throughout a distance, such as moving an object against an opposing force. An example would be when a lawnmower is pushed, and the force applied is sufficient to start the motion of the mower in opposition to the friction produced by the wheels and the thick grass. According to Newton's law, once the mower is moving, it will continue to move until another force stops it. You know that as soon as you cease pushing on the mower it stops shortly because there is another force (heavy grass and wheel friction) opposite to your push.

Work is measured in *Foot-Pounds (ft-lbs.)* in the English system and *Meter-Newton (M-N)* in the Metric system. The foot or meter is the distance the object is moved, and the Pounds or Newton is the force or effort applied. Mathematically, work can be defined by the equation: W = F x D, where the W represents work which is the product of the Force (F) times the Distance (D).

Torque

Torque is the turning or twisting force exerted by the crankshaft. The pressures developed by the combustion of the air-fuel mixture are transmitted to the piston and connecting rod to the crankshaft. The amount of torque depends on the pressure applied to the piston and the length of the crankarm. The measurement unit is in ft-lbs. A 10-pound weight hanging at the end of a wrench that is 1-foot long produces a torque, or twisting, on a bolt of 10 pounds at 1-foot length, or 10 ft-lbs.

Torque should not be confused with work. Work is when an object is actually moved through a distance and is measured in ft-lbs, while torque is an instantaneous twisting effort and does not necessarily result in an object being moved through a distance. An example would be when a screw is attempted to be loosened with a screwdriver. The twisting force you apply with the tool is the torque; however, if the twisting effort results in motion, then work is accomplished. Torque generally leads to work being accomplished.

In all engines, torque becomes greater as the engine speed increases, up to a point. After that point is reached, the torque begins to decrease at a rate depending on the engine design. The increase in torque is caused by the wide open throttle plate in the carburetor, which allows a greater amount of fuel charge to enter the cylinder. Torque begins to decrease at a certain point even though the engine accelerates because the speed of the piston movement becomes so fast that fuel is unable to enter the cylinder as freely as at lower engine speeds. As a result, there is less fuel for combustion causing less force to be applied by the piston to the crankshaft, as shown in figure 2-47.

When torque decreases, despite an increase in engine speed, the engine power output will stabilize and then drop. Even though internal engine friction becomes greater as engine speed increases, little effect is realized until the engine torque drops. At this point, the effect of the loss of power from the friction prevents the total power output from rising any further and eventually causes a decline in total power.

At low engine speeds, the engine develops only enough torque to keep it running. The net torque—the torque beyond that which is required to keep the engine turning over—is practically zero. If a load is applied, the engine will stall out.

Most manufacturer's have their optimal operating speed somewhat higher than the speed at which the maximum torque is produced. The reason is that when an engine is suddenly loaded, the operating speed drops immediately until the governor system reacts. By having the operating speed higher than the peak torque speed, a torque reserve is created so that when the engine speed decreases, the torque will increase and adjust quickly to the increased load.

Power

Power is the amount of work that is done in a period of time. If it takes you one hour to mow the

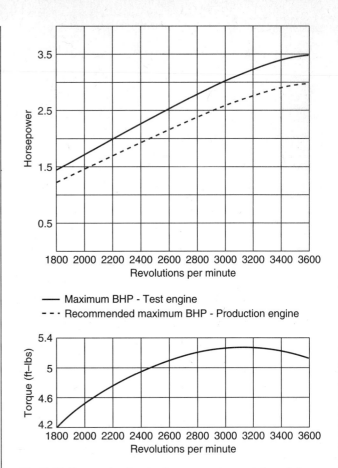

— Maximum BHP - Test engine
- - - Recommended maximum BHP - Production engine

Fig. 2-47 Torque begins to decrease at a certain point even though the engine accelerates.

lawn, but it takes your daughter two hours, you are producing more power than she is because the work accomplished is the same, but your rate of time for the work is faster. Power is measured in Horsepower in the English system and Watts in the Metric system.

Engine power output can be rated by different measurements in different systems, but the most common method in the small engine industry is to use the term Horsepower or Watt.

HORSEPOWER

Engine power is the amount of work performed per unit of time. It can be expressed by the equation:

$$\text{Power} = \frac{\text{Work}}{\text{Time}}$$

If a one-pound weight is lifted one foot, the amount of work is one foot-pound

(Work = Distance x Force).

If this work is done in a one-second time period, then it can be expressed as

$$\text{Power} = \frac{1 \text{ ft-lb}}{1 \text{ sec}}$$

James Watt, the inventor of the steam engine, found that an English work-horse was able to lift a 550-pound block at a rate of one-foot per second. He defined this power unit as,

$$\frac{550 \text{ ft-lb}}{1 \text{ sec}}, \text{ or } \textbf{one horsepower}.$$

Horsepower has been the standard unit of power in the English measurement system, while the **Watt** is the unit in the metric system. 1000 watts (1 kilowatt) is equal to 1.34 horsepower, or 1 horsepower is equal to 746 watts.

Different horsepower ratings can be assigned to the air-cooled engine and are defined below. The Brake horsepower is the most common method of rating net engine power.

BRAKE HORSEPOWER OR NET HORSEPOWER

Brake horsepower (BHP) is the actual horsepower output by the engine to the driven device. It is the net result of the theoretical horsepower output that can be calculated mathematically minus the loss of power from the internal friction components.

Horsepower is directly related to engine speed. Engine speed is measured in revolutions per minute, or RPM. The faster the engine turns, the higher the RPM. An engine operating at 1000 RPM will produce less horsepower than when it is operating at 3000 RPM. In general, the faster the engine speed, the more horsepower it will develop. However, at a certain RPM limit, the power is just enough to overcome the greater internal friction that has developed, and any increase in engine speed beyond that point will be unable to overcome the increased friction with a resultant decrease in horsepower.

HORSEPOWER CURVE CHART

The relationship of horsepower output to engine speed is illustrated on the horsepower curve chart which engine manufacturers prepare for each different engine design, as shown in figure 2-37. The bottom of the chart is scaled to indicate engine speed in RPM, while the left hand vertical scale reveals horsepower. The curved line across the chart demonstrates how the horsepower increases with engine speed. Horsepower is measured in accordance with the appropriate SAE (Society of Automotive Engineers) test so that when a buyer is comparing a power curve of a Briggs & Stratton engine against a Tecumseh engine, the units are standard.

TEST ENGINE VS. PRODUCTION ENGINE HORSEPOWER

The maximum BHP curves are corrected to standard conditions of a sea level barometer and an atmospheric temperature of 60° F. The results are from *test engines* that are equipped with a standard muffler and air cleaner. *Production engines* will develop approximately 85 percent of the maximum BHP when tested after the break-in period (to reduce friction) and after the combustion chamber and valve ports have been cleaned of any carbon. The carburetor and ignition system must be properly adjusted. For practical operation, BHP load and speed should be under the recommended maximum BHP curve.

Brake horsepower derives its name from the apparatus used to measure the output horsepower, called a prony brake.

DYNAMOMETER

The working ability of an engine can be measured using a device called a dynamometer. It actually "loads" the engine, causing it to work. The dynamometer contains instrumentation which tells us how well the engine is performing under varying conditions. The dynamometer is used in the type of analysis that would be done by engine rebuilding shops to make sure an engine is performing according to specifications, a racing engine builder to achieve maximum power output, or an engine manufacturer for engine design research and reliability testing.

Three of the most common types of dynamometers are the:

1. *Prony brake*

2. *Water absorption (Hydraulic)*

3. *Electric*

Prony brake dynamometer

The prony brake is a simple device that is used to measure the torque, or turning, of an engine, as shown in figure 2-48. The torque measurement is inserted into the formula:

$$BHP = \frac{F \times L \times 2\pi \times RPM}{3300}$$

The "F" is the scale reading in pounds, the "L" is the distance of the brake arm, the "RPM" is the speed of the engine.

A prony brake can be constructed by placing an adjustable collar (brake) around the output shaft of the engine and measuring the distance from the center of the shaft to the end of the arm which is connected to a spring scale. The engine is operated at a set speed while the friction adjusting wheel is tightened to increase the load on the engine. The governor system on the engine keeps the speed constant while it opens the carburetor's throttle plate to increase the engine power. When the throttle plate is fully opened, a reading in pounds on the spring scale is observed and inserted into the formula.

As an example, let us imagine that the reading on

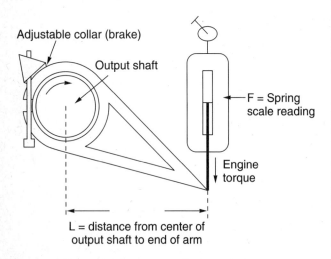

L = distance from center of output shaft to end of arm

Fig. 2-48 The prony brake is a device used to measure the torque and horsepower of an engine.

the spring scale was 25 pounds for an engine operating at 3600 RPM with the throttle plate wide open. The distance from the shaft to the end of the arm was set at one foot. Inserting the numbers into the formula, the brake horsepower rating would be 17.1 horsepower.

Water absorption dynamometer

The water absorption, or hydraulic type, is the most common type used and is found in many industrial applications, as shown in figure 2-49. It consists of a finned rotor mounted in a cradled housing. The engine being tested is attached to and drives this finned rotor section. During the testing sequence, a liquid (usually water) is introduced into the dynamometer under pressure. The engine being tested must pump this liquid and, consequently, this places a load on the engine. The load is varied by increasing or decreasing the liquid placed in the system. As the liquid is pumped by the rotor, it moves out toward the sides of the housing, causing the housing to rotate in its cradle. By attaching a scale to the housing, this force can be measured. The Mechanical energy of the engine is converted to heat energy in the form of hot water. The hot water is then pumped out of the dynamometer, so cooling is not a problem.

Electric dynamometer

A dynamometer is a device used to determine the brake horsepower of an engine. One type of dynamometer converts Mechanical energy of the engine into electricity. An electrical generator is connected to the output shaft and the current produced by the generator is measured. The electrical current is converted to watts, which is then converted to horsepower. This type of dynamometer is the most sophisticated and accurate, but is very expensive when compared to the water absorption type.

THERMAL EFFICIENCY

Thermal efficiency of an internal combustion engine compares the heat developed into useful work to the heating value of the fuel. The heat energy of the fuel goes into:

Fig. 2-49 Water absorption dynamometer.

1. *Heat dissipated by the cooling system*
2. *Heat carried away by the exhaust gases*
3. *Mechanical work on the piston to overcome friction*
4. *Mechanical work on the piston at crankshaft output*

Heat dissipated by cooling system

Since the metals of the combustion chamber would warp and burn if allowed to reach the temperature of the gases contained in the cylinder, provision must be made to cool the metals and limit the exposed surfaces to a safe temperature. Cylinder walls are cooled with a water jacket or air flow on their outer surface in order to keep a safe temperature at the inner surface. This is a necessary consumption of energy, which amounts to 20 percent of the energy in the fuel in the average engine, as shown in figure 2-50.

Heat carried away by exhaust gases

Fifty percent of the energy of the fuel is lost in the exhaust gases as heat energy. The exhaust tem-

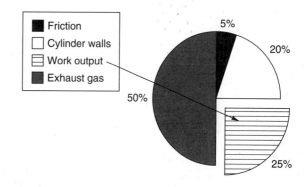

Fig. 2-50 The energy of combustion is converted to heat and mechanical energy. The percentage of the total energy released is represented by the pie graph.

peratures can be 2000° F above the outside air temperature.

Mechanical work to overcome friction

Five percent of the energy of the fuel is converted to mechanical work that is used to overcome the internal engine friction.

Mechanical work at crankshaft output

Twenty-five percent of the energy of the fuel can be harnessed at the crankshaft output. Engineering modifications to the combustion chamber and other engine designs attempt to make the internal combustion engine more efficient.

SUMMARY

It is important to understand the basic principles of engine operation and terms associated so that a common language used in explaining procedures in the following chapters will be more meaningful.

The internal combustion engine is the most common design used in the small air-cooled engine and can be found in a four-stroke-cycle or two-stroke-cycle design. In the "four stroke," the five events of combustion (power, exhaust, intake, compression, and ignition) occur in four strokes of the piston, while in the "two stroke," the five events of combustion occur in only two strokes of the piston.

The cylinder and valve arrangement determine the style of the engine. The "L" head engine has been used extensively, but the overhead valve engine is gaining in popularity because of better efficiency and emission and longevity.

The power of an engine is dependent upon many items that include the piston displacement, compression ratio, amount of air-fuel intake, valve timing, degree of compression, ignition timing, type of fuel, engine friction, and atmospheric conditions.

Engine power can be measured and rated in horsepower. Horsepower is the rate at which the engine work is done. The higher the horsepower, the faster the work can be accomplished. An engine dynamometer can be used to measure the horsepower of an internal combustion engine. Horsepower and Torque charts are a method of comparing engines and are used by the manufacturers to support their power output claims.

Questions (Multiple Choice)

1. **Five events are necessary for proper combustion. Technician A says that the sequence they follow is exhaust, compression, intake, ignition, and power. Technician B says that the sequence is intake, compression, ignition, power, and exhaust. Who is correct?**

   A. Only Technician A
   B. Only Technician B
   C. Both Technician A and B
   D. Neither Technician A or B

2. **Two common types of internal combustion engines used on the air-cooled application are the four-stroke-cycle and the two-stroke-cycle engine. Technician A says that one of the differences between the two is the number of power strokes for every revolution of the crankshaft. Technician B says that one of the differences between the two is how the engine is lubricated. Who is correct?**

   A. Only Technician A
   B. Only Technician B
   C. Both Technician A and B
   D. Neither Technician A or B

3. **There are advantages of an air-cooled vs. a water-cooled engine. Technician A says that the air-cooled engine removes the combustion heat better and has a longer potential life. Technician B says that the air-cooled engine has more components which make the engine heavier**

compared to a water-cooled engine of
the same horsepower rating.
Who is correct?

A. Only Technician A
B. Only Technician B
C. Both Technician A and B
D. Neither Technician A or B

4. **An engine operating at 3000 RPM is
producing less than the rated brake
horsepower. Technician A states that
since engine power is affected by the
compression ratio, changing to a higher
ratio will lower the power output
(all other factors constant). Technician B states that changing to a
carburetor with a larger venturi will
decrease the power output (all other
factors constant).**
Who is correct?

A. Only Technician A
B. Only Technician B
C. Both Technician A and B
D. Neither Technician A or B

5. **When the throttle plate on the carburetor
is brought from the wide open position
to near the closed position, the engine
speed decreases. Technician A says that
it is true because the volume of atmospheric oxygen available to the
combustion chamber is reduced, while
the fuel flow remains the same.
Technician B says that it is true because
the fuel flow into the combustion
chamber decreases as well as the
amount of available oxygen.**
Who is correct?

A. Only Technician A
B. Only Technician B
C. Both Technician A and B
D. Neither Technician A or B

6. **The "L" head engine is a popular design
for the small air-cooled engine, but the
OHV engine is gaining in popularity.
Technician A says that this is because
the discharged gases from the OHV
engine are cleaner and better for the
environment. Technician B says that the
OHV engine is able to remove the
combustion heat more efficiently, so it
has a longer engine life.**
Who is correct?

A. Only Technician A
B. Only Technician B
C. Both Technician A and B
D. Neither Technician A or B

7. **Technician A says that the push on the
piston top during the power stroke is
transferred to the crankshaft through
the connecting rod. Technician B states
that the power is transferred from the
piston through the valve train to the
crankshaft.**
Who is correct?

A. Only Technician A
B. Only Technician B
C. Both Technician A and B
D. Neither Technician A or B

8. The brake horsepower is the measurement of power available at the crankshaft. Technician A says that the maximum horsepower will most likely increase if the muffler and air cleaner are removed. Technician B says that the maximum brake horsepower will always remain the same for any engine because it is determined by a mathematical calculation of the bore, stroke, compression ratio, etc.
Who is correct?

A. Only Technician A
B. Only Technician B
C. Both Technician A and B
D. Neither Technician A or B

9. Two engines are operating at 3000 RPM. The torque produced by engine #1 is greater than engine #2. Technician A says that engine #1 would have a higher maximum brake horsepower rating. Technician B says that even though there is a relation between the output torque and horsepower, it is impossible to determine the engine with the higher maximum brake horsepower from the information given.
Who is correct?

A. Only Technician A
B. Only Technician B
C. Both Technician A and B
D. Neither Technician A or B

10. When combustion occurs, heat is produced. Technician A says that most of the heat is removed in the exhaust gases. Technician B says that most of the heat is removed by the cylinder fins around the cylinder.
Who is correct?

A. Only Technician A
B. Only Technician B
C. Both Technician A and B
D. Neither Technician A or B

11. Combustion products may be deposited on the exhaust valve face or seat so that the valve never completely closes. Technician A says that the flywheel will be harder to spin when starting the engine. Technician B says that the engine will lose power.
Who is correct?

A. Only Technician A
B. Only Technician B
C. Both Technician A and B
D. Neither Technician A or B

12. The amount of fuel that enters an engine combustion chamber varies with the position of the throttle plate. Technician A says this is true because the throttle plate directly controls the volume of air flowing through the venturi in the carburetor. Technician B says that this statement is false because the main adjustment screw on the carburetor is solely responsible for the fuel that enters the combustion chamber.
Who is correct?

A. Only Technician A
B. Only Technician B
C. Both Technician A and B
D. Neither Technician A or B

Questions (Matching)

Match the correct letter to the number by placing the letter into the box provided.

☐	1. "L" head engine	A. The discharge of products of the burned fuel.
☐	2. OHV engine	B. A cool air mass is brought into the combustion chamber.
☐	3. Combustion chamber	C. Rapidly expanding air creates a high pressure on the piston.
☐	4. Compression stroke	D. The intake gases are packed together.
☐	5. Connecting rod	E. The burning of the fuel is outside the cylinder.
☐	6. Dynamometer	F. The burning of the fuel is inside the cylinder.
☐	7. Exhaust stroke	G. The five parts of combustion occur for every four strokes.
☐	8. External combustion	H. The five parts of combustion occur for every two strokes.
☐	9. Flywheel	I. Both the intake and exhaust valves are located on the side.
☐	10. Four-stroke-cycle	J. Intake and exhaust valve are located in the cylinder head.
☐	11. Horsepower	K. 9.025 cubic inches.
☐	12. Intake stroke	L. Formed by the cylinder and cylinder head.
☐	13. Internal combustion	M. Connects the crankshaft to the piston.
☐	14. Main bearings	N. To support the crankshaft.
☐	15. Mechanical work	O. Control the intake and exhaust of gases.
☐	16. Piston displacement	P. A mass needed to maintain inertia.

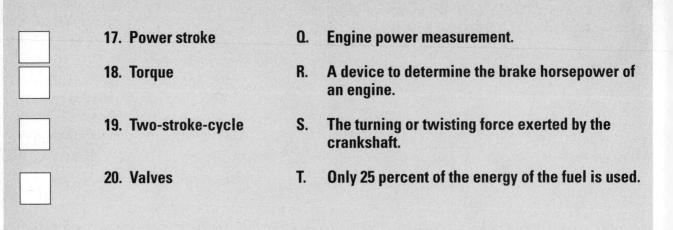

17. **Power stroke**

18. **Torque**

19. **Two-stroke-cycle**

20. **Valves**

Q. **Engine power measurement.**

R. **A device to determine the brake horsepower of an engine.**

S. **The turning or twisting force exerted by the crankshaft.**

T. **Only 25 percent of the energy of the fuel is used.**

CHAPTER 3

Fuel and Lubrication System

INTRODUCTION

The fuel system performs many functions, which include:

1. *storing the fuel*

2. *delivering the fuel to the carburetor*

3. *changing the liquid gasoline to a vapor*

The fuel tank holds a quantity of fuel until it is needed by the carburetor, as shown in figure 3-1. The stored fuel is then delivered to the carburetor by the pull of gravity or by a fuel pump.

Since only vaporized gasoline will burn, the fuel system must change the liquid fuel to a gaseous state. This is accomplished in the carburetor and the intake port of the engine.

The carburetor mixes air with the vaporized fuel in different proportions of air to gasoline to meet different operating conditions. When starting an engine, the fuel system must deliver a very rich mixture (rich in gasoline) of about nine parts air to one part gasoline. After the engine has warmed up, the mixture must be "leaned out" (made less rich) to about 15 parts air to one part gasoline.

The fuel system is responsible for most engine problems. A good understanding of its principles and designs is necessary for proper troubleshooting and servicing.

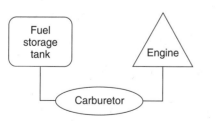

Fig. 3-1 The fuel tank holds the fuel until it is needed by the carburetor.

TYPES OF FUEL SYSTEMS

The type of fuel system is determined by how the fuel is delivered to the carburetor. The most common types are:

1. *the gravity fuel system*

2. *the suction fuel system*

3. *the fuel pump system*

4. *the pressurized fuel system*

GRAVITY FUEL SYSTEM

In the gravity fuel system, the fuel tank is located above the carburetor, and fuel is delivered to the carburetor by the force of gravity, as shown in figure 3-2. When the engine is not running, the flow of fuel is stopped by the needle valve and seat in the carburetor or a shut-off valve in the fuel line or fuel tank. The carburetor inlet valve, consisting of a needle and seat, may not be able to shut off the flow of fuel if the fuel tank is located too far above the carburetor. The force of gravity on the fuel will force fuel into the carburetor.

Normally, a fuel tank shut-off valve is installed and closed when the engine is not in operation. A dangerous situation can arise when fuel leaks past

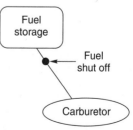

Fig. 3-2 In the gravity fuel system the fuel tank is located above the carburetor.

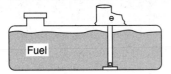

Fig. 3-3 In the suction system the fuel tank is attached to the carburetor, as in the Briggs & Stratton Vacu-Jet type.

the inlet and overflows the float bowl so that it drains into the crankcase of the engine. If the crankcase oil level rises from time to time, check for a leaking inlet needle and seat.

SUCTION FUEL SYSTEM

The suction fuel system is an unsophisticated system wherein the fuel tank is located below the carburetor and gasoline is sucked upward, as shown in figure 3-3. The suction is caused by the downward movement of the piston in the cylinder while the exhaust valve is closed and sealing, and the intake valve open. The resulting low cylinder pressure causes the outside atmospheric pressure to push air into the carburetor, across the venturi, and into the cylinder.

The Briggs & Stratton's Vacu-Jet carburetor is the most common application of this system and is found on many "low-powered" engines. The fuel tank must be very close to the carburetor and the fuel pick-up tube.

FUEL PUMP SYSTEM

The fuel pump system is very versatile because the fuel tank can be placed in a variety of positions, either above or below the carburetor. The flow of fuel from the fuel tank to the carburetor does not depend on gravity. When the fuel pump is used in conjunction with a diaphragm carburetor, a two-stroke-cycle chain saw engine becomes an all-position engine. The fuel pump can be operated by the mechanical action of the engine, an electric motor, or by using the crankcase pulsations of the engine.

PRESSURIZED FUEL SYSTEM

Fuel can be forced to travel long distances to the carburetor in a pressurized fuel system. This method is used with older outboard marine engines where there are two lines between the fuel tank and the engine: One line brings gasoline to the carburetor, and the other conducts pressurized air from the crankcase to an air-tight fuel tank.

GASOLINE

Gasoline is one of the products produced when crude oil is refined. It is a volatile liquid used to power the internal combustion engine. An ideal gasoline has the following properties: easy starting, permits the engine to run smoothly with only a small amount of detonation, resists vapor locking, minimizes deposits in the fuel system and combustion chamber, provides good economy, and is inexpensive.

HYDROCARBONS

Gasoline is a hydrocarbon (HC) made up of hydrogen and carbon compounds. These compounds react with oxygen to release energy and produce water and carbon dioxide, as shown in figure 3-4. Perfect combustion is not possible, so compounds of carbon monoxide and unburned hydrocarbons are also present in the exhaust gases. Carbon monoxide and unburned hydrocarbons are causes of pollution, as shown in figure 3-5.

One method used to reduce the the by-products of carbon monoxide and more efficiently burn all the hydrocarbons is to operate an engine at a higher temperature. When the combustion chamber temperatures are increased, the nitrogen gas found in the air is combined with oxygen molecules to form different compounds of nitrogen oxide, a major contributor to atmospheric smog, as shown in figure 3-6.

BLENDING

Gasoline is a blending of different hydrocarbons, each with its own set of characteristics. By mixing

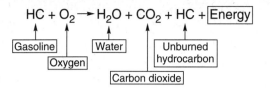

Fig. 3-4 Gasoline combines chemically with oxygen gas to form energy, water and carbon dioxide (non-poisonous gas).

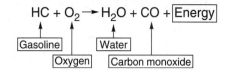

Fig. 3-5 Since perfect combustion is not possible, Carbon monoxide (a poisonous gas) is produced.

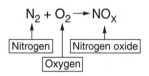

Fig. 3-6 Atmospheric air combines with oxygen to for different compounds of nitrogen oxide.

various basic fuels, a gasoline is obtained that provides satisfactory engine performance under the many different operating conditions that the engine will meet. Volatility, antiknock value, and freedom from harmful chemicals are factors that must be considered when blending gasolines.

VOLATILITY

Gasoline is a blended hydrocarbon with a varying ratio of light, short-chained, volatile molecules and heavier, long-chained molecules. The volatile components provide easy vaporization and starting capabilities, and the longer molecules provide proper combustion in a warm-running engine. The amount of lighter molecules blended in the fuel will vary from season to season, depending on the climate. In the northern states, a winter blend of fuel has a higher ratio of lighter molecules than a summer blend. Problems arise when fuel bought in one season is used in another. Winter gas used in the summer may vaporize easily and cause a vapor lock condition, while summer fuel used in the winter may cause starting problems.

STORAGE LIFE

The organic substances in gasoline are constantly changing and forming new compounds that alter the characteristics of the fuel over a period of time. The same molecules that make up the volatile part of gasoline can react with oxygen and other elements in the environment. New molecules are formed, which constitute gummy residues and varnish that can clog up the passages in the fuel lines and carburetor. Most fuels are blended with oxidation inhibitors that allow the gasoline to withstand

three months of storage without substantial breakdown. When oil is mixed with the fuel, the life of the stored gasoline is reduced to *one month.*

Fuel additives can be bought and added to the fuel to decrease the oxidation process. These additives are recommended for all fuels if they are stored seasonally or for a long period of time. The additive is only effective if it is added to the fuel, and the engine is then operated for a period of time to bring the protected fuel into the fuel lines and carburetor.

Some of the additive manufacturers claim that the fuel can be protected from oxidation for up to two years. Throughout this extended period, a vented fuel cap will cause many of the lighter molecules to be lost and may cause starting problems even though the fuel is protected from oxidation breakdown. Briggs & Stratton markets a gasoline additive (part #5041) that is a four-part solution which includes:

1. *Antioxidants*, *which reduce the formation of gums and varnishes and maintain the octane rating.*

2. *Metal deactivators*, *which neutralize the effect of copper alloys that can cause a gel which can plug gasoline filters. The deactivators are usually effective for thirty days.*

3. **Rust inhibitors,** *which reduce rusting and corrosion of steel parts within the engine.*

4. **Detergents**, *which reduce the build-up of varnish and carbon deposits within the engine.*

ANTI-KNOCK VALUE

Gasoline blends are rated with an anti-knock value (octane rating), with common values such as 86, 87, 89, and 92. The higher the number, the more resistant the fuel is to combustion problems, such as detonation and preignition.

The compression ratio is determined by the engine design and is expressed by the comparison of the volume of the combustion chamber when the piston is at BDC (bottom dead center) to the volume with the piston at TDC (top dead center), as shown in figure 3-7. The higher the compression ratio, the more the air-fuel mixture is "squeezed" during the compression stroke.

The amount of power an engine develops and the efficiency with which the fuel is burned is directly related to the degree of compression achieved in the cylinder. As the compression ratio increases, the need for controlling the burn rate by the addition of anti-knock compounds to the fuel also increases.

If an engine operates adequately with an octane rating of 87, then no increase in performance will be achieved by using fuel with a higher number. Use the fuel octane rating recommended by the engine's manufacturer.

The type of combustion desired in an engine is a progressive burning of the fuel and air mixture where the flame starts at the spark plug and travels in all directions in a controlled type of action, as shown in figure 3-8. As the flame travels, heat is generated and gaseous products are released.

Detonation

The pressure and temperature of the flame front may reach the point where the mixture may be ignited by itself to cause detonation. Detonation produces an opposing flame front that collides with the normal flame front to produce an uncontrolled explosion that may damage engine components, as shown in figure 3-9. The burn rate of the fuel can be controlled chemically by blending tetra-ethyl lead or alcohol to create an octane rating. Most small air-cooled engines can operate on fuels with low octane ratings because of their low compression ratios.

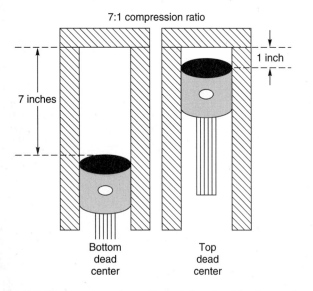

Fig. 3-7 The compression ratio is determined by the engine design.

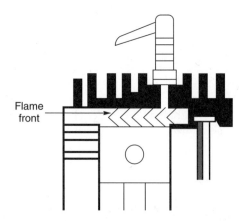

Fig. 3-8 Combustion is a controlled burning starting at the spark plug and continuing throughout the combustion chamber.

Preignition

Another form of uncontrolled combustion is caused by some spot in the cylinder remaining hot enough after the exhaust stroke to ignite the incoming mixture before the normal spark ignition. This condition, called preignition, will cause an abnormal buildup of pressure, rough sounding operation, and loss of power, as shown in figure 3-10. All forms of uncontrolled combustion result in excessive pressures on rings, piston, cylinder head, and bearings, which may lead to mechanical failure of some engine part.

ALCOHOL ADDITIVES (GASOHOL)

Since the EPA (Environmental Protection Agency) issued an order to remove lead from gasoline, fuel quality has become a major problem. Tetra-ethyl lead was added to fuel to improve the anti-knock quality of gasoline. Lead also acts as a lubricant by coating the valves, valve seats, and combustion areas with lead oxide. The modern engine, with its hardened valves and its need for low octane fuel, has not been hurt by the removal of lead.

In recent years, the octane ratings of gasoline have dropped dramatically, and unleaded gasolines now occupy the majority of pump space at fuel stations. Most automobile engines were not designed to run on low octane gasoline, so gasoline manufactur-

ers have raised the octane rating by introducing alcohol. The most commonly used alcohol additives for gasoline are ethanol (ethyl alcohol), methanol (methyl alcohol, originally called wood alcohol, but now largely produced from methane gas), or MTBE (Methyl-Tertiary-Butyl-Ether), as shown in figure 3-11. The percentage of alcohol additives used in gasoline will vary throughout the year, both geographically and from supplier to supplier. This addition of alcohol has been responsible for many problems in two- and four-stroke-cycle engines. The use of MTBE and ethanol are less damaging to engine components than methanol. MTBE is the octane enhancer of choice by most manufacturers because of its stability and compatibility with gasoline and also because it is not as sensitive to moisture as the other octane enhancers.

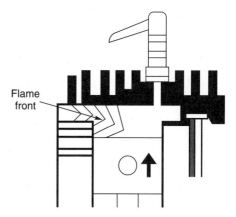

Fig. 3-10 Preignition is abnormal combustion where the fuel-air mixture is ignited by another hot spot instead of the spark plug.

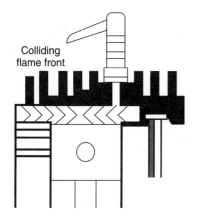

Fig. 3-9 Detonation is abnormal combustion that produces an opposing flame front that collides with the normal flame front.

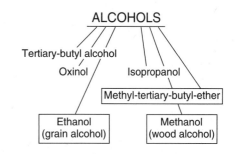

Fig. 3-11 Some of the components of the alcohol family are used as a fuel additive.

Many manufacturers are changing the materials used in engine construction to allow the use of alcohol fuels. Engines of the future must be able to withstand the use of 40–60 percent alcohol. Engines will contain fewer and fewer metal parts, and the use of ceramics and plastics will be more prevalent.

Pros and cons of alcohol addition

Ethanol (grain alcohol) has been the main ingredient added to gasoline to restore the loss of anti-knock quality provided by lead. Alcohol is a renewable re-source that is distilled from farm grains and is not dependent upon importation in North America. It is cleaner burning and less polluting than petroleum-based engine fuels.

Some of the undesirable effects of alcohol are as follow:

1. *Alcohol acts as a solvent. It counteracts the lubricating properties of the fuel-oil mist in a two-stroke-cycle engine. Gasoline with a high alcohol content can break down an oil's ability to remain suspended in gasoline, which is necessary for two-stroke-cycle engine lubrication, and may leave internal moving parts unprotected.*

2. *The water absorbed by the alcohol may cause a separation of the gas and oil mixture used in a two-stroke-cycle engine. Since water is absorbed by alcohol, the presence of water in the fuel mixture will result in the settling of the "alcohol-water layer" to the bottom of the tank—the area from which the fuel is drawn.*

3. *Alcohol can destroy plastic, brass, rubber, and fiberglass materials. These substances are used in fuel tanks, fuel lines, filters, and carburetor float bowls.*

4. *Methanol corrodes aluminum, zinc, and magnesium castings of the carburetor and other engine components.*

5. *Since alcohol will remove corrosion from the inside of the fuel tank and the particles will remain suspended in the fuel, the fuel filter should be changed often.*

Test for alcohol

The following is one way to test for alcohol in fuel:

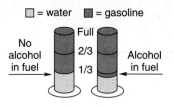

Fig. 3-12 The water layer will decrease if alcohol is present because some of the water is absorbed by the alcohol in the fuel.

1. *Fill a graduated glass cylinder or container one-third full of water.*

2. *Mark or note the water level.*

3. *Fill the container to the top with gasoline.*

The water layer below the fuel layer will remain undisturbed if no alcohol is present in the fuel mixture. If ethanol or methanol is present, however, some of the alcohol will be absorbed by the water layer. The original water layer will appear to increase, as shown in figure 3-12.

Briggs & Stratton offers an alcohol tester (part # 100023) that can be purchased through your local supplier, as shown in figure 3-13.

Alcohol protector

Fuel additives are available that will prevent damage caused by alcohol in gasoline and are used to improve the alcohol/gasoline solubility even when water is present. These additives will prevent the reaction of alcohol with the metal engine parts and protect the fuel system from rust and corrosion.

LEADED VS. UNLEADED GASOLINE

Engines are constructed so that lead is not required in the fuel for proper combustion. When leaded gasoline is used, it forms white, hard, scaly deposits which break off with time and cause engine wear. When unleaded gasoline burns, it leaves a soft black powder residue that is easily discharged from the engine without causing abrasive wear.

If you are already using leaded gasoline, it is recommended that you continue doing so. If the switch is made to unleaded fuel, the engine's cylinder head

Fig. 3-13 Briggs & Stratton offers an alcohol tester (#100023).

must first be removed and all deposits cleaned from the combustion chamber. Unleaded gasoline burns hotter than leaded fuel. This greater heat causes the lead deposits to break off faster, and if they become lodged in the valve area, they will cause engine failure. The preferred fuel is unleaded; it should be used as much as possible.

CLEAN AIR STANDARDS

Legislation limiting harmful emissions from automobile engines has been in effect for many years. The small air-cooled engine industry will also be affected by certain future standards. An estimated 4 percent of all the total pollutants from piston engine sources can be linked to the air-cooled engine. Some states are mandating that total emissions be lowered by 55 percent.

The proposed levels are far below the current emissions of any two-stroke-cycle utility engine, and the regulations will cause the fatality or redesigning of this engine. The four-stroke-cycle engine can more easily meet the proposed emission levels with many of the new technology engines. The older "L" head engine's combustion chamber design is inefficient and will not meet the proposed requirements. To comply with the new emission levels, the four-stroke-cycle engines will have to be an overhead valve design that will run very lean. New or larger cooling systems will have to be used to dissipate the additional heat produced, and the horsepower per cubic inch will have to be reduced.

RECOMMENDED FUELS

Use fresh, clean, *unleaded,* low-octane, non-alcohol gasoline for all four-stroke-cycle engines. When it is not available, exceptions may be made for:

Tecumseh

Leaded gasoline or gasohol containing no more than 10 percent ethanol (grain alcohol) can be utilized. Never buy gasoline containing as little as 1 percent methanol, gasohol containing more than 10 percent ethanol, premium gasoline, or white gas.

Briggs & Stratton

Unleaded gasoline containing not more than 10 percent ethanol (grain alcohol) may be used, but the fuel must be removed from the engine during storage. The minimum octane rating should be 77, except for the Vanguard engine, which requires 85.

Kohler

Regular grade, leaded gasoline can be used as a substitute; however, the combustion chamber and cylinder head may require more frequent cleaning. The use of alcohol-blended gasoline is allowed, provided the maximum concentration of ethanol in the fuel is 10 percent. No methanol-blended gasoline should be used.

OTHER FUELS

LIQUIFIED PETROLEUM GAS (LPG)

Some advantages are attained when engines are made to run on LPG (Liquified Petroleum Gas, such as propane, butane, or natural gas).

1. *The byproducts of combustion are cleaner. This enables engines using LPG to meet strict*

emission standards. LPG engines can also be operated indoors.

2. Reduction in combustion deposits from LPG in the cylinder extends the life of the mechanical parts, which results in reduced maintenance costs.

3. LPG cannot wash down the cylinder walls and remove lubricants because it enters the engine as a vapor. This decreases wear on cylinder walls, pistons, and piston rings.

LPG and gasoline are both highly volatile and flammable, but LPG presents one additional danger. It is heavier than air and tends to settle in low areas, where the slightest spark may ignite the gas. Extra precautions must be taken to ensure safety when using LPG as an engine fuel. Other disadvantages include the higher cost for the modifications to the engine for the use of the alternative fuel and the increased energy cost per BTU. Since gaseous fuels burn very clean and dry, they provide no lubricating properties to the valve train (valve rotators are not recommended). The spark plug gap should be reduced to 0.018-inch. This tighter gap setting provides better starting in cold temperatures, when the fuel does not vaporize as readily.

KEROSENE

Small gas engines can be converted to run on kerosene or fuel oil with special modifications. First, the compression ratio must be decreased by installing an extra head gasket. The engine must be started on gasoline and then switched over to the kerosene when the operating temperature has been reached. The advantage or disadvantage of this arrangement would be the cost of kerosene compared to gasoline.

LUBRICATION SYSTEM

Engine oil is used in the lubrication system to performs five basic functions:

1. Lubrication of components
2. Sealing of power
3. Cooling
4. Corrosion protection

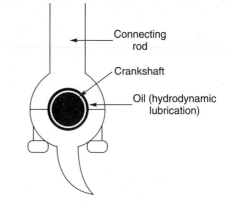

Fig. 3-14 The oil layer between the connecting rod and the crankshaft provide hydrodynamic lubrication.

5. Cleaning
6. Reducing engine noise

LUBRICATE COMPONENTS

The lubrication system must protect the working components by reducing friction. All machined metal surfaces have a great amount of roughness that is not visible to the naked eye. A smooth metal surface that appears glasslike may actually have many peaks and valleys when viewed through a magnifying glass. A sufficiently thick oil layer must separate the moving metal components so that all metal-to-metal contact is eliminated.

Hydrodynamic lubrication

Hydrodynamic lubrication occurs when two mating metal surfaces are separated by a layer or film of oil, as shown in figure 3-14. Hydrodynamic lubrication is not difficult to attain at constant engine speeds with a light work load, but under a heavily loaded condition the oil must adequately separate the metal surfaces. The oil viscosity or thickness must be sufficient so that even at high temperatures of operation, the oil film will be thick enough to protect the surfaces.

Boundary lubrication

When it is impossible to sustain a constant oil film between the metal parts, a brief metal-to-metal contact between the high spots on the sliding sur-

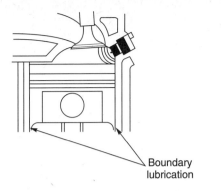

Fig. 3-15 Boundary lubrication is the coating of oil on the metal parts to reduce friction.

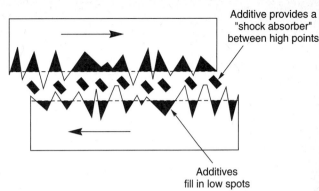
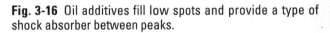

Additive provides a "shock absorber" between high points

Additives fill in low spots

Fig. 3-16 Oil additives fill low spots and provide a type of shock absorber between peaks.

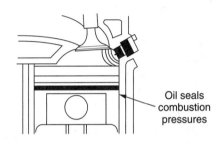

Oil seals combustion pressures

Fig. 3-17 Oil seals the area around the piston, confining combustion to the area above the piston.

faces, such as the piston skirt to cylinder wall, can produce heat that will cause the parts to wear prematurely, as shown in figure 3-15. This type of partial lubrication is called boundary lubrication and is always present during engine start-up, initial operation of a remanufactured engine, and in places where the oil supply is limited and temperatures from the combustion are high (such as the top piston ring and the cylinder wall).

To protect the moving parts against friction, high-quality engine oils have chemicals added that react with metal surfaces to form a coating that serves as a boundary lubricant. The additives are added to permanently fill in the low spots between the two metal surfaces and also provide a type of shock absorber between the surface peaks so that at the moment of contact, the peaks are deformed to a lower level rather than breaking off or welding together, as shown in figure 3-16. Effective boundary lubrication provided by the special additives reduces the rough areas and decreases the friction and heat that causes wear and failure.

SEAL ENGINE POWER

Oil seals the area around the piston, piston rings, and cylinder so that the pressure of combustion is confined to the space above the piston, as shown in figure 3-17. Immediately following the burning of the fuel in the combustion chamber of the engine, high pressure gases are formed. These pressures on the piston top push the piston down, causing the engine to operate. It is necessary to keep these gases

from leaking past the piston rings and cylinder wall, and since piston rings alone cannot prevent leakage of combustion gases, the engine oil must provide the sealing necessary to prevent unacceptable loss of power from the "blow-by." The oil must have the proper thickness, or viscosity rating, to seal properly at high operating temperatures.

COOL THE ENGINE

Although the cooling system removes the major portion of the unwanted heat, the oil must transport it from some critical areas. The oil assists the cooling system by absorbing heat when it passes over the parts and flows back to the crankcase, where the heat is transferred to the surrounding air.

CORROSION PROTECTION

Corrosive byproducts of incomplete combustion continually enter the oil when some of the combustion gases pass the seal between the piston rings and the cylinder wall. The corrosive combustion agents include unburned fuel, soot, water, and acids which remain in the oil until the next oil change. It is important that the engine oil contain additives that will coat the metal components to prevent the corrosion caused by the byproducts over a period of time.

CLEAN INTERNAL PARTS

Detergents and dispersants are added to engine oil to keep most of the contaminants suspended as microscopic particles so that they do not form sludge and damaging deposits. The detergent additive cleans the metal surfaces as it circulates by preventing the formation of sludge or varnish deposits which are formed when the combustion byproducts react with oxygen at high temperatures. The oil temperature in an air-cooled engine is higher than a liquid-cooled engine and thus accelerates sludge formation.

The dispersant additive reacts with the removed sludge to keep it from settling on the internal components, which might create a barrier to the dissipation of the crankcase heat. Since most small air-cooled engines do not have an oil filtering system to remove the large contaminants, it is imperative that the recommended interval between the oil changes be followed. The contaminants are removed only when the oil is changed.

REDUCE ENGINE NOISE

As the air-fuel mixture is ignited near the end of the compression stroke, the combustion pressures in the cylinder quickly increase. A heavy load of as much as several tons is suddenly imposed on the piston, piston wrist pin, connecting rod, and on their bearing surfaces. This load attempts to squeeze out the oil between the component surface, but the ability of the oil to resist the shock load helps quiet the engine and prevents any metal-to-metal contact, which would increase undesirable wear.

FOSSIL BASE LUBRICANT

Lubricants are refined from a chemically complex fluid called petroleum or crude oil. Crude oil is extracted from the ground. Geologists and petroleum scientists are not sure how it was originally formed, but the prevalent theory is that it was formed, long ago from a huge number of animals and plants living in shallow coastal waters and who suddenly died. The plant and animal remains did not decompose immediately because of a lack of oxygen. Eventually the land shifted over the ooze, subjecting it to intense pressure, and at the same time providing it with the needed oxygen required for decomposition. This chain of events happened in many places on earth at about the same time, but the chemical composition of each field was different, which explains why no two crude oils are exactly alike.

Crude oil is extracted from many places in the world. Texas, Alaska, Venezuela, the Middle-East, and Pennsylvania are the dominant areas where crude oil is extracted from the ground. High-quality engine oils, formulated from the refined base oil, have a higher resistance to thickening when cold or to thinning when hot.

SYNTHETIC BASE LUBRICANT

The base stock of a synthetic base lubricant is a synthesized hydrocarbon or organic ester. Two of the most outstanding characteristics of the synthetic base are oxidation stability (which allows longer operation in an environment above 220° F without breaking down) and the ability to lubricate at low temperatures without slowing the flow.

The major difference between a synthetic and the conventional lubrication stock is the cost. The synthetics are more expensive, but since they are man-made, they are more consistent and uniform in their structure.

Synthetic lubricants are not very compatible with other synthetic brands or with fossil base lubricants and should never be mixed. The additives used to enhance both of the base oils will deteriorate with time, so both must be changed at the usual interval time to validate the warranty and enhance engine life.

ENGINE OIL ADDITIVES

The base oil stock which is produced at the refinery is not adequate to meet the demands of the lubrication system. Many chemicals must be added to the base stock to meet the minimum requirements, such as:

1. anti-wear additives
2. friction modifier additives
3. oxidation inhibitor additive
4. rust and corrosion inhibitor additive
5. pour point depressant additive
6. foam inhibitor additive
7. viscosity index improver additive

ANTI-WEAR ADDITIVES

Anti-wear additives may also be referred to as Extreme Pressure (EP) additives that are added to form a protective coating of the metal surfaces at the boundary lubrication areas. The additives usually contain sulfur, phosphorous, and fatty materials.

FRICTION MODIFIER ADDITIVES

Friction modifier additives allow an oil to be classified as "energy conserving." The friction modifiers lower the friction between moving parts of the engine to increase the potential for improved fuel economy.

OXIDATION INHIBITOR ADDITIVE

When oil gets hot, it can react with oxygen, which creates a new molecule that thickens the oil. Anti-oxidants prevent atmospheric oxygen from combining with the molecules in the oil so that the oil remains at the proper viscosity for an extended period.

RUST AND CORROSION INHIBITOR ADDITIVE

One of the combustion products is water which, mixed with the engine oil along with other contaminants, forms a corrosive acid that can cause rust and damage to internal engine components. The additive works primarily by neutralizing the acids in the oil and by surrounding the corrosive molecules, effectively isolating them from the metal engine parts.

POUR POINT DEPRESSANT ADDITIVE

Pour point depressant additives are important in cold weather because they work like an antifreeze for the oil by preventing wax, contained in the base oil, from coming out of the solution. The pour point depressants effectively lower the temperature at which the oil will pour or flow adequately by surrounding the wax crystals and preventing them from growing.

FOAM INHIBITOR ADDITIVE

Foam inhibitors weaken the surface tension of the oil so that air bubbles are released quicker when they are mixed with the oil. Without the foam inhibitors, oil would become an emulsion of air and oil that would severely reduce its ability to function as a good lubricant.

VISCOSITY INDEX IMPROVER ADDITIVE

The operation of the internal combustion engine involves both high and low oil temperatures. When the oil cools, it thickens, becomes heavy (viscous), and resists flow. It may not circulate properly as a result. When the oil is heated, it thins out, becomes light, and may cause high oil consumption. It is necessary to blend an oil that will not become too thick when cold or too thin when heated.

Viscosity index improvers are "polymers," or plastic ingredients, that decrease the rate at which an oil becomes thinner or less viscous with temperature. The viscosity improver polymers can be imagined as little plastic solids that are small when the oil is cool, but expand or uncoil when they are heated. The expansion creates more friction between the oil molecules and thickens the fluid.

The oil can be blended to maintain a stable flow rate over a wide range of temperature. Viscosity index improvers are a key ingredient to multi-grade or all season engine oils.

OIL GRADES

In order to choose a correct engine oil, the technician must know how the lubricating oils are rated.

The oil should be selected that meets the manufacturer's viscosity recommendations and performance classification, as shown in figure 3-18. The American Petroleum Institute (API) is a voluntary collaboration of the automotive and oil industries working together to assure certain standards. The API symbol found on the oil container gives the technician some consistent standards of information among oil brands and formulations, as shown in figure 3-19. A container of engine oil bearing the API and SAE symbols conforms to the standards and specifications of the API and the Society of Automotive Engineers (SAE). The symbols contain information about the areas of:

1. *viscosity grade*

2. *oil performance classification*

3. *energy conservation rating*

SINGLE AND MULTI-VISCOSITY GRADES

The viscosity grade rating is found in the center area of the API service symbol, as shown in figure 3-20. Viscosity is the measurement of the oil's resistance to flow as temperature changes. In cold weather, oil thickens, and in the presence of heat, it flows more freely.

The SAE formulated a system for classifying the viscosity of engine oils to enable the consumer to choose the correct grade. The familiar SAE designations for the grades of oil refer to the viscosity of the oil at a specified temperature, such as the single weight or grade of oils, SAE 5W, SAE 10W, SAE 20W, SAE 30, SAE 40, and SAE 50. The oil can be designated as a single grade, or viscosity, or a multi-viscosity oil with a number such as SAE 5W-30, SAE 10W-30, or SAE 10W-40.

Higher viscosity oils, which are thicker and heavier, are identified by higher numbers than the thinner oils. The higher the number, the thicker the oil.

The "W" that follows some of the numbers originally stood for "winter," but its current meaning is to indicate the oil's cold start capabilities. The lower the "W" number, the easier the engine will start in cold temperatures. For instance, a SAE 5W-30 oil

Fig. 3-18 Viscosity rating of motor oil.

will allow an engine to start much easier when the temperatures are below 0° F than a SAE 10W-30 oil.

To have the "W" attached to a number, the oil must meet requirements for cold starting. Numbers without the "W" indicate that they are tested for their performance above 212° F, which is the average engine oil temperature of a water-cooled engine, but much lower than the 300° F operating temperature of the air-cooled engine.

A 5W-30 oil has the viscosity of 5 in cold weather and in a cold engine for easy starting. When the engine gets hot, then the oil has a viscosity of 30.

A single weight oil will become thick at low temperatures, which will make starting a concern in the winter months. If a lawnmower engine is operated only in the summer months, then a SAE-30 oil is recommended because it contains more base oil than the multi-viscosity and less viscosity improvers that tend to break down in high temperature conditions. A single weight oil will also control

Fig. 3-19 The API symbol found on oil containers.

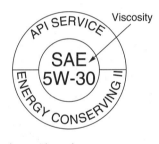

Fig. 3-20 Viscosity grade rating.

the loss of oil from blow-by between the piston rings and the cylinder. Air-cooled engines do not dissipate the heat of combustion evenly, and certain areas of the cylinder are hotter than others. This imbalance of heat will distort the cylinder slightly, causing some loss of oil.

Multi-viscosity oils show a lesser change in viscosity over a wide range of temperatures than single viscosity oils. This difference is accomplished by using viscosity index improver additives.

Consumers often specify an engine oil only by its viscosity. However, the viscosity has nothing to do with additional important testing relating to rust and corrosion protection, deposit control, oxidation inhibition, or other performance features in a modern engine oil.

Fig. 3-21 Oil performance classification section of the API symbol.

OIL PERFORMANCE CLASSIFICATION

The oil performance classification section of the API symbol is as important, or more important, than the viscosity rating in determining if the engine oil is of a quality suitable for the air-cooled or automobile engine, as shown in figure 3-21. The quality of each oil category (SG, for example) is determined in carefully controlled engine tests to measure the ability of an oil to reduce wear, sludge, varnish, oil thickening, rust, corrosion, and piston deposits.

The first letter "S" designates a quality suitable for gasoline engines, while if the first letter is a "C," it is a quality oil for diesel engines. The second letter denotes the rating of the additives, from the lowest letter "A" to the highest. If the oil is rated as SG/CC, then it can be used in a diesel or gasoline engine.

The highest rating presently is the SG rating for the modern auto engine. The highest rating should be used for the air-cooled engine, but the minimum rating recommended by the manufacturer can be used.

API rating: SA

The SA rating denotes a straight mineral oil with no additives.

API rating: SB

The SB rating indicates a non-detergent type of oil with very few additives that should not be used in the modern engine.

API rating: SC

The SC rating designates an oil blended to control the loss of oil and the formation of low temperature deposits and rust. It was formulated for use in the automobile produced from 1964 through 1967.

API rating: SD

The SD rating conveys an oil which was designed for use in gasoline engines from 1968 through 1970, and it provided more protection against high and low temperature engine deposits. The SD rating could be used in all engines that required a SC rating.

API rating: SE

The SE rating indicates an oil used in gasoline engines produced from 1971 through 1980. The oil was designed to provide even greater protection from oil oxidation, engine deposits, rust, and corrosion. The SE category should be used in all engines that require SC or SD ratings.

API rating: SF

The SF rating denotes an oil specified for gasoline engines from 1980 through 1990. The engines were downsized, and the oil was blended to withstand the additional stresses and heat. The SF category should be used in all engines that require SC, SD, or SE ratings.

API rating: SG

The SG oil rating is used in engines from 1991. Additional additives were used to increase the oil's performance. This is the oil rating that should be used in an air-cooled engine. Most air-cooled engines require a minimum of a SC rating, but any SD, SE, SF, or SG rating may be used.

ENERGY CONSERVATION RATING

The bottom part of the API circle contains the "energy conserving" rating, as shown in figure 3-22. Oil labelled "Energy Conserving II" offers the greatest benefits of reduced friction and improved efficiency.

RECOMMENDED LUBRICATION

When determining the general lubricating oil to use in the small air-cooled engines, the manufacturer's recommendation, found in the service manual, should be followed. Generally, in order to select the correct oil, three factors should be considered:

1. *The air temperature that the engine will operate.*
2. *The SAE viscosity grades.*
3. *The API engine service classifications.*

WHEN AIR TEMPERATURE IS ABOVE 40° F

The recommended viscosity for an engine that will be operated in temperatures above 40° F is a straight grade SAE 30, as shown in figure 3-23. SAE 10W-30 is an acceptable substitute that can be used, but engine oil consumption during operation will increase, so it is important to check the oil level more frequently.

NOTE: NEVER USE SAE 10W-40.

The API classification should have a minimum service classification of SC, but SD, SE, SF, and SG can also be used. The higher the classification, the better the additives to the basic oil. The highest classification is recommended.

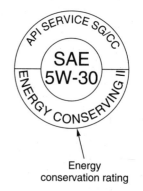

Energy
conservation rating

Fig. 3-22 Energy conserving rating of the API symbol.

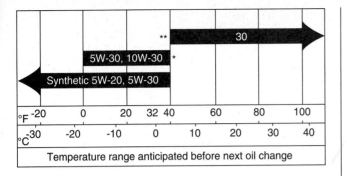

		**		30			
	5W-30, 10W-30	*					
Synthetic 5W-20, 5W-30							

| °F | -20 | 0 | 20 | 32 40 | 60 | 80 | 100 |
| °C | -30 | -20 | -10 | 0 | 10 | 20 | 30 | 40 |

Temperature range anticipated before next oil change

Fig. 3-23 Briggs & Stratton recommended oil viscosity rating.

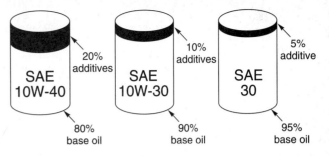

Fig. 3-24 Percentage of oil and additives of various viscosities of oil.

WHEN AIR TEMPERATURE IS BELOW 40° F

The recommended viscosity for an engine that will be operated in temperatures below 40° F is a multi-grade SAE 10W-30 or SAE 5W-30. If temperatures are below 0° F, then SAE 5W-20 or a synthetic oil base is recommended.

The API classification should have a minimum service classification of SC, but SD, SE, SF, SG can also be used. The higher the classification, the better the additives to the basic oil. The highest classification is recommended.

LUBRICATION PROBLEMS

Because of the increased operating engine temperature in the air-cooled engine, certain lubrication problems can occur.

SAE 10W-40 MULTI-VISCOSITY OIL

SAE 10W-40 viscosity oil should not be used in the small air-cooled engine. The SAE 10W-40 oil is formulated from about 80 percent lubricating stock and 20 percent additives, as shown in figure 3-24. A large portion of the additives are viscosity improvers that permit the wide range of ambient temperatures. SAE 10W-30 viscosity oil is blended with about 90 percent base stock and 10 percent of additives because less of the volume is used by the viscosity improvers to achieve this rating. A straight SAE-30 oil has about 95 percent base stock, and the

remaining 5 percent of the volume are additives necessary to attain the API SG rating without any addition of viscosity improvers. SAE-30 oil is recommended because there is more lubricating oil present per quart and no viscosity improvers to break down in the extreme operating temperatures.

The viscosity improver additive contains a chemical polymer that breaks down quickly at temperatures above 280° F. Under heavy load and high engine operating temperatures, a SAE 10W-40 oil's viscosity improvers break down to form a sludge that reduces the lubricating quality of the oil and coats the internal crankcase walls. The sludge is composed of a new substance that is thicker and absent of any lubricating qualities. Tests have shown that in extreme conditions, air-cooled engines have failed from lack of lubrication in as little as 10 hours when a SAE 10W-40 is used.

LOW OIL RESERVOIR

Check the oil level and fill to the "full" mark. An engine that operates with the oil in the maximum level position will protect the engine from overheating more so than an engine with a low crankcase supply. One of the functions of the engine oil is to cool the engine. As the oil is splashed or pumped to the engine components, the engine heat is absorbed and returned to the crankcase. The heat then dissipates to the metal crankcase, which radiates it to the atmosphere. The closer the oil level is to the "full" mark, the greater amount of heat can be absorbed by the oil and dissipated to the metal crankcase.

EXTRA ADDITIVES

Many engine oil additives are available that claim better lubrication and performance. With a proper maintenance schedule and frequent oil changes, there is no need for any additives. The incorporating of any extra additives many times does more harm than good.

"If a little is good, then a lot is better" is a belief that stimulates the introduction of large amounts of additives into the engine oil. Most additives are formulated to be used in the automobile (even though all the manufacturers discourage their use), and the mixing directions are for a large capacity oil reservoir. Too much additive is added to the engine oil, and with the additional heat from the air-cooled engine, the additive breaks down and forms a sludge in the crankcase.

TWO-CYCLE OIL

Two-cycle engines are lubricated with oil that has been pre-mixed or injected into the gasoline. When this mixture reaches the carburetor, two things happen: The gasoline is vaporized, and the oil is suspended as tiny droplets. As the air, oil, and fuel mixture flows through the crankcase, the oil droplets adhere to all the interior surfaces supplying lubrication to the bearings, piston, rings, and cylinder walls. The remaining air-fuel and some oil then flows to the combustion chamber, where it is burnt and its stored energy is released to help operate the engine.

Most engines have injection oilers that are timed to inject a calibrated amount of oil into the carburetor or intake manifold.

TWO-CYCLE VS FOUR-CYCLE OIL DEMANDS

The two-stroke-cycle engine can develop almost twice the power of a comparable-sized four-stroke-cycle engine. This is possible because a *two-stroke-cycle engine* has a power stroke *every time* the piston reaches the top of its travel, while in a *four-stroke-cycle engine,* there is a power stroke *every other time.* This feature tends to make the two-stroke-cycle engine run hotter and puts a tremendous stress on the lubricating oil. A two-stroke-cycle lubricating oil is blended with different additives than those found in ordinary four-stroke-cycle engine oils. The oil used in a four-stroke-cycle engine oil should not be used in a two-stroke-cycle engine because special additives were not meant to be burned and can cause the formation of undesirable deposits in the combustion chamber.

WATER-COOLED VS. AIR-COOLED

Outboard engines run cooler because they are water-cooled. Since the engines operate cooler, they require less lubricating oil than does an air-cooled engine. The water-cooled manufacturers are able to reduce the amount or oil used per gallon and can easily operate their equipment at a 50:1 ratio. The air-cooled engine normally must have a higher ratio of oil to each gallon of fuel, such as 32:1. The precise ratio differs from engine to engine.

SUMMARY

The fuel system performs many functions, which include storing the fuel, delivering the fuel to the carburetor, changing the liquid gasoline to a vapor, and mixing the vapor with air in the proportions for proper combustion.

The lubrication system must lubricate, seal, cool, clean, and provide corrosion resistance to the internal engine parts. The primary purpose of the lubricating system is to provide a flow of lubricating oil to internal engine parts to reduce wear and friction. Many additives are used in the lubricating oil to give the oil certain desirable properties. These include detergents, dispersants, extreme pressure agents, viscosity index improvers, and agents to inhibit oxidation, corrosion, rust, and foam.

Questions

1. The fuel system is necessary to engine operation. Technician A states that the fuel system performs the functions of delivering the fuel to the carburetor and changing the liquid gasoline to a vapor. Technician B states that the fuel system performs the functions of removing the burned combustion materials from the cylinder and regulating the water vapor in the incoming fuel mixture.
   Who is correct?

   A. Only Technician A
   B. Only Technician B
   C. Both Technician A and B
   D. Neither Technician A or B

2. A rich air fuel mixture is necessary for a cold engine to start. Technician A states that this is approximately nine parts gasoline to one part air. Technician B states that this is approximately nine parts air to one part gasoline.
   Who is correct?

   A. Only Technician A
   B. Only Technician B
   C. Both Technician A and B
   D. Neither Technician A or B

3. The anti-knock value is a rating of the octane rating of gasoline. Technician A says that the lower the compression ratio of an engine, the higher the anti-knock value of the gasoline. Technician B says that alcohol can be mixed with the gasoline to increase the anti-knock value rating.
   Who is correct?

   A. Only Technician A
   B. Only Technician B
   C. Both Technician A and B
   D. Neither Technician A or B

4. An uncontrolled combustion that is caused by a hot spot in the cylinder causes the incoming mixture to ignite early. Technician A says that this is called detonation. Technician B says that it is preignition.
   Who is correct?

   A. Only Technician A
   B. Only Technician B
   C. Both Technician A and B
   D. Neither Technician A or B

5. Alcohol can cause undesirable effects. Technician A says that that one of these effects can be that it can destroy rubber materials. Technician B says that it can counteract the lubricating properties of the fuel-oil mist in a two-stroke-cycle engine.
   Who is correct?

   A. Only Technician A
   B. Only Technician B
   C. Both Technician A and B
   D. Neither Technician A or B

6. In addition to lubricating the components in an engine, Technician A says that it will also reduce engine noise. Technician B says that it will also cool the engine parts.
   Who is correct?

   A. Only Technician A
   B. Only Technician B
   C. Both Technician A and B
   D. Neither Technician A or B

7. The correct viscosity of oil should be used. Technician A says that the most commonly recommended is SAE 10W-40. Technician B says that it should be SAE-30.
   Who is correct?

   A. Only Technician A
   B. Only Technician B
   C. Both Technician A and B
   D. Neither Technician A or B

8. If the oil meets the viscosity requirements of the engine and the API rating is also adequate, Technician A says that it does not matter what brand of oil is used. Technician B says that only the oil that is sold under the manufacturer's name should be used, eg., Briggs & Stratton oil for B&S engines.
   Who is correct?

   A. Only Technician A
   B. Only Technician B
   C. Both Technician A and B
   D. Neither Technician A or B

9. Fresh gasoline is necessary for proper engine operation. Technician A says that gasoline can have a stabilizer added to it to keep it fresh for one year. Technician B says that it is best to buy fuel like it is purchased for the automobile, which is one tank at a time, and to not store gasoline for more than three months without a stabilizer added.
   Who is correct?

   A. Only Technician A
   B. Only Technician B
   C. Both Technician A and B
   D. Neither Technician A or B

10. Gasoline is a blended hydrocarbon. Technician A says that the lighter, short-chained molecules are increased during the summer months. Technician B says that the lighter, short-chained molecules are decreased during the summer months.
    Who is correct?

    A. Only Technician A
    B. Only Technician B
    C. Both Technician A and B
    D. Neither Technician A or B

CHAPTER 4

Carburetors

INTRODUCTION

The carburetor measures the correct quantity of fuel to be supplied to the engine, breaking it down into small droplets called atomization and mixing the fuel with air. The carburetor must correctly proportion the amounts of fuel and air, regardless of the speed and load at which the engine is operating. Due to the differences in weight between fuel and air and the changing pressures to which they are subjected, the problem of maintaining correct mixture proportions is very difficult. For this reason, good carburetors have complicated construction.

The carburetor breaks the fuel into various sizes of drops suspended in an air stream so that the droplets vaporize easily before they reach the combustion chamber. The completeness of the vaporization depends on the following factors:

1. *Volatility of the fuel. The more volatile fuels vaporize more rapidly.*

2. *Temperature of the air. Higher temperatures increase the rate of vaporization.*

3. *Degree of atomization. A greater degree of fuel atomization presents the maximum amount of surface, and the fuel is more readily vaporized.*

In the time allowed for mixing the vapors of fuel with air, it is not possible to obtain a thorough mixing so that all the fuel particles will react with all of the oxygen particles. In order to consume all of the fuel or oxygen, there must be an excess of one or the other. Maximum power will occur when all the oxygen is consumed, which necessitates extra fuel particles. Maximum power fuel will be a rich mixture. The most economical mixture will have all the fuel particles consumed by supplying a lean fuel mixture with excessive air.

A carburetor must be able to provide different air-fuel ratios between 11 to 1 and 16 to 1 to provide the maximum power and greatest economy.

CARBURETOR CONSTRUCTION

A carburetor is a metering device for mixing fuel and air, as shown in figure 4-1. The correct mixture in the combustion chamber is essential for the engine to run properly. Two conditions must be met for proper carburetion. The fuel must be introduced to the incoming air stream and it must be vaporized.

VENTURI

The venturi, marked with #1 in figure 4-1, is the narrowed part of the carburetor tube where a suction is created and the velocity of the incoming air is increased (see figure 4-2). If you open the choke and throttle plate of the carburetor and look through the channel or air horn, you will notice that the venturi is the constricted section, as shown in figure 4-3.

MAIN DISCHARGE TUBE

The main discharge tube, marked with #2 in figure 4-1, is the part through which the fuel travels from the fuel container to the air stream in the venturi during high speed operation.

The main discharge tube of the carburetor is a tube, one end of which is connected to the venturi and the other in the fuel container found below. While air is flowing through the venturi, the effect will be the same as putting a straw in one's mouth and sucking on it. If one end of the straw is placed into a liquid, the fluid will be drawn up the straw. The fuel container can be the float bowl, part of the fuel tank, or the entire fuel tank. The fuel is actually pushed through the main jet tube by the difference

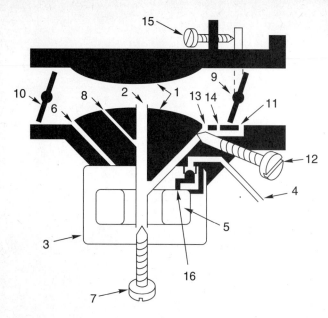

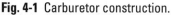

Fig. 4-1 Carburetor construction.

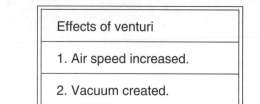

Effects of venturi
1. Air speed increased.
2. Vacuum created.

Fig. 4-2 Effects of venturi.

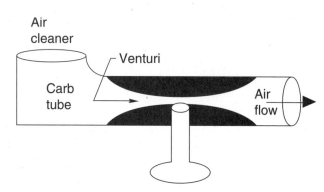

Fig. 4-3 Venturi is the constricted section in the air horn.

in pressure between the atmosphere and the venturi throat, as shown in figure 4-4. Normal atmospheric pressure pushes on the fuel in the reservoir and moves the fuel to the low pressure area in the venturi.

The greater the volume of air passing through the venturi, the higher the vacuum and the larger the amount of fuel that will be sprayed into the air stream at the main discharge tube.

FUEL CONTAINER

The fuel container, marked with #3 in figure 4-1, holds the fuel for use by the different metering circuits in the carburetor.

FUEL SUPPLY INLET

The fuel supply inlet, marked with #4 in figure 4-1, is where fuel enters the fuel container from the engine's fuel tank.

FLOAT

The float, marked with #5 in figure 4-1, is used to control the level of fuel in the fuel container. An

essentially unchanging level of fuel must be maintained. Proper metering of fuel-to-air ratios is dependent on a constant distance from the venturi to the surface of the fuel in the container.

When the engine speed or load increases, fuel is rapidly pulled out of the fuel bowl and into the venturi. This makes the fuel level and float drop in the bowl. Fuel enters the bowl and the float rises.

INLET NEEDLE

The inlet needle, marked #16 in figure 4-1, opens as the float level drops and fuel is allowed to enter the bowl area. As the fuel level rises, the float pushes the inlet needle back and shuts off the incoming fuel. When the engine is operating, the float and inlet needle regulate the incoming fuel flow to maintain the proper fuel level in the fuel container.

BOWL VENT

The bowl vent, marked with #6 in figure 4-1, allows atmospheric air pressure to enter the carburetor system. The difference in atmospheric pressure (relatively high) and the venturi pressure (relatively low) pushes the fuel from the fuel container into the

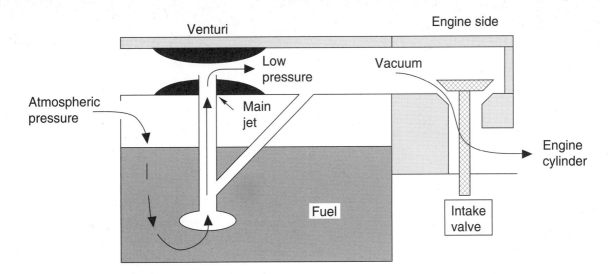

Fig. 4-4 Fuel is pushed through the main jet tube.

Fig. 4-5 External bowl vent.

Fig. 4-6 Internal bowl vent.

venturi while the engine is operating. The vent tends to maintain the air pressure above the surface of the fuel in the bowl at atmospheric levels.

The bowl vent may be an external or internal type. An external vent can be found on the outside of carburetor body, as shown in figure 4-5. An internal vent is commonly found in the carburetor air horn near the choke side of the carburetor, as shown in figure 4-6. An advantage of an internal vent is that a larger venturi can be used in order to maximize the horsepower of the engine because the pressure sup-

plied to the surface of the fuel in the fuel container is higher than the atmospheric pressure since air moving through the carburetor is forced into the vent opening.

HIGH SPEED MIXTURE ADJUSTMENT NEEDLE

The high speed mixture adjustment needle, marked with #7 in figure 4-1, is used to control the amount of fuel entering the air stream at high speed. It can be turned in to decrease the amount of fuel, which makes the air-fuel mixture lean (in fuel), or turned out for a rich (in fuel) air-fuel mixture.

HIGH SPEED AIR BLEED

The high speed air bleed, marked with #8 in figure 4-1, allows air to break up the fuel before entering the air stream in the venturi. When the air enters the carburetor, it forms a slight pressure near the venturi as the molecules are backed up while waiting to enter the venturi. This backup of air molecules increases the pressure slightly on the choke side of the venturi. With the high speed air bleed located in this high pressure area, some of the air moves through this channel to mix with the fuel in the main discharge tube.

THROTTLE PLATE

The throttle plate, marked with #9 in figure 4-1, controls the air flow through the venturi, thereby controlling the fuel flow to the engine. When the throttle plate is closed, all the air flow to the engine would cease, therefore, the throttle plate must be held open slightly by the idle speed screw, marked with #15 in figure 4-1. When the throttle plate is opened fully, the air flow into the engine is limited by the size of the venturi.

CHOKE PLATE

The choke plate, marked with #10 in figure 4-1, partially blocks off air flow, creating low pressure throughout the carburetor to provide a rich in-fuel mixture for cold starting. The fuel is drawn into the limited air flow from the main discharge tube and the idle passages.

IDLE PASSAGE

The idle passage, marked with #11 in figure 4-1, connects the carburetor's bowl to the engine side of the throttle plate. Fuel is forced through this passage when the throttle plate moves to the idle position.

LOW SPEED MIXTURE ADJUSTMENT SCREW

The low speed mixture adjustment screw, marked with #12 in figure 4-1, is used to meter the precise amount of fuel for engine operation at idle.

IDLE AIR BLEED

The idle air bleed, marked with #13 in figure 4-1, allows air to atomize the fuel before entering the air stream while the engine is idling. This premixing of the fuel and air increases the efficiency of engine combustion. When the throttle plate is in the idle position, the transitional fuel passage, marked with #14 in figure 4-1, also allows air to bubble into the idle passages.

TRANSITIONAL FUEL PASSAGES

The transitional fuel passage, marked with #14 in figure 4-1, provides a temporary fuel supply to the engine during the transition from idle to high speed operation. As the throttle plate begins to open, both the transitional fuel passages and the idle passage provide the air-fuel mixture.

CARBURETOR FUNCTIONS

EVAPORATION OF GASOLINE

Liquid gasoline will not support rapid combustion without being changed to a mist or broken down or atomized into tiny drops. When liquid gasoline is introduced to the venturi, it is gasified by its collision with the rapidly moving air mass. The atomized fuel then flows to the intake manifold where the pressure is low, and any remaining liquid gasoline evaporates further. Vaporization is finally completed with the heat generated by the compression of the air-fuel mixture in the cylinder.

FUEL CONTROL

The size of the venturi is matched with the power of the engine. If the engine is operated constantly in a full load condition, a relatively large venturi can be used for maximum efficiency and power, but since most engines are designed to run from no-load to full-load (idle to high speed), the air-to-gasoline ratio must be changed for different operating conditions. Heavy loads and fast acceleration necessitate a richer mixture. The venturi is narrowed enough to allow good suction at mid range speeds, and an idle circuit is added so that the engine will still draw

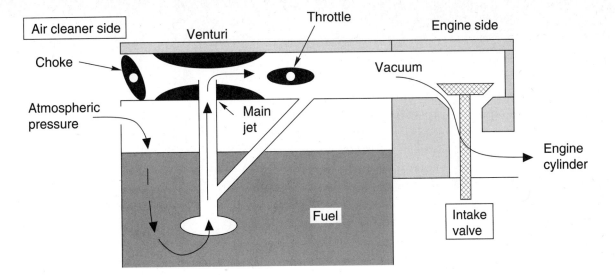

Fig. 4-7 Choke system.

fuel at low speeds. Either a fuel adjustment screw regulates the amount of fuel going up the main nozzle or a non-adjustable metering jet is used.

The venturi size directly affects the volumetric efficiency of the engine, which determines the horsepower rating of the engine.

EFFICIENT COMBUSTION

If the engine is warm, the air-fuel mixture enters the combustion chamber in a vaporized state. In an engine that has not yet reached operating temperature, some of the gasoline condenses before it goes into the cylinder. The ratio of air-fuel for proper combustion is dependent upon engine and intake air temperatures as well as many other factors.

During combustion the air-fuel mixture burns very rapidly. Combustion is not an uncontrolled explosion, but rather a regulated burn with a rapid release of heat energy, causing the air molecules in the cylinder to expand and force the piston downward.

STARTING THE ENGINE

When the air-fuel mixture is drawn into the cylinder on the intake stroke of the piston and the engine

is cold, the gasoline vapors tend to condense into large drops on their way to the cylinder. Because all the gasoline supplied to the cylinders does not vaporize, it becomes necessary to supply a rich mixture in order to have enough vapor for combustion to occur. This is accomplished by a choke, enrichment system, or primer.

Choke system

When the choke is tilted in the carburetor to restrict the amount of air entering the air horn, greater suction is created and a larger amount of fuel is drawn into the combustion chamber from the idle passages and the main discharge tube, as shown in figure 4-7.

Enrichment system

An enrichment system is an air-fuel metering circuit separate from other carburetor circuits. Fuel for the circuit is drawn from the float bowl through an enrichment jet. The enrichment jet is activated by the user through a special lever. When the plunger is down, the enrichment circuit is blocked and does not operate. When the plunger is raised, a rich air-fuel mixture is discharged from the enrichment port, which is located directly behind the venturi.

Fig. 4.8 Pneumatic primer system.

Primer system

The primer system provides a rich mixture by increasing the pressure in the top of the fuel bowl to force extra fuel into the venturi. Another way is by pushing down on the float to force a higher than normal float bowl level and create a rich condition.

In a diaphragm carburetor, the diaphragm may be lifted mechanically or pneumatically, as shown in figure 4-8.

HIGH SPEED OPERATION

The high speed circuit is provided by the main discharge tube, marked with #2 in figure 4-1, which is centered in the venturi. When the throttle plate is opened sufficiently, the air passing through creates a pressure difference in the venturi, which causes a discharge of fuel from the nozzle, as shown in figure 4-4. Throughout the intermediate and high speed ranges, this discharge increases with the volume of air passing through it so that a uniform air-fuel mixture ratio is maintained.

An adjusting needle or preset main jet controls the amount of fuel that enters the air stream. Maximum power is obtained with a mixture of from 12-to-14-parts of air to 1-part fuel. An excessively lean or rich fuel mixture will not burn properly and will not produce maximum power. If an engine is operating on too lean a mixture for maximum power and carburetor adjustments are made to increase the amount of fuel in the mixture, the horsepower output will also increase.

At partially open throttle plate position, it is possible to obtain maximum power at a leaner fuel mixture wherein the best air fuel ratio is 16 to 1.

IDLE SPEED OPERATION

When the throttle plate is closed or slightly open, only a small amount of air can pass through the air horn and flow around the throttle plate, as shown in figure 4-9. The air speed is so low, and there is such a small amount of air passing through, that hardly any vacuum develops in the venturi; thus, the main discharge tube in the venturi will not feed any fuel. For this reason, the carburetor must have an idle circuit.

When the throttle plate is in the idle position, the amount of air and fuel is greatly reduced as compared with wide-open throttle plate conditions. The mixture tends to burn more slowly and a richer fuel mixture is necessary.

Idling speed of the engine is controlled by the air flow past the throttle plate, and the volume of the air flow is controlled by the setting of the idle speed adjusting screw.

INTERMEDIATE SPEED OPERATION

When the throttle plate is placed at a position that is less than the high speed position but greater than the idle position, not only does the vacuum formed in the main venturi draw fuel into the air stream, but the idle circuit also delivers fuel at the secondary low pressure area around the idle fuel inlets. Both the high speed and the low speed mixture adjustments affect the air-fuel ratio.

MAINTENANCE AND TROUBLESHOOTING

Many engine performance problems are mistaken for carburetor malfunctions.

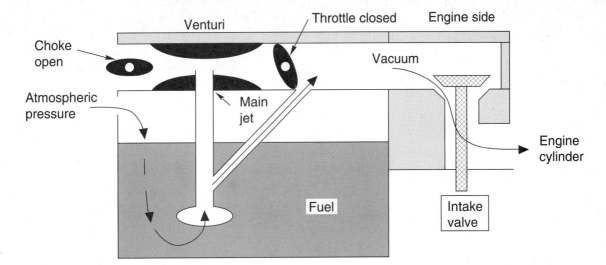

Fig. 4-9 Idle speed operation.

PRELIMINARY CHECKLIST

Good carburetor adjustments require reasonable integrity of all other engine systems.

Fuel tank

The interior of the tank must be free of corrosion or other contaminants that may settle in the bottom and block the flow of fuel.

Fuel lines

The fuel lines must not be bent so that the flow of fuel is obstructed in any way. Some fuel lines may disintegrate and collapse, causing the flow to the carburetor to stop.

Clean fuel

Check the fuel to determine if it is stale or dirty. White scale in the carburetor is evidence of water in the fuel. Once this type of oxidation has developed, there is no stopping it. The carburetor must be replaced.

Filters

The filter must be clean and installed in the correct direction. Arrows are stamped on the filter's cover showing the direction of the flow. The filter should be changed often.

Ignition

The symptoms of poor running carburetor adjustment may stem from a faulty ignition. A faulty ignition will cause an intermittent symptom of carburetor misadjustment. The proper spark plug must be used, with the gap adjusted to the manufacturer's specifications.

Compression

Without proper compression, the engine will be difficult to adjust. Insure that the compression is adequate.

Exhaust

A restricted exhaust port will cause a loss of power and difficulty in adjusting the carburetor. Make sure that the exhaust port is clear of any carbon obstructions.

Manifold gaskets

The manifold gasket must be in good condition and not leak. An air leak will cause the engine to operate in a lean condition. An intake manifold leak can be detected by putting oil in the area while the engine is idling and then listening for a change in engine speed. The oil will temporarily block the leaking gasket, and the mixture and engine sound will be affected.

Vapor lock

Vapor lock is a term used to describe the action of liquid fuel transforming into a gaseous "vapor" state, typically caused by heat. During this transformation, air bubbles are formed. Always use clean, fresh fuel that has been blended for the appropriate season.

Air bubbles from a vapor lock condition can become trapped at a high point in the fuel line or at a fuel filter, stopping the flow of fuel. Vapor lock may occur when the equipment is being operated or when restarting a warm engine.

Vapor lock during operation can be caused from insufficient cooling air. The exhaust gases may be trapped by the equipment design and recirculated to the cooling system. The warm cooling air may be misdirected near or at fuel lines, carburetor, and fuel supply.

Vapor lock occurs when a hot engine is shut down. Since the engine is hottest at about one minute after it has been turned off, it is best to cool the engine by letting it run at a speed just above idle for about 30 seconds before stopping it.

AIR CLEANER

Observe how the unit has been maintained. Check to see if the air cleaner is plugged or missing. The poly foam element may have been oiled incorrectly. A dirty element can cause cylinder bore wear that will affect performance. Dirt may be brought into the engine with air that is sucked around a worn choke or throttle plate shaft, and the engine may be difficult to adjust properly. Tecumseh furnishes dust seals on chrome plated shafts on their current carburetors to prevent this type of wear.

INITIAL SETTINGS

Make certain the governor linkage is connected correctly and adjust the mechanical governor. Make the initial settings on the carburetor and let the engine warm up for 3 to 5 minutes.

PROPER ADJUSTMENT

The home technician may have fine-tuned his machine, and the repair job may be as simple as correcting his work by adjusting the carburetor. Find the lean and rich points of the main adjustment screws and put the setting at half-way between the two points. It is best to find the lean point first, then the rich point, to make a proper adjustment.

Bring the engine to idle and adjust the idle screw in the same manner as the main mixture adjustment. Even though the idle mixture screw may be adjusting either air or fuel flow, the procedure will apply to both systems.

Take the engine back to high speed and reset the main mixture adjustment. Return the engine to idle again and reset the idle mixture adjustment. It is important to make the adjustment between the main and idle at least twice in order to ensure proper operation.

AIR LEAKS

Air leaks at the carburetor gaskets can cause a hunting condition. Anytime a gasket is disturbed, a new one should be used in its place. Make sure the float is adjusted properly.

FUEL SYSTEM FAILURES

Even though the fuel system is very simple, pin-pointing the difficulty can be frustrating. Fuel system problems can be separated into five categories:

1. *Incorrect fuel*
2. *Bad fuel*
3. *Too little fuel*
4. *Too much fuel*
5. *No fuel supplied to the engine*

Incorrect fuel

Some possible ways incorrect fuel may make its way into the system are:

- Fuel with an excessive amount of alcohol
- Diesel fuel put into the tank by mistake
- Fuel blended for one season, but used in another

Bad fuel

Some possible ways bad fuel may make its way into the system are:

- Fuel deteriorated
- Fuel with sand, dirt, grass, etc., in it
- Fuel left out in the rain without a cap

Too little fuel

The following may prevent the engine from receiving enough fuel:

- A partially plugged fuel filter
- A restricted fuel tank vent
- A pinched fuel line
- A collapsed fuel line that has separated internally
- A fuel pump with damaged, worn, or dirty parts
- An engine with a poor crankcase vacuum caused by leaking gaskets, loose oil-fill caps, loose oil filters, leaking piston rings, or bad seals
- A sticking or misadjusted carburetor inlet needle
- Damaged, dirty, or misadjusted carburetor mixture jets
- The wrong mixture jets in the carburetor
- The wrong carburetor
- An incorrect or missing air filter
- An excess amount of air being forced through the carburetor by the air intake system
- An incorrectly oiled air filter pre-cleaner
- A loose air cleaner housing
- A worn throttle plate shaft or carburetor body
- Loose, warped, or damaged connections in the intake manifold, which allows air to be drawn in

Too much fuel

Check for these problems that can result in too much fuel being supplied to the engine:

- Misadjusted or leaking carburetor fuel inlet needle
- Misadjusted or damaged carburetor fuel mixture jets
- Leaking internal welch plugs
- Leaking, missing, or incorrectly installed gaskets
- The wrong mixture jets installed in the carburetor
- A fuel pump that has a high output pressure
- The wrong carburetor

No fuel supplied to the engine

Look for the following possible problems:

- A plugged fuel filter
- An incorrectly positioned or missing fuel tank pick-up hose
- A pinched-off fuel line
- An internally collapsed fuel line
- A disconnected fuel line
- A failed fuel pump
- A sticking or plugged carburetor inlet needle
- Misadjusted or plugged carburetor fuel mixture jets

MISCELLANEOUS PROBLEMS

If the engine dies in between idle and high speed, the idle circuit must be cleaned with the welch plugs removed.

If the carburetor floods, it is generally because the inlet seat is leaking. The new inlet seat must be installed in the carburetor body in proper direction.

A plugged atmospheric vent hole will cause a lean setting in the carburetor. The external vents must be kept free of debris.

BRIGGS & STRATTON CARBURETORS

INTRODUCTION

The basic purpose of a carburetor is to produce a mixture of fuel and air in the correct proportions so

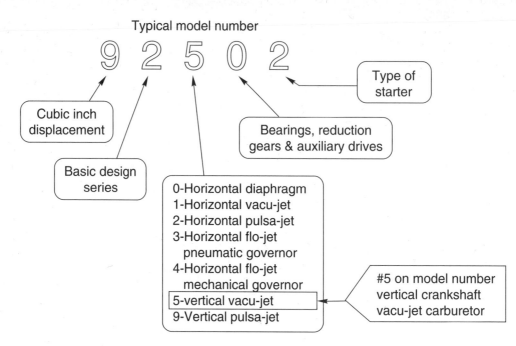

Fig. 4-10 Briggs & Stratton Vacu-Jet identification.

that the engine will operate properly at different speeds and loads. Briggs & Stratton manufactures three different basic types, designated as: Vacu-Jet, Pulsa-Jet, and Flo-Jet. Because environmental safety demands stricter emission and performance controls, many manufacturers realize that it is more cost effective to contract another company (which specializes in carburetors) to design and build them. Briggs & Stratton has entered into a business relationship with Walbro carburetors, and together they are producing a Flo-Jet type of carburetor with both of their names on it.

Prior to removing any carburetors for repair, note and record the position of governor springs, governor link, remote control, or other attachments in order to facilitate re-assembly.

VACU-JET CARBURETOR
Vacu-Jet construction

The Vacu-Jet is a simple carburetor which is used in many small displacement Briggs & Stratton engines and which functions adequately. The manu-facturing cost of this carburetor has helped the corporation stay competitive. It is adjusted differently than are other carburetors because of its unique features, as described below.

Vacu-Jet identification

The easiest way to identify the Vacu-Jet design is to check the engine's model number, as shown in figure 4-10. Another method is by recognizing its appearance. This is sometimes difficult to do because of changes in outward features from time to time, as shown in figures 4-11 and 4-12. The most reliable method of identification is when the carburetor is separated from the fuel tank and only one pick-up tube is present.

Vacu-Jet design

Fuel has to be lifted to the venturi. The atmospheric pressure enters the vent in the fuel cap and pushes on the mass of fuel. A single pick-up tube extends into the fuel tank through the bottom of the carburetor, figure 4-13. When air passes through the

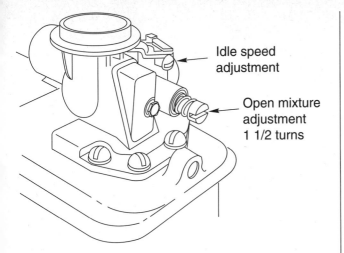

Fig. 4-11 Vacu-Jet all temperature carburetor with automatic choke.

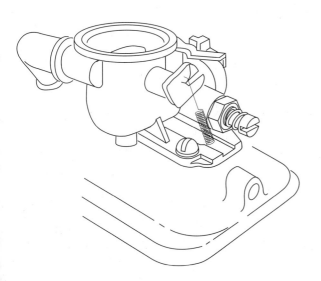

Fig. 4-12 Vacu-Jet Choke-a-matic.

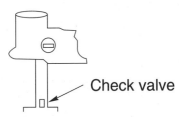

Check valve

Fig. 4-13 Vacu-Jet check valve.

venturi of the carburetor, the pressure in this area is reduced.

FUEL TANK BELOW CARBURETOR The fuel from the tank is pushed up through the tube to the low pressure area in the venturi. As the fuel is used up in the tank and the level is lowered, the greater the distance the fuel must be lifted. The change in the distance that the fuel must be lifted can cause a variation in the air-fuel mixture in the engine.

SINGLE FUEL PICK-UP TUBE This carburetor is identified by a single fuel pick-up tube incorporating a check valve, as shown in figure 4-13. The one-way valve helps to keep the fuel in the tube when the engine is not operating. Hard starting may indicate that the check valve is not working.

SINGLE MIXTURE ADJUSTMENT SCREW The amount of fuel flow through the single tube is regulated and adjusted by a single screw and a limiting orifice which affects both the high and low speed mixture adjustment. The limiting orifice, as shown in figure 4-14, does not allow excess fuel to enter and wash down the oil from the cylinder wall.

NO ACCELERATION PUMP Many of these carburetors are used on lawnmowers, where rapid acceleration is needed. The carburetor is adjusted to run "richer" to compensate for this.

Vacu-Jet adjustment

INITIAL CARBURETOR VACU-JET ADJUSTMENT The air cleaner must be assembled to the carburetor before running the engine. The best carburetor adjustment is obtained with the following steps:

1. *Make sure that the level of fuel in the tank is approximately 1/4 full.*

2. *Gently turn the needle valve clockwise until it just seats. The valve may be damaged by turning it in too far.*

3. *Open the needle valve 1-1/2 turns counter-clockwise. This initial adjustment will permit the engine to be started.*

4. *Next, with a warmed-up engine (3–5 minutes operating time), make the final adjustment.*

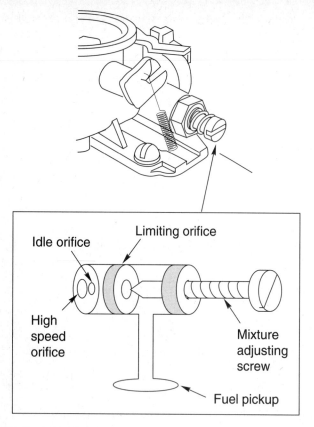

Fig. 4-14 Limiting orifice.

NOTE: ORIGINALLY, BRIGGS & STRATTON RECOMMENDED THAT THE FUEL TANK BE HALF FULL OF FUEL WHEN ADJUSTING. MANY WARM RESTART PROBLEMS HAVE PROMPTED A CHANGE IN THE INITIAL FUEL TANK LEVEL. THE DISTANCE THE FUEL HAS TO BE LIFTED WILL AFFECT THE MIXTURE ADJUSTMENT. AT 1/4 FULL, WE HAVE AN AVERAGE OPERATING CONDITION, AND THE ADJUSTMENT WILL BE SATISFACTORY EVEN IF THE ENGINE IS RUN WITH THE TANK FULL OR NEARLY EMPTY.

FINAL CARBURETOR VACU-JET ADJUSTMENT

1. *Place the speed control lever in the "fast" position.*

2. *Turn the needle valve in clockwise until the engine speed just starts to slow. Now open the needle valve 3/8-turn counter-clockwise, as shown in figure 4-15.*

3. *With your fingers, rotate the throttle plate counter-clockwise and hold it against the throttle plate stop while adjusting idle RPM. Turn the idle speed adjusting screw to obtain 1750 RPM, as recommended by the factory specifications. Use a tachometer to set the idle speed. If the idle speed is lower than recommended, the engine may not perform properly.*

4. *Release the throttle plate. The engine should accelerate smoothly. If the engine does not accelerate properly, the carburetor should be readjusted, usually to a slightly richer mixture, by opening the needle valve an additional 1/8-turn (counter-clockwise).*

5. *Check the adjustment by moving the engine control from "slow" to "fast" speed. The engine should accelerate smoothly. If the engine tends to stall or die out, increase the idle speed or re-adjust the carburetor, usually to a slightly richer mixture.*

NOTE: PREVIOUSLY, BRIGGS & STRATTON RECOMMENDED THAT THE CARBURETOR BE ADJUSTED TO THE "MID-POINT," BETWEEN TOO RICH AND TOO LEAN. THE ADJUSTMENT WORKED WELL WITH AN ENGINE BEING STARTED FOR THE FIRST TIME, BUT NOT IN THE CASE OF A HOT ENGINE. THE MID-POINT PROVED TO BE TOO RICH AND RESULTED IN AN ENGINE THAT WAS HARD TO RESTART. AS A CONSEQUENCE, THE COMPANY ADOPTED NEW RECOMMENDATIONS.

STARTING A FLOODED VACU-JET Flooding can occur if the engine is tipped at an angle for a prolonged period of time, if the engine is cranked repeatedly when the spark plug wire is disconnected, or if the carburetor mixture is adjusted too rich.

If flooding occurs in a Vacu-Jet with an automatic choke, move the governor control to "STOP" and pull the starter rope at least six times; in this situation the governor spring holds the throttle plate in a closed (idle) position. (Cranking the engine with a closed throttle creates a higher vacuum which opens the choke rapidly, permitting the engine to clear itself of excess fuel.) Then, move the control to "FAST" position and start the engine. If the engine continues to flood, lean the carburetor needle valve 1/8-turn clockwise.

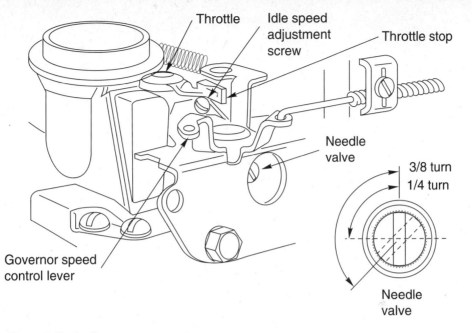

Fig. 4-15 Final adjustment.

PULSA-JET CARBURETOR

The Pulsa-Jet is a carburetor which incorporates a diaphragm fuel pump and a constant level fuel chamber, as shown in figure 4-16. The fuel tank, fuel pump, and constant level fuel chamber serve the same functions as the gravity feed tank, the float, and the float chamber of the conventional "float type" carburetors.

With this carburetor, the fuel level stays constant in the small chamber no matter what fuel level exists in the main tank. Very little "lift" is required to draw the gasoline into the venturi. The venturi can be made larger, as shown in figure 4-17, permitting a greater volume of air-fuel mixture to flow into the engine, enhancing the volumetric efficiency of the engine, and increasing the horsepower rating.

Pulsa-Jet construction

When the carburetor is removed from the fuel tank, notice that there are two different sized fuel pipes. The long one transfers fuel from the large tank to the small fuel cup by means of a fuel pump, while the short tube transfers the fuel from the fuel cup to the venturi.

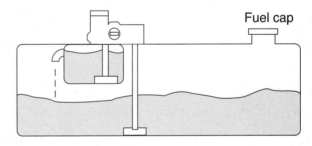

Fig. 4-16 Briggs & Stratton Pulsa-Jet carburetor with constant level fuel chamber.

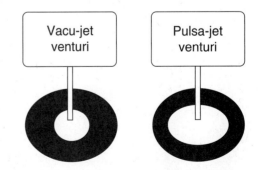

Fig. 4-17 Pulsa-jet uses a larger venturi than the Vacu-Jet.

Carburetor identification

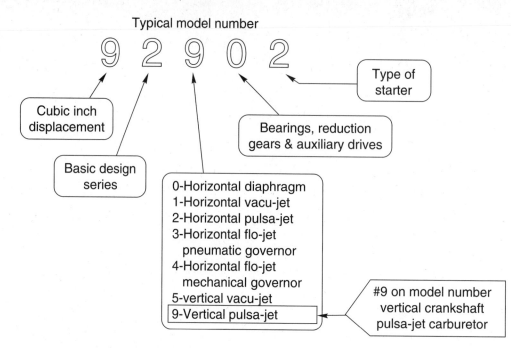

Typical model number

9 2 9 0 2

- Type of starter
- Cubic inch displacement
- Basic design series
- Bearings, reduction gears & auxiliary drives

0-Horizontal diaphragm
1-Horizontal vacu-jet
2-Horizontal pulsa-jet
3-Horizontal flo-jet
  pneumatic governor
4-Horizontal flo-jet
  mechanical governor
5-vertical vacu-jet
9-Vertical pulsa-jet

#9 on model number vertical crankshaft pulsa-jet carburetor

Fig. 4-18 Briggs & Stratton Pulsa-Jet identification.

Pulsa-Jet identification

The easiest way to identify the Pulsa-Jet design is to check the engine's model number, as described in figure 4-18. Another method is by recognizing its appearance. This is sometimes difficult to do because of changes in outward features from time to time, as shown in figures 4-19 to 4-24.

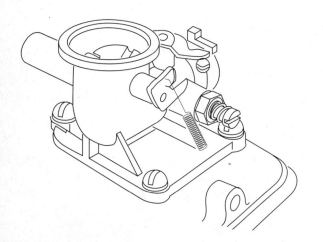

Fig. 4-19 Pulsa-Jet with Choke-A-Matic.

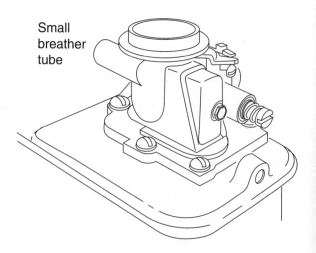

Small breather tube

Fig. 4-20 Pulsa-Jet with automatic choke.

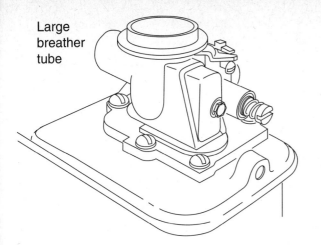

Large breather tube

Fig. 4-21 Pulsa-Jet all temperature with automatic choke.

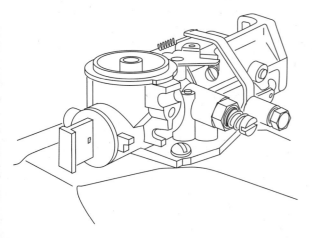

Fig. 4-22 Pulsa-Jet with slide choke.

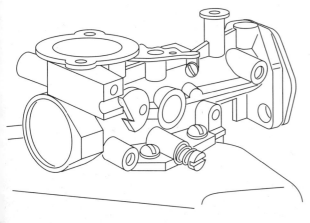

Fig. 4-23 Pulsa-jet with fixed main jet.

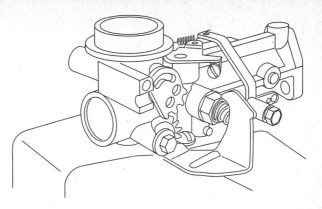

Fig. 4-24 Pulsa-Jet with rotary choke.

Pulsa-Jet fuel pump

The built-in fuel pump is actuated by the changes in pressures in the carburetor air horn or throat. When the piston is moving downward on the intake stroke, a low pressure area builds on the engine side of the venturi, as shown in figure 4-25. The low pressure is transferred to the pump diaphragm through a hole in the bottom of the air horn. The diaphragm is raised upward against the tension of the spring, and the area below the diaphragm develops a lower pressure or suction as it attempts to expand. The space does increase as fuel is sucked through valve #1 (a one-way flap permitting fuel to flow into the chamber), but not through valve #2 (also a one-direction flexible flap), which allows fuel to exit the compartment only.

When the piston is not on the intake stroke, the engine's intake valve is closed, as shown in figure 4-26, and the engine vacuum in the venturi area disappears. The spring that is located in the upper pump area pushes the diaphragm downward, causing an increase in the pressure in the lower chamber. The added pressure attempts to drive the fuel out of the area, but only valve #2 is allowed to open, as valve #1 is held shut by the pressure. Fuel is pushed into the small fuel cup of the carburetor.

This vacuum–no vacuum condition creates a pulsating action of the fuel pump's diaphragm which, along with the two pump valves, moves the fuel from the pick-up tube to the small fuel cup. Keep in mind that the diaphragm separates the air compartment and spring from the fuel compartment!

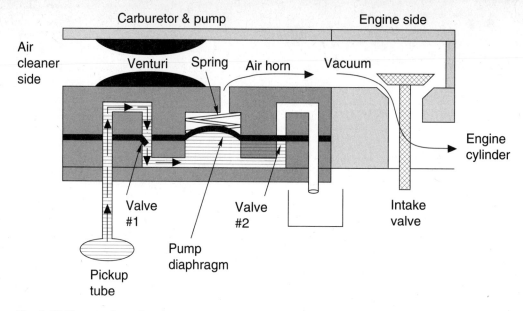

Fig. 4-25 Vacuum from the piston movement moves the pump diaphragm upward.

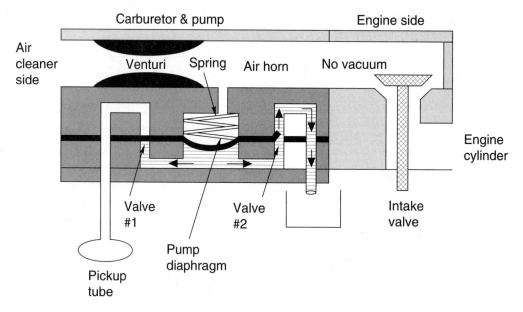

Fig. 4-26 Spring pushes pump diaphragm downward when no vacuum is present.

A common mistake in assembling the carburetor is the placement of the fuel pump spring and spring cover. The inexperienced person will place it into the fuel tank side of the diaphragm. The correct position, as shown in figure 4-27 (page 114), is in the recess on the carburetor side of the diaphragm. This is a calibrated spring and should never be stretched. If it is bad, it should be replaced.

Pulsa-Jet variations

The newest version of the Pulsa-Jet aluminum carburetor, as shown in figure 4-23, has a fixed main fuel jet with an adjustable idle fuel circuit. A venturi is cast as an integral part of the carburetor body, improving the engine's starting, idling, acceleration, and response to load. The fuel tank is vented through a passage within the carburetor. These new aluminum body carburetors are equipped with a throttle plate shaft dust seal.

A 0.046-inch diameter hole has been placed in the fuel tank's reservoir cup. When an engine runs out of fuel, both the tank and reservoir cup are nearly empty of fuel. When the fuel tank is filled to a level above this hole, fuel will transfer into the reservoir cup. This reduces the number of pulls required to "prime" the reservoir cup, allowing for quicker and easier starting.

Pulsa-Prime variation

Another new Pulsa-Jet family carburetor is the "Pulsa-Prime" carburetor. The main body of the carburetor is injection-molded glass reinforced nylon polymers. By using plastic, no machining is required. A wet bulb primer has been added to the carburetor. This eliminates the need for the automatic choke. The wet bulb primer will inject fuel directly into the throat of the carburetor for quick start with a cold engine through the high speed nozzle. The primer bulb will also assist when starting an engine after refueling. The Pulsa-Prime carburetor has no idle system and no air or fuel adjustments. The governor system will allow for a 600 RPM decrease in speed from top no-load speed, but the engine will never reach a true idle speed. The carburetor has a fixed high-speed jet that will provide fuel throughout the operating speed of the engine.

Initial Pulsa-Jet carburetor adjustment

The air cleaner must be assembled to the carburetor before running the engine.

1. *Gently turn the needle valve clockwise until it just seats. The valve may be damaged by turning it in too far.*

2. *Open the needle valve 1-1/2 turns counter-clockwise. This initial adjustment will permit the engine to be started.*

3. *Next, with a warmed-up engine (3–5 minutes operating time), make the final adjustment.*

Final carburetor Pulsa-Jet adjustment

1. *Place the speed control lever in the "fast" position.*

2. *Turn the needle valve inward, clockwise, until the engine speed just starts to slow. Now open the needle valve 3/8-turn counter-clockwise, as shown in figure 4-16.*

3. *With your fingers, rotate the throttle plate counter-clockwise and hold it against the throttle plate stop while adjusting idle RPM. Turn the idle speed adjusting screw to obtain 1750 RPM, as recommended by the factory specifications. Use a tachometer to set the idle speed. If the idle speed is lower than recommended, the engine may not perform properly.*

4. *Release the throttle plate. The engine should accelerate smoothly. If the engine does not accelerate properly, the carburetor should be readjusted, usually to a slightly richer mixture, by opening the needle valve an additional 1/8-turn (counter-clockwise).*

5. *Check the adjustment by moving the engine control from "slow" to "fast" speed. The engine should accelerate smoothly. If the engine tends to stall or die out, increase the idle speed or readjust the carburetor, usually to a slightly richer mixture.*

NOTE: PREVIOUSLY, BRIGGS AND STRATTON RECOMMENDED THAT THE CARBURETOR BE ADJUSTED TO THE "MID-POINT," BETWEEN TOO RICH AND TOO LEAN. THE ADJUSTMENT WORKED WELL WITH AN ENGINE BEING STARTED FOR THE FIRST TIME, BUT NOT IN THE CASE OF A HOT ENGINE. THE MID-POINT PROVED TO BE TOO RICH AND RESULTED IN AN ENGINE THAT WAS HARD TO RESTART. AS A CONSEQUENCE, THE COMPANY ADOPTED NEW RECOMMENDATIONS.

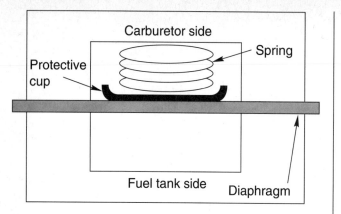

Carburetor side

Protective cup

Spring

Fuel tank side

Diaphragm

Fig. 4-27 Correct placement of the fuel pump spring.

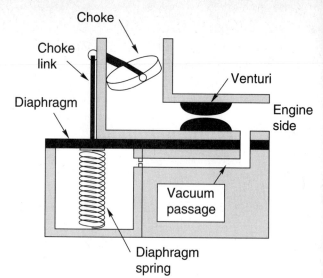

Choke

Choke link

Diaphragm

Choke

Venturi

Engine side

Vacuum passage

Diaphragm spring

Fig. 4-28 Briggs & Stratton automatic choke.

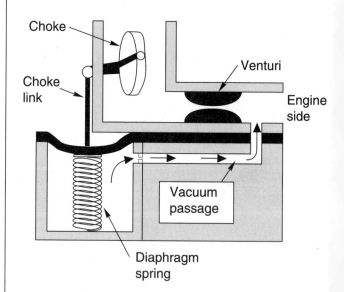
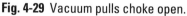

Choke

Choke link

Choke

Venturi

Engine side

Vacuum passage

Diaphragm spring

Fig. 4-29 Vacuum pulls choke open.

AUTOMATIC CHOKE (VACU-JET/PULSA-JET)

Theory of operation

The automatic choke provides a rich mixture condition when starting the engine. A diaphragm under the carburetor is connected to the choke shaft by a rigid link. A calibrated spring under the diaphragm holds the choke valve closed, as shown in figure 4-28, when the engine is not running. When the engine starts, the choke opens as the vacuum created in front of the venturi is transferred via a calibrated passage to the area below the choke diaphragm. As the diaphragm and choke link moves downward, shown in figure 4-29, the choke is pulled open.

This system has the ability to respond in a fashion similar to an acceleration pump. As speed decreases during heavy loads, the choke valve partially closes, resulting in a richer mixture and improved acceleration performance.

Pre-loading diaphragm

Before tightening the carburetor mounting screws in a staggered sequence, move the choke plate to an over-center position and hold it there. This pushes the choke link downward to the bottom of its travel and pre-stretches the diaphragm, as shown in figures 4-30 and 4-31. Tighten the screws in a crossing sequence to 35 inch-pounds.

Automatic choke inspection

1. *Remove the air cleaner and replace the stud. Observe the position of the choke valve. It should be fully closed.*

2. *Move the speed control to the stop position. The governor spring should be holding the throttle*

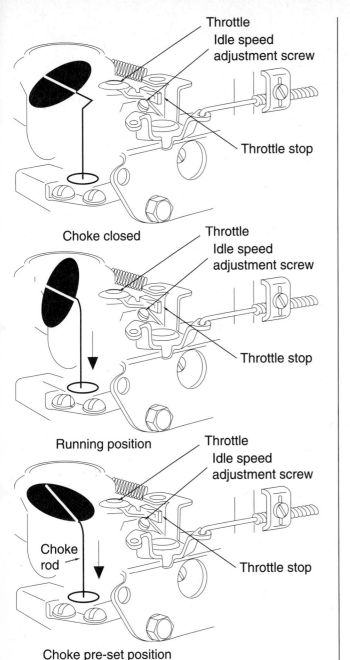

Choke closed

Running position

Choke rod

Choke pre-set position

Fig. 4-30 Off closed, Running position, Adjust position.

Throttle
Idle speed
adjustment screw

Throttle stop

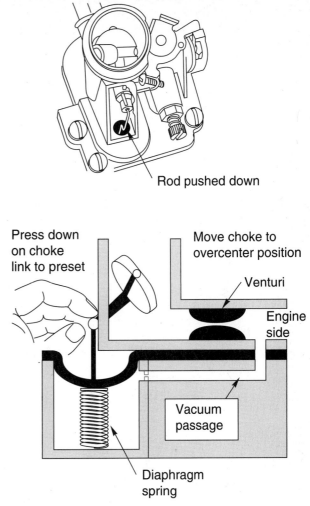

Rod pushed down

Press down on choke link to preset

Move choke to overcenter position

Venturi

Engine side

Vacuum passage

Diaphragm spring

Fig. 4-31 Press down on choke link to preset.

plate in a closed position. Pull the starter rope rapidly. The choke valve should alternately open and close.

3. *If the engine can be started, run for 2 or 3 minutes at a normal operating speed. Close the needle valve (turning clockwise) enough to make the mixture lean, then adjust needle valve 3/8-turn open (counter-clockwise).*

4. *Allow the engine to run at idle speed for 3 to 5 minutes. Again, close the adjusting speed needle; the mixture should become so lean that the engine will stop. If the engine continues to run at idle with the needle valve closed, a fuel leak is occurring at the diaphragm.*

Bi-metal choke

An engine equipped with a safety stopping mechanism can be expected to be stopped, and then restarted, on a frequent basis. This means that the engine must start easily each time. The bi-metal automatic choke carburetor is used in these situations.

The diaphragm springs that are used in the automatic choke carburetors may look alike, but will vary in function, depending on usage. Therefore, utilize only the spring recommended in the parts book for the engine you are tuning. The springs have been color-coded according to their strength. The stronger the spring, the longer the engine will be kept in the choke condition. The springs from the strongest to the weakest are: no color, red, blue, and green.

The operation of the automatic choke remains the same at moderate engine starting temperatures. However, at "hot" or "cold" starting temperatures, the bi-metal choke, as shown in figure 4-32, carburetor automatically compensates for temperature changes by influencing the choke opening time.

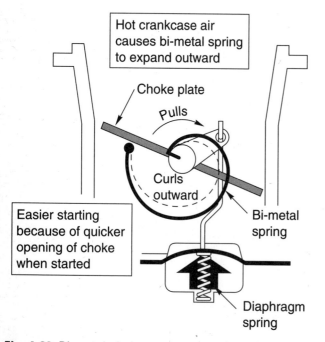

Hot crankcase air causes bi-metal spring to expand outward

Choke plate

Pulls

Curls outward

Easier starting because of quicker opening of choke when started

Bi-metal spring

Diaphragm spring

Fig. 4-32 Bi-metal choke compensates for temperature changes.

At higher engine starting temperatures, the "hot" crankcase air passing through the breather tube causes the bi-metal spring to expand and curl outward. This action causes the inner end of the bi-metal spring to "pull" the choke shaft/choke plate open, but since the diaphragm spring is opposing this force, the choke plate continues to remain closed. This action does, however, make it easier for starting vacuum to open the choke plate quickly and improve "hot" starting problems.

At lower engine starting temperatures, the "cold" ambient air causes the bi-metal spring to contract and curl inward. This action causes the inner end of the bi-metal spring to "push" the choke shaft/choke plate closed, which assists the diaphragm spring in holding the choke plate closed longer. The result is a slightly longer period of a rich air-fuel mixture, assuring improved "cold" temperature starting.

Automatic choke inspection areas

When a problem is suspected in the automatic choke system, certain areas should be inspected to determine whether:

1. *The carburetor is adjusted too lean or too rich*
2. *The fuel pipe check valve is inoperative (Vacu-Jet only)*
3. *There is a bent air cleaner stud*
4. *There is a sticking choke shaft due to dirt, etc.*
5. *The choke spring is damaged or too short*
6. *The diaphragm is not pre-loaded*
7. *The diaphragm is ruptured*
8. *The vacuum passage is restricted*
9. *Gasoline or oil is in the vacuum chamber*
10. *There is a leak between link and diaphragm*
11. *The diaphragm folded during assembly*

Auto choke problems and solutions

1. *Flooding, during hot restart, has been improved by a change in the fuel tank. The "V" notch channel in the fuel tank top regulates the necessary amount of vacuum underneath the*

choke diaphragm. To obtain a more precise method of metering this vacuum, a drilled hole (0.014-inch diameter) has replaced the "V" notch, as shown in figure 4-33.

2. *An engine with a hot restart condition will have the following symptoms: the engine starts and runs well, but when shut off, it is difficult or* impossible to restart until the engine cools off. Too much fuel added to a warm engine equipped with a tank mount-style carburetor assembly will cause a rich condition, resulting in difficult hot restart. The flooded condition will be even worse if the engine is bumped, tipped on an angle, or if the fuel tank cap is not venting properly.

3. A damaged diaphragm can cause a rich condition. If the diaphragm has been perforated, fuel may bypass the carburetor's metering system and be pulled directly into the engine through the air compartment and hole in the air horn.

4. The condition of the diaphragm link, diaphragm, and diaphragm spring can affect proper operation of the choke. If any of these components are damaged, a hot restart problem can occur. The use of non-original parts, which have not been manufactured to the original manufacturer's specifications, can cause a hot restart condition.

5. The "plastic" Minlon carburetor is a mineral-filled nylon material used in the production of the Vacu-Jet carburetor and was introduced in the late 1970s. The injection molding process created passageways that were smooth and efficient, and the adjustments and "pre-loading" procedure are identical to the metal version of the carburetor. In the metal carburetor, the air cleaner stud threads into the carburetor body, while in the Minlon carburetor, the air cleaner stud extends through the carburetor body and threads into the fuel tank top. Before "pre-loading" the Minlon carburetor diaphragm, the air cleaner stud hole in the carburetor body must first be properly aligned with the tapped air cleaner stud hole in the fuel tank top. A misalignment between the air cleaner stud hole in the carburetor body and the tapped air cleaner stud hole in the tank top can cause the choke plate to bind on the air cleaner stud. This situation will prevent the choke plate from opening completely (engine running too rich), or the choke plate from closing completely (engine hard to start). Obviously, for these reasons, the alignment is most important for trouble-free automatic choke operation.

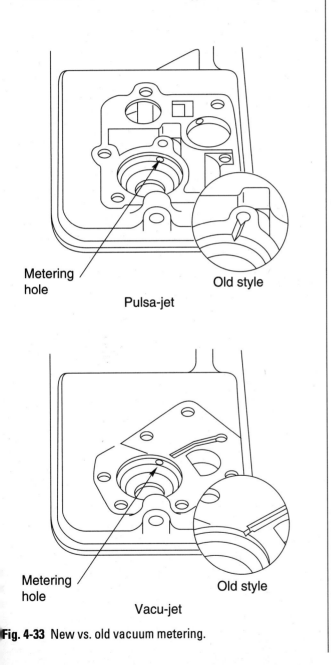

Metering hole

Old style

Pulsa-jet

Metering hole

Old style

Vacu-jet

Fig. 4-33 New vs. old vacuum metering.

FLO-JET CARBURETOR

Flo-Jet construction and design

The Flo-Jet resembles the internal workings of the carburetors generally found on automobiles. The fuel tank is mounted higher than the carburetor, and gasoline flows into it by the pull of gravity. A float inside the carburetor bowl regulates the flow of gasoline, similar to the float inside your toilet tank that regulates the level and flow of water.

Flo-Jet identification

The easiest way to identify the Flo-Jet design is to check the engine's model number. Another method is by recognizing its appearance, as shown in figures 4-34 to 4-36.

Flo-Jet carburetor adjustment

FLO-JET DISSASSEMBLY Prior to the separation of the upper and lower half of the carburetor, the fuel nozzle, as shown in figure 4-37, must be removed by using Briggs & Stratton tool #19280, or by using a screwdriver without a taper, as damage can result to the screw section. Failure to remove the nozzle will result in a carburetor that may leak when the engine is not operating. The fuel nozzle is visible after the high speed adjusting screw and packing nut are removed.

FLO-JET CARBURETOR CLEANING Particles of dirt and debris in the fuel system will create erratic engine operation and affect carburetor adjustments. If this problem is suspected, clean the entire fuel system, including the carburetor. Install a fuel filter in the fuel line and adjust the float level, as described below:

FLO-JET FLOAT ADJUSTMENT With the body gasket in place on the upper body and the float valve and float installed, the float should be parallel to the body mounting surface, as shown in figure 4-38. If

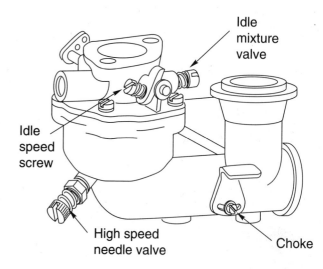

Fig. 4-35 Medium Briggs & Stratton Flo-Jet carburetor.

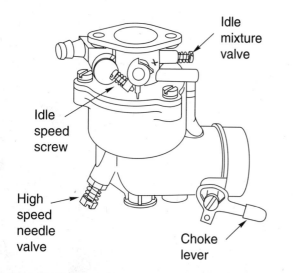

Fig. 4-34 Small Briggs & Stratton Flo-Jet carburetor.

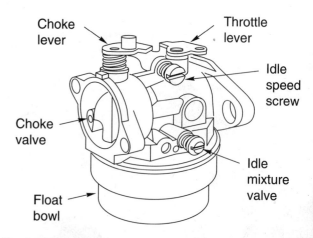

Fig. 4-36 Briggs & Stratton-Walbro carburetor.

Fig. 4-37 Removal of the fuel nozzle.

not, bend the tang on the float with a screwdriver until it is parallel. Do not press on the float.

INITIAL FLO-JET ADJUSTMENTS Before starting the engine, the air cleaner should be clean and assembled to the carburetor. If the mixture valves are adjusted without the air cleaner in place, difficulties will be encountered after the air cleaner is attached. Some resistance to the air intake is caused by the cleaner, and this will result in added suction in the air horn (which was not present when the adjustments were made).

1. Gently turn the mixture valves clockwise until they just seat. Valves may be damaged by turning them too far and too tightly.

2. Open the high speed needle valve, as shown in figure 4-39, 1-1/2 turns counter-clockwise and the idle valve one turn. This initial adjustment will permit the engine to be started.

3. Start engine and allow it to warm up (approximately 5 minutes) prior to final adjustment.

FINAL FLO-JET ADJUSTMENT

1. When the engine is warm and running, place the speed control lever in the "fast" position.

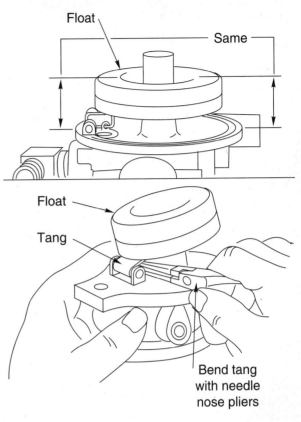

Fig. 4-38 Flo-Jet float adjustment.

2. Turn the high-speed needle valve inward until the engine slows (clockwise-lean mixture). Then turn it outward, past smooth operating point (counter-clockwise-rich mixture). Now turn the needle valve to mid-point between rich and lean.

3. Adjust the idle RPM to 1750 RPM, as specified by the manufacturer, by rotating the throttle plate counter-clockwise and holding it against the stop and turning the idle mixture valve screw.

4. While holding the throttle plate against the idle stop, turn the idle mixture needle valve clockwise until the engine slows. Then turn it counter-clockwise, past the smooth operating point. Set it at a midpoint between rich and lean.

5. Recheck the idle RPM and release the throttle plate. If the engine does not accelerate properly, readjust the high speed mixture valve approximately 1/8-turn counter-clockwise (rich).

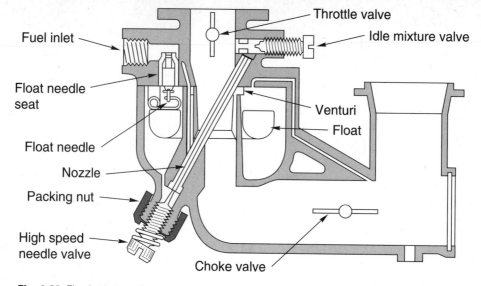

Fuel inlet

Float needle seat

Float needle

Nozzle

Packing nut

High speed needle valve

Throttle valve

Idle mixture valve

Venturi

Float

Choke valve

Fig. 4-39 Flo-Jet internal parts.

Flo-Jet problems

LEAKAGE Carburetor leakage is not only irritating, but also a safety concern. The first step toward finding the cause of the problem is to identify when the leak occurs.

Float bounce is one cause for the Flo-Jet carburetor to leak. Float bounce is a condition that typically occurs when the engine/equipment is transported. The use of a fuel shut off valve is recommended for use on all float-style carburetor systems.

The fuel supply (fuel tank) may be located too far above the carburetor, resulting in excessive pressure at the fuel inlet needle and a leak. The maximum tank height recommended for gravity feed fuel systems is forty-five inches. The use of an in-line filter is recommended for all float-style carburetor systems.

If the carburetor leaks shortly after the engine is turned off, it may be due to a long "coast down" period (prolonged spinning of the engine). A long "coast down" period will cause accumulation of unburned fuel. Whenever possible, slow the engine's speed to an idle before shutting off the engine. Using the choke as a means to shut off the engine will only aggravate the leakage condition and is not recommended.

The remaining causes for Flo-Jet carburetor leakage involve parts which are loose, missing, assembled/adjusted incorrectly, damaged, or affected by contaminants in the fuel system such as dirt, water, or additives. Understanding the design of the carburetor is necessary to efficiently isolate and identify the actual cause of the problem.

A common cause of fuel leakage in a two-piece Flo-Jet carburetor is an improper seal between the main nozzle and the lower half of the carburetor. There are three ways to correct this problem: First, take an old nozzle and grind all the threads off the outside. Do not leave any sharp edges that could damage the threads inside the carburetor. Next, put a small amount of fine lapping compound on the shoulder of the nozzle. Using a screwdriver as a lapping tool, lap the nozzle into the carburetor. This will remove corrosion and restore a sealing surface. Be sure to thoroughly clean the carburetor before reassembly. Another way to correct the problem is to use a part from the #391413 carburetor repair kit for servicing a Pulsa-Jet carburetor. Using a Teflon washer from the kit, force the washer over the end of the nozzle. The washer will act as a gasket, stopping any leakage between the nozzle and carburetor body. The third option is to

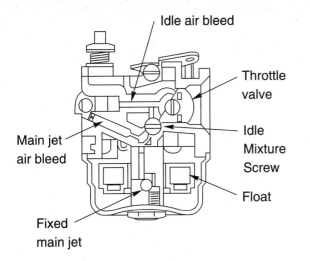

Fig. 4-40 Briggs & Stratton-Walbro carburetor-internals.

replace the lower half of the carburetor or the entire carburetor assembly.

Damage at the float valve seat/bushing may be the result of abrasives, corrosion, or careless use of tools. If the condition is limited to the seat/bushing, the problem can be corrected quick and easily. Pressed-in float valve seat/bushings are replaceable on all float-style carburetors.

HARD HOT RESTART When the warm engine is difficult to start, the primary cause is a rich or flooded condition. If an engine has a hard hot restart symptom, first check the engine's spark plug to determine if a flooded condition exists. This is accomplished by removing the plug and observing the tip to see if it is covered with fuel. Hard hot restart can be caused by an improperly adjusted engine and/or equipment controls, or by a partially restricted air filter. Perform all initial adjustments and recheck for a hard restart condition before attempting more repairs.

A damaged adjusting needle, O-ring, or loose needle/seat will cause a rich condition which contributes to hard hot restart problem. Inspect for missing, damaged, or loose parts.

Another critical factor is the engine's idle speed. When the engine's idle RPM is set too low, hesitation occurs during acceleration. To eliminate this situation, some technicians make the air-fuel mixture richer, which will cure the hesitation, but also will cause a hot restart condition.

Winter grade fuel used in warm weather may vaporize too readily and cause a flooding (hot restart) condition.

Briggs/Walbro one-piece Flo-Jet

Walbro now produces a carburetor, as shown in figure 4-40, in association with Briggs & Stratton. Hot start problems plagued the industry for many years, and many manufacturers decided to use the high technology of the Walbro Corporation. Briggs & Stratton has an agreement whereby the Walbro Flo-Jet carburetor and parts can be sold only to Briggs & Stratton Corporation. Briggs & Stratton then uses their distribution system to supply the repair parts and information.

These carburetors are made in two basic types: the fixed high-speed main jet and the adjustable high-speed main jet.

BRIGGS/WALBRO FLO-JET ADJUSTMENT

1. *Turn the idle mixture screw inward until it just seats, and then back out the screw one turn. This will permit the engine to start.*

2. *On carburetors with an adjustable high speed needle, turn the high speed adjustment needle clockwise until it just touches the needle seat, then back off 1-1/4-turn.*

3. *Start and run the engine for five minutes at a moderate speed to bring the engine up to operating temperature.*

4. *Move the equipment speed control to the idle position. Turn the idle speed screw to obtain a 1750 RPM minimum.*

5. *Turn the idle mixture screw clockwise slowly until the engine just begins to slow. Next, turn the screw in the opposite direction until the engine begins to slow. Turn the screw back to the midpoint.*

6. *On carburetors with an adjustable high speed needle, move the speed control to the fast position. Adjust the high speed mixture needle in the same manner as the idle mixture screw.*

7. *Move the equipment speed control from idle to high-speed position. The engine should accelerate smoothly. If it doesn't, open the idle mixture needle screw 1/8-turn.*

TECUMSEH CARBURETORS

TYPES

Series I Tecumseh carburetor

This carburetor, as shown in figure 4-41, comes in a variety of styles for use in the four-stroke-cycle engine ranging from two through seven horsepower. This float-style, small venturi carburetor may have an adjustable low and high speed air-fuel mixture screws, or a fixed main with an adjustable idle.

Series II Tecumseh carburetor

This carburetor was designed similar to the Series I for outboard boat engines where the external fuel tank would be below the carburetor, as shown in fig-ure 4-42. The carburetor had an additional fuel pump circuit built into it along with a special emulsion idle adjust that could be manipulated from the operator's control panel.

Series III & Series IV Tecumseh carburetors

These carburetors, as shown in figure 4-43, are generally used on eight through ten horsepower four-stroke-cycle engines and have a larger venturi size. The quickest way to identify these carburetors is by the bosses on each side of the idle mixture screw, which were placed there for the possibility of attaching different types of control panels. Series III has one screw holding the choke plate in place, and Series IV has two screws. Two screws are added to the Series IV because of its larger size and the tendency of the choke plate to warp if held only by one screw.

These carburetors can have either an external or internal atmospheric float vent and either a solid fuel or an emulsion adjust idle circuit. The operation of these carburetors duplicates the Series I.

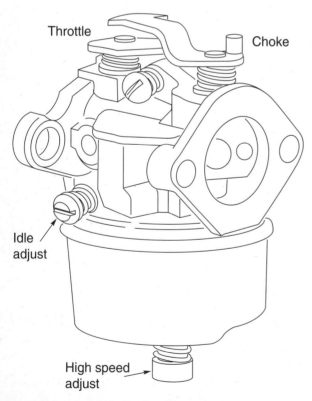

Fig. 4-41 Tecumseh Series I carburetor.

Fig. 4-42 Tecumseh Series II carburetor.

Series V Tecumseh carburetors

This carburetor was never produced. Since the engineering drawings were made, the number was skipped.

Series VI carburetor

This carburetor, as shown in figure 4-44, is used on many recent engines. It is similar to the Dual System carburetor and is equipped with a fixed main jet as well as a fixed idle circuit, as shown in figure 4-45. The primer is designed to pressurize the float bowl area to obtain a better delivery of fuel to the venturi when starting a cold engine. The bowl nut on some models has a left hand thread.

Dual System Tecumseh carburetor

Originally, the non-adjustable carburetor produced by Tecumseh was called AutoMagic™, and it had one fuel orifice leading to the main jet. When the engine was used on a lawnmower, it was designed to stop the engine from running when mowing a steep incline, as shown in figure 4-46. This feature protected the engine from lubrication failure when the oil pump was "starved" for oil. This chokeless carburetor was also difficult to start in temperatures below 40° F. The carburetor was updated with a primer system, as shown in figure 4-47, for easier starting, and an extra fuel orifice was added to the main jet area, originating the name "Dual System."

The easiest way to identify the dual system carburetor is by the large primer bulb on the side. The absence of adjustment needles helps to identify the carburetor as well. The dual system carburetor is used on four-stroke-cycle vertical crankshaft engines used on a lawnmower.

Tecumseh diaphragm carburetor

This carburetor, as shown in figure 4-48, is noticeably different by its lack of a float area. It has a rubber-like diaphragm, which is exposed to intake manifold pressure on one side and to atmospheric pressure on the other. As the intake manifold pressure decreases, the diaphragm moves against the inlet needle, allowing the inlet needle to move from

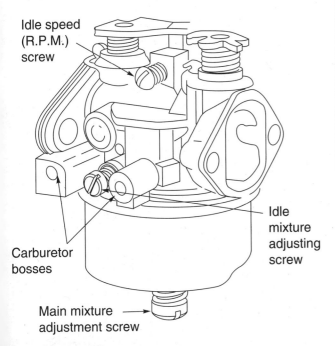

Fig. 4-43 Tecumseh Series III carburetor.

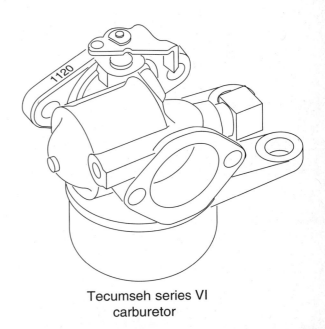

Tecumseh series VI carburetor

Fig. 4-44 Tecumseh Series VI carburetor.

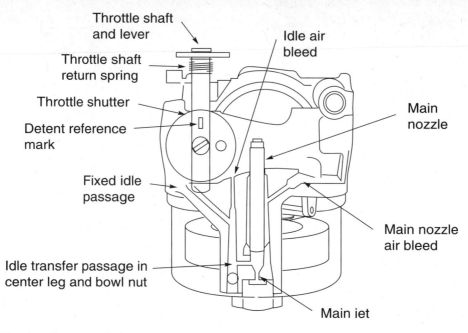

Fig. 4-45 Tecumseh Series VI carburetor-internals.

its seat. This permits the fuel flow through the inlet valve to maintain the correct fuel level in the fuel chamber. An advantage of the diaphragm carburetor over the float feed system is that the diaphragm system will allow the engine to operate at a greater angle and on bumpy terrain.

Tecumseh Vector carburetor

This unique carburetor, with its fixed main and idle jets, is a recent addition to the Tecumseh carburetor family, as shown in figure 4-49. The upper part is constructed of extruded aluminum with a pressed-in plastic venturi. Long aluminum bars are

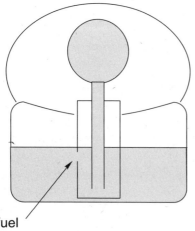

Fig. 4-46 Tecumseh AutoMagic fuel orifice.

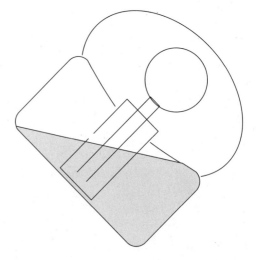

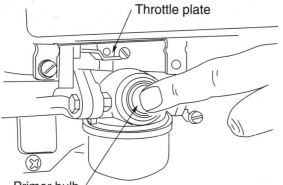

Fig. 4-47 Tecumseh Dual system carburetor with large primer.

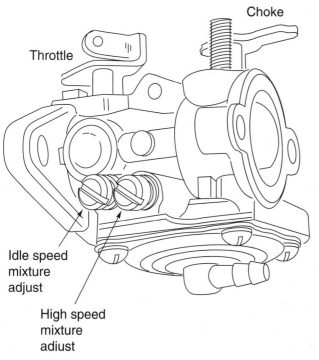

Fig. 4-48 Tecumseh diaphragm carburetor.

formed and then cut into two-inch sections. The extrusion eliminates porosity found in the conventional casted carburetor body and reduces the machining operations to a few drillings. The lower area is an all plastic bowl and float system that contains the main and idle jets.

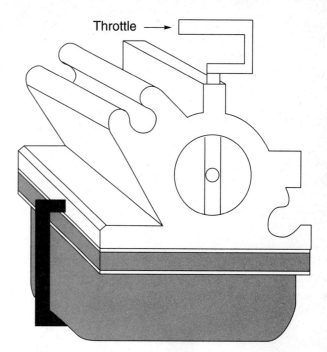

Fig. 4-49 Tecumseh Vector carburetor.

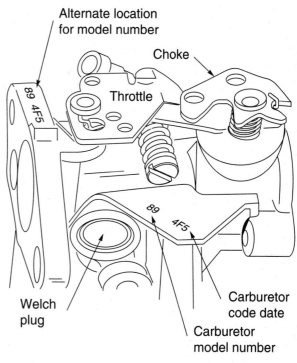

Fig. 4-50 Carburetor model and code date location.

Fig. 4-51 Carburetor model 688 manufactured on October 16, 1978.

The advantages of this carburetor is the simplicity of construction and its resistance to alcohol fuels. It has a different mounting system, which will not permit its use in retrofitting earlier engines. The float bowl is held on with a "fruit jar" bale that easily snaps off with a screwdriver to allow quick cleaning. The aluminum top section will rarely need servicing as a result of special fuel restriction on top of the idle system. If severe deterioration has been caused by fuel, the bottom plastic components can be replaced easily and economically.

The replaceable main nozzle is made of plastic and has a spring below it that holds it in place in the upper half. The plastic float, hinge, and inlet snap together and require no adjustments.

TECUMSEH CARBURETOR CODING

The carburetor model and code date number are found stamped in the carburetor body, as shown in figures 4-50 and 4-51.

Prior to 1988

The model number, as shown in figure 4-52, expresses the variations of a basic carburetor design

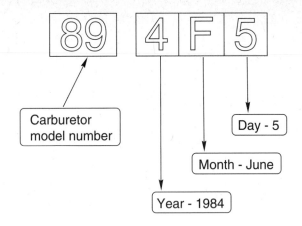

Fig. 4-52 Model number interpretation.

specified for a certain engine. Care must be taken when interchanging carburetors that have different model numbers from one engine to another. Even though the carburetor may be a typical Tecumseh series, internal differences such as venturi size, idle circuit, etc., may cause poor performance.

The date code can be interpreted as shown in figure 4-52. The chart in figure 4-53 converts the letter code to the month.

1988. . . .

The model and date code are found in the same place, but the date code is interpreted differently.

The month and day are determined by the same coding as pre-1988, as shown in figure 4-54, but the

Letter → Month coding			
Letter	Month	Letter	Month
A	January	G	July
B	February	H	August
C	March	I	September
D	April	K	October
E	May	L	November
F	June	M	December
Note: The letter "J" is not used because it looks like the letter "I"			

Fig. 4-53 Letter to month conversion.

year is not given directly. The day of the week code is converted by using the chart in figure 4-55. By knowing the day of the week, the date of the day of the month, and the month, one can derive the year of manufacture of a carburetor by referring to a perpetual calendar.

In the example above the:

Month C	=	March
Date	=	9
Day of week is D	=	Thursday

Since the 9th day of March falls on a Thursday in 1989, the deduction is that this is the date it was built.

THEORY OF OPERATION (SERIES I)

Tecumseh carburetors, as shown in figure 4-56, have seven basic passage ways: fuel inlet, main jet, main nozzle, main nozzle air bleed, atmospheric vent, idle circuit, and primer circuit.

Fuel inlet passage

FLOAT BOWL The fuel is stored in the float bowl. Sediment that bypasses the fuel filter will settle in the bowl. The bowl should be drained often to remove the foreign particles. Most float bowls have a drain to accomplish this.

FLOAT The float-type carburetors use a hollow copper metal, cork, or plastic float to maintain the operating level of fuel in the carburetor. As the fuel is used, the fuel level in the carburetor bowl drops and the float moves downward. This actuates the inlet needle valve to allow fuel to flow into the fuel bowl. As the fuel level in the bowl again rises, it raises the float. This float motion adjusts the fuel flow at the proper rate and keeps the fuel at the proper mixture level.

FLOAT NEEDLE AND SEAT The entry of fuel to the float bowl is controlled by the needle and seat. Fuel entry is halted when the needle contacts the seat. The fuel must pass by the needle and seat to enter the float bowl. The inlet needle contacts a viton seat in the carburetor body, as shown in figure 4-57. The seat can be removed with a #5 crochet

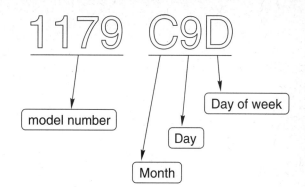

Fig. 4-54 Post-1988 dating code.

Letter	Day
A	Monday
B	Tuesday
C	Wednesday
D	Thursday
E	Friday
F	Saturday
G	Sunday

Fig. 4-55 Letter to day of week conversion.

hook or a short piece of hooked wire. It is also possible to pack the inlet passage with wheel bearing grease and expel the seat and grease with a short blast of compressed air. To reinstall, moisten the seat with engine oil, and insert with smooth side toward inlet needle or grooved side facing the carburetor. Press it into the cavity using an old plunger from a Briggs & Stratton ignition kit or a flat punch close to the diameter of the seat, making sure it is firmly seated. Do not use a sharp object as it will damage the seal.

The inlet needle hooks onto the float tab by means of a spring clip. To prevent the float from binding during the up and down movement, the long, straight end of the clip should face the choke end of the carburetor, as shown in figure 4-58.

FLOAT PIVOT PIN The pivot pin transfers movement of the float to the needle and seat. Friction in this area would cause poor fuel level control.

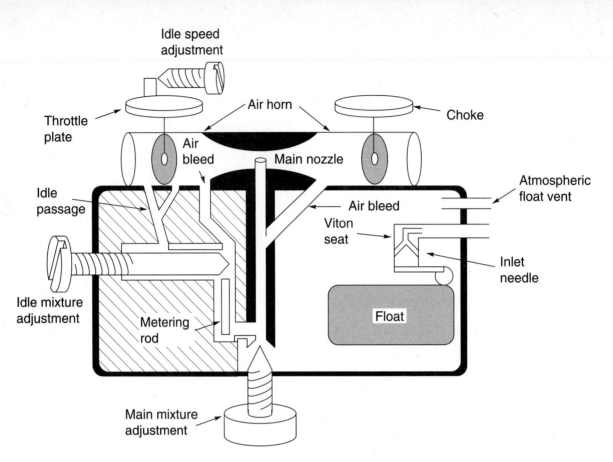

Fig. 4-56 Tecumseh Series I carburetor with Idle air adjust.

FLOAT ADJUSTMENT The float will be raised or lowered by the level of the fuel in the float bowl. The maximum quantity of fuel in the float bowl is controlled by the attachment of the float to the needle and seat. The float will stabilize the fuel level in the bowl by closing the needle and seat. As the carburetor draws fuel from the float bowl, the float drops and retracts the needle from the seat, allowing more fuel to enter.

With the float setting tool, part #670253A, as shown in figure 4-59, check the position of the float. The toe of the float must be within the tolerances for the float setting tool. The toe must be under step, as in figure 4-59-1, and can touch step, as in figure 4-59-2, without a gap. If the float is too high or too low, adjust the height by removing the float and bending the tab or tang accordingly.

NOTE: WHEN THE TOOL IS NOT AVAILABLE, A #54 DRILL BIT (0.055-INCH) CAN BE PLACED BETWEEN THE CARBURETOR BODY AND THE FLOAT FOR PROPER ADJUSTMENT.

Sometimes there is a dimple pressed into the bottom of the float bowl. The bowl should be installed so that the dimple lines up at the farthest end or drop side of the float. If no dimple is present, line the crease in the float bowl parallel to the hinge pin so that the deepest part is facing away from the hinge.

Main nozzle passage

During high speed operation, the throttle plate is fully open and air flows through the air horn at high speed. The venturi, which decreases the size of the

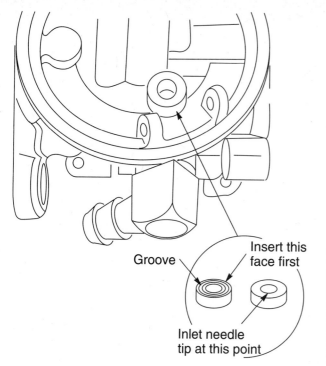

Fig. 4-57 Viton inlet seat.

Groove

Insert this face first

Inlet needle tip at this point

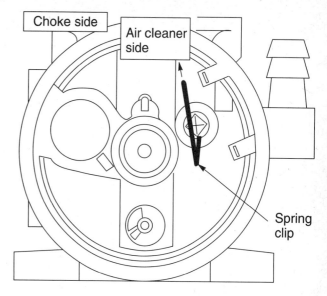

Choke side

Air cleaner side

Spring clip

Fig. 4-58 Inlet needle and clip installed correctly.

air passage through the carburetor, further accelerates the air flow. This high speed movement of air decreases the air pressure, and fuel is drawn into the air stream through the main nozzle, as shown in figure 4-60, that opens into the venturi. The amount of fuel in the main nozzle is controlled by the main mixture adjustment screw located on the bottom of the float bowl.

Main nozzle air bleed passage

Air is sucked or forced into the main nozzle through the air bleed located in the air horn. This allows the fuel to atomize into small packets and helps lift the fuel in the main nozzle.

Float atmospheric vent passage

The vent is needed to prevent a build-up of vacuum or pressure in the air space above the fuel in the float bowl. The carburetor may be vented either internally or externally.

EXTERNAL VENT A passage runs from the float bowl cavity to the outside of the carburetor, as

shown in figure 4-61. The float level can be affected by dirt in the outside vent on the carburetor. When the vent blocks the creation of a pressure difference between the venturi and the pressure on the fuel in the float area, a lean condition or a "no run" condition occurs. This hole, if present, will be located above the idle mixture screw.

INTERNAL VENT When the vent passage terminates inside of the air horn of the carburetor, as shown in figure 4-62, it is labeled as an internal vent. The internal vent has an advantage over the external because of its use of clean filtered air and lesser tendency to become blocked. An added advantage is the air ram effect, which lifts the fuel to the venturi for easier starting and operation. While the air is rushing through the air horn, it pushes air into the vent hole, resulting in additional pressure on the fuel surface in the float bowl. This vent can be found by looking into the air horn from the choke side at the 9 o'clock position.

Idle circuit passage

Tecumseh Series I carburetors may regulate the flow of the fuel/air mixture by using an air adjust or the solid fuel adjust method.

IDLE AIR ADJUST CARBURETOR The easiest way to identify the idle air adjust carburetor system is by

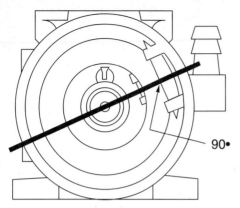

Pull at 90 degrees to hinge pin

Fig. 4-59-1 Carburetor main nozzle.

TECUMSEH
670253 A

Pull →

No higher
than here

1

TECUMSEH
670253 A

Pull →

Can touch
here without
gap

2

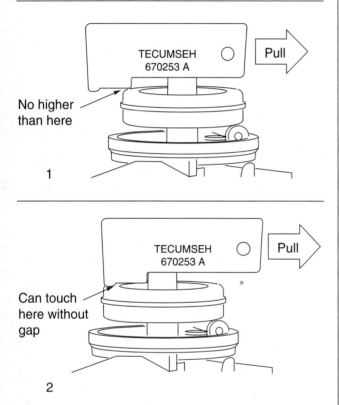

Fig. 4-59 Tecumseh float setting.

Fig. 4-59-2 Carburetor main nozzle.

removing the idle mixture screw and noticing the length, as shown in figure 4-56. The longer screw indicates the air adjust system. If you have the long screw, you will also notice an additional hole at the 7 o'clock position in front of the venturi on the throttle plate side. Because it is difficult to drill a

long small diameter passage in the carburetor block during the manufacturing process, a larger conduit is made, and then the passage is reduced by inserting a metering rod in the space. Proper operation of the metering rod can be detected by a rattle sound when the carburetor is shaken. Avoid damage to this rod by installing the idle adjustment screw only with the carburetor in the normal operating position (not upside down), as shown in figure 4-63.

At idle, a relatively small amount of fuel and air is required to operate the engine. The throttle plate is almost closed, shutting off the fuel supply from all openings except the primary idle passage, so that the suction created by the engine draws fuel only

Fig. 4-61 External vent (cutaway).

Fig. 4-62 Internal vent (cutaway).

from that passage near the side of the carburetor air horn.

NOTE: THE IDLE CIRCUIT IS AN AIR ADJUST SYSTEM. YOU ARE ADJUSTING THE AMOUNT OF AIR THAT WILL MIX WITH THE FUEL. WHEN THE IDLE ADJUSTMENT NEEDLE IS TURNED OUT, MORE AIR IS ALLOWED INTO THE IDLE CIRCUIT THROUGH THE IDLE AIR BLEED. TURNING THE NEEDLE IN DECREASES THE AMOUNT OF AIR IN THE CIRCUIT, WHICH RICHENS THE MIXTURE.

Fig. 4-63 Metering rod.

During intermediate operation, a second idle passage is uncovered as the throttle plate opens, and more fuel is allowed to mix with the air flowing into the engine.

SOLID FUEL ADJUST CARBURETOR This system can be identified by a short idle mixture screw, as shown in figure 4-64. The screw will adjust the amount of fuel that will mix with air for the closed throttle plate, idle position. At the closed throttle plate position, air enters the idle secondary passage and flows to the idle passage and then into the engine side of the throttle plate. Fuel is allowed to enter this air stream through the idle adjustment screw passage.

This system does not have a metering rod in the fuel transfer passage like the air adjust models. During intermediate operation, a second idle passage is uncovered as the throttle plate opens, and more fuel is allowed to mix with the air flowing into

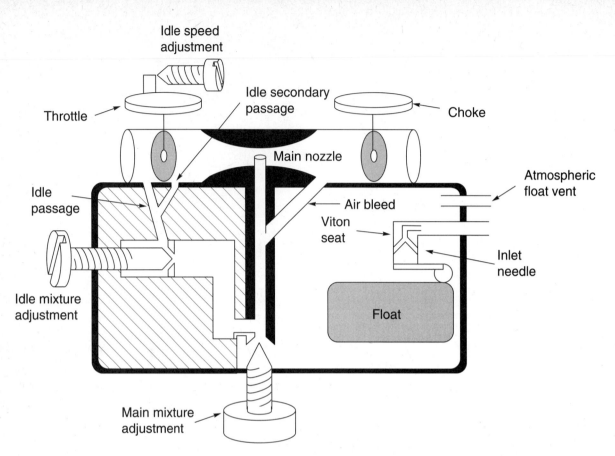

Fig. 4-64 Tecumseh Series I carburetor with solid fuel adjust.

the engine. The idle passages can be cleaned by removing the welch plug cover and blowing out all the passages with compressed air in the opposite direction of normal fuel flow.

THEORY OF OPERATION (DUAL SYSTEM)

This carburetor is a float-feed, non-adjustable, without an idle circuit or a choke, as shown in figure 4-65, and is used only on rotary lawnmowers applications.

Prime well

In the before-start position, fuel fills the prime well, as shown in figure 4-66, in the bowl screw, to the level maintained by the float. This will provide the rich mixture required to start a cold engine. It takes about five seconds for the prime well to refill after each crank or when the engine is stopped.

If the engine is used under cool weather conditions, the primer can be used to force the fuel through the main nozzle to provide a richer mixture. In order to prime a dual system carburetor properly, the operator must wait approximately three seconds between each prime or push on the primer bulb. When the operator presses the bulb the required number of times in fast succession, only one prime is accomplished because the fuel has not had the time necessary to refill the bowl screws between primes.

Fixed main jet

When the engine starts and runs, the fuel level in the bowl and prime well stabilize, as shown in figure 4-67. The quantity of fuel is controlled by the main jet shown in figures 4-68 and 4-69 (page 135). The air from the air bleed and fuel from the main jet are pulled up the main nozzle for engine operation.

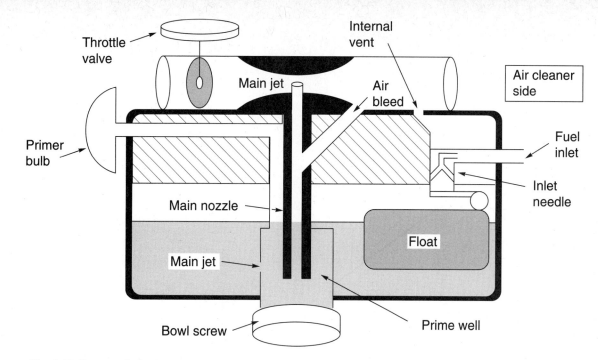

Fig. 4-65 Tecumseh dual system carburetor.

Do not exceed the original equipment manufacturer's (OEM) recommended RPM since excess speed will cause the carburetor to lean out, resulting in overheating and failure of the engine.

High altitude adjustment

Carburetors that have a non-adjustable main jet may experience starting and operating problems when used in areas of the country where the altitude is four thousand feet above sea level or higher. A leaner air-fuel mixture is more desirable at higher elevations to sustain good engine performance. Different bowl nuts are available from the manufacturer that can be interchanged for better operation. Do not install the leaner jetted bowl nuts on engines used below four thousand feet.

TECUMSEH CARBURETOR ADJUSTMENTS
Idle speed adjustment

This screw is located on top of the carburetor and contacts the throttle plate, as shown in figure 4-70.

To initially adjust this speed, back out the screw, turn in until the screw just touches the throttle plate lever, then turn the screw inward one turn. When the engine is running, the final idle RPM can be adjusted with the use of a tachometer.

Idle and main mixture adjustment

Some carburetors have fixed-main and/or idle jets and are identified by the absence of adjustment screws. These carburetors have no adjustment except for the float level.

Carburetors that have idle and main mixture adjustment screws can be adjusted as described below:

1. *Preset the mixture needles by turning inward the idle adjusting screw, as shown in figure 4-71, and the main adjusting screw, as shown in figure 4-72, finger-tight, then back them out to 1-1/4 turn for the main and 1 turn for the idle, (page 136).*

2. *Allow the engine to warm up to a normal running temperature for five minutes at a moderate speed.*

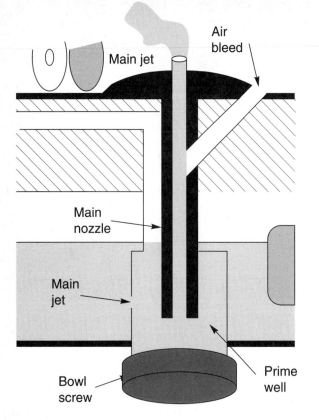

Fig. 4-66 Dual system starting operation.

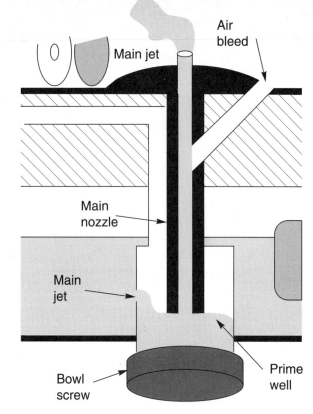

Fig. 4-67 Dual system running operation.

3. *Increase the engine running speed to the maximum recommended RPM and tighten (turn clockwise) the main adjustment screw until the engine starts to run rough. Then turn the screw in the opposite direction until the engine starts to run rough again. Turn the screw in to the mid point between the two points.*

4. *After the main system is adjusted, move the speed control lever to the idle position and follow the same procedure for adjusting the idle system. The idle speed should be kept about 1800 RPM while adjusting.*

5. *Repeat the adjustment between the main screw and idle at least three times. This ensures that the carburetor will work properly under load after it leaves the service area.*

6. *Test the engine by running it under normal load. The engine should respond to load pick-up immediately. An engine that "dies" is too*

lean. An engine which ran rough before picking up the load is adjusted too rich. If the adjustment seems too "touchy," check the float for proper setting and for "sticking."

Speed control adjustment

Vertical shaft Tecumseh engines commonly have a control mounted above the carburetor similar to the one shown in figure 4-73 (page 137). To adjust the speed control panel for proper operation:

1. *Loosen the two screws on the top of the panel.*

2. *Move the control lever to full high-speed position.*

3. *Install a wire or aligning pin through the:*

   • Hole in the top panel.

   • Hole in the choke actuating lever.

Fig. 4-68 Main jet in bowl bolt.

Fig. 4-69 Automagic main jet.

Fig. 4-70 Idle speed adjustment.

design on their engines until 1987. Following 1987 a new fixed main jet, side draft carburetor, has been used on most Kohler engines from 8 through 23 HP models, as shown in figure 4-74 (page 138). These carburetors are built to Kohler specifications by the Walbro Corp., and the benefits gained include:

- Elimination of main fuel needle adjustment, which provides a consistent air/fuel mixture that results in longer valve and engine life.

- Improved acceleration.

- Alcohol resistant fuel inlet needle valve, float, and carburetor body.

- Improved access to mounting bolts.

KOHLER CARBURETOR SERVICE
Kohler adjustable carburetor

CARBURETOR DISASSEMBLY

1. *Remove the bowl retaining screw, retaining screw gasket, and fuel bowl.*

- Hole in the choke lever.

4. *Tighten the two adjusting screws on the panel.*

KOHLER CARBURETORS

INTRODUCTION

In the early 1960s, the Kohler corporation purchased Carter carburetors and used this carburetor

Fig. 4-71 Idle mixture adjustment.

Fig. 4-72 High-speed mixture adjustment.

2. *Remove the float pin, float, fuel inlet needle, baffle, and bowl gasket.*

3. *Remove the fuel inlet seat and inlet seat gasket. Remove the idle and main fuel adjusting needles and springs. Remove the idle speed adjusting screw and spring.*

4. *The throttle plate and choke shafts should only be disassembled if these parts are to be replaced.*

CARBURETOR CLEANING All parts should be carefully cleaned with a carburetor cleaner such as acetone. Do not submerge the carburetor in a cleaner or solvent when the fiber and rubber seals are installed as they may be damaged. Be sure all gum deposits are removed from the following areas:

- carburetor body and bore, especially the areas where the throttle plate, choke plate, and shafts are seated.

- float and float hinge.

- fuel bowl.

- idle fuel and "off-idle" ports in the carburetor bore, ports in the main fuel adjusting needle, and main fuel seat. These areas can be cleaned by using a piece of fine wire in addition to cleaners. Be careful not to enlarge the ports or break the cleaning wire within the ports.

- blow out all passages with compressed air.

CARBURETOR INSPECTION Carefully inspect all components and replace those that are worn or damaged:

- Inspect the carburetor body for cracks, holes, and other wear or damage.

- Inspect the float for dents or holes. Check the float hinge for wear and missing or damaged float tabs.

- Inspect the inlet needle and seat for wear or grooves.

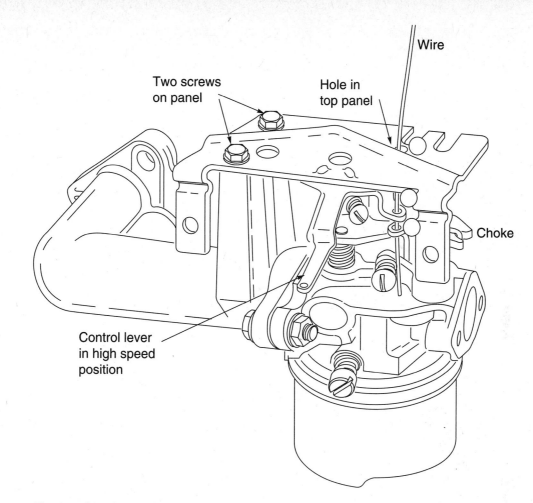

Two screws
on panel

Hole in
top panel

Wire

Choke

Control lever
in high speed
position

Fig. 4-73 Speed control adjustment.

- Inspect the tips of the main fuel and idle fuel adjusting needles for wear or grooves.

- Inspect the throttle plate and choke shaft and plate assemblies for wear or excessive play.

CARBURETOR ASSEMBLY Always use new gaskets when servicing and reinstalling the carburetor, as shown in figure 4-75A.

1. *Install the fuel inlet seat gasket and fuel inlet seat into the carburetor body. Torque the seat to 35–45 inch-pounds.*

2. *Install the fuel inlet needle into the inlet seat. Install the float and slide float pin into the carburetor body.*

3. *Set the float level, as shown in figure 4-75B, by bending the float tab or tang with a small screwdriver. Next, adjust the float drop by turning the carburetor over to its normal operating position and allow the float to drop to its lowest level, as shown in figure 4-76. The float drop should be limited to 1-1/32 inches between the machined surface of the body and the bottom of the free end of the float.*

4. *Install the bowl gasket and baffle gasket. Position the baffle gasket so that the inner edge is against the float hinge towers.*

5. *Install the fuel bowl so that it is centered on the baffle gasket and the inner edge is against the float hinge towers.*

Fig. 4-74 Kohler/Walbro carburetor.

6. Install the bowl retaining screw gasket and bowl retaining screw. Torque the screw to 50–60 inch-pounds.

7. Install the idle speed adjusting screw and spring. Install the idle fuel and main fuel adjusting needles and spring. Turn the adjusting needles clockwise until they bottom lightly.

8. Reinstall the carburetor to the engine using new gaskets.

9. Make preliminary adjustments.

10. Make final adjustments.

KOHLER CARBURETOR PRELIMINARY ADJUSTMENTS Turn the idle fuel adjusting screw 1-1/2 turns and main adjusting screws two turns counter-clockwise from lightly bottomed, as shown in figure 4-77.

KOHLER CARBURETOR FINAL ADJUSTMENTS

1. Start the engine and allow it to warm up.

2. After it has warmed up, operate at full throttle plate and under load, if possible.

3. Turn the main fuel mixture adjustment screw inward until the engine slows down (lean side), then outward until it slows down again from an over-rich setting. Note the positions of the screw at both settings, then set it about halfway between the two.

4. Bring the engine to an idle speed above 1200 RPM and adjust the idle fuel screw in the same manner for smoothest idle.

Kohler /Walbro fixed jet carburetor

IDENTIFICATION This carburetor is designed to deliver the correct air-fuel mixture to the engine under all operating conditions. The main fuel jet is calibrated at the factory and is not adjustable. Engines that are operated at high altitudes require a special main jet.

The idle fuel adjusting needle is set at the factory and normally does not need adjustment. If the engine is hard starting or does not operate correctly at idle speeds, it may be necessary to adjust the idle fuel needle.

KOHLER FIXED JET CARBURETOR INITIAL SETTINGS

1. Turn the idle fuel adjusting needle inward (clockwise) until it lightly seats. Be careful not to damage the critical dimensions of the tapered needle.

2. Turn the needle outward (counter-clockwise) 1-1/4 turns.

KOHLER FIXED JET CARBURETOR FINAL SETTINGS

1. Start the engine and run at a moderate speed for 5 to 10 minutes to warm up the engine before making the final settings.

2. Place the throttle plate into the idle or slow position and set the idle speed to 1200 RPM by turning the idle speed adjusting screw inward or outward. Check the speed using a tachometer.

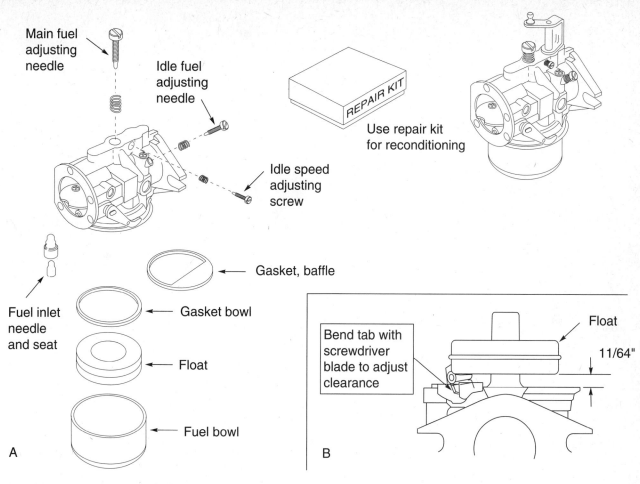

Fig. 4-75 Carburetor float adjustment (Kohler).

Use repair kit for reconditioning

Main fuel adjusting needle

Idle fuel adjusting needle

Idle speed adjusting screw

Fuel inlet needle and seat

Gasket, baffle

Gasket bowl

Float

Fuel bowl

A

Bend tab with screwdriver blade to adjust clearance

Float

11/64"

B

3. *Place the throttle plate at the idle or slow position.*

4. *Turn the idle fuel adjusting needle outward (counter-clockwise) to "richen" the mixture from the preliminary setting until the engine speed decreases. Note the position of the needle.*

5. *Now turn the adjusting needle inward (clockwise) to lean the engine. The engine speed may increase, then it will decrease as the needle is turned inward. Note the position of the needle.*

6. *Set the adjusting needle midway between the rich and lean settings noted.*

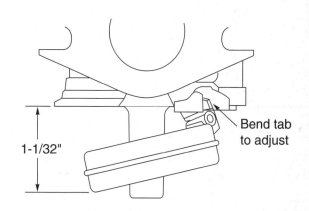

1-1/32"

Bend tab to adjust

Fig. 4-76 Float drop adjustment (Kohler).

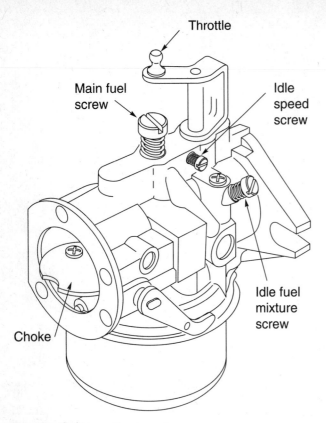

Throttle

Main fuel
screw

Idle
speed
screw

Idle fuel
mixture
screw

Choke

Fig. 4-77 Kohler carburetor adjustment screws.

Fig. 4-78 Diaphragm carburetor used on a chain saw.

DIAPHRAGM CARBURETOR WITH FUEL PUMP

INTRODUCTION

The diaphragm carburetor iswidely used, especially in chain saws where the engine may be used in many different positions, as shown in figure 4-78. Most diaphragm carburetors have a built-in fuel pump, as shown in figure 4-79, to insure that the fuel supply is available to the carburetor regardless of the position of the fuel tank.

With small variations to the basic overhaul technique, you can easily and correctly tune adjustable carburetors not only on chain saws, but also on string trimmers, snowmobiles, lawnmowers, outboards, and probably any two-stroke engine.

FUEL PUMP OPERATION

The fuel pump side of the carburetor is identified by the presence of a fuel inlet fitting. When the pump cover is removed, a fuel gasket and diaphragm are found. Each revolution of the engine produces two changes in crankcase air pressure. The downward movement of the piston creates a positive pressure in the crankcase, while the upward movement creates a negative pressure (vacuum), as shown in figure 4-80. These impulses are channeled to the fuel pump diaphragm through an impulse channel hole. The impulses actuate the diaphragm just above the fuel reservoir in the pump chamber by moving it up and down to pump fuel from the tank. The pump's "one-way-flap-valves" work in conjunction with the crankcase pressure variations to keep the fuel moving in one direction. A negative impulse, as shown in figure 4-81, brings fuel in from the fuel line through valve #1 and closes valve #2. A positive impulse, as shown in figure 4-82, closes valve #1 and pushes the fuel through valve #2 into the metering side of the carburetor.

The impulse channel hole in the carburetor may be external or internal. The external channel is connected atop the fuel pump cover, while the internal connects through the mounting against the

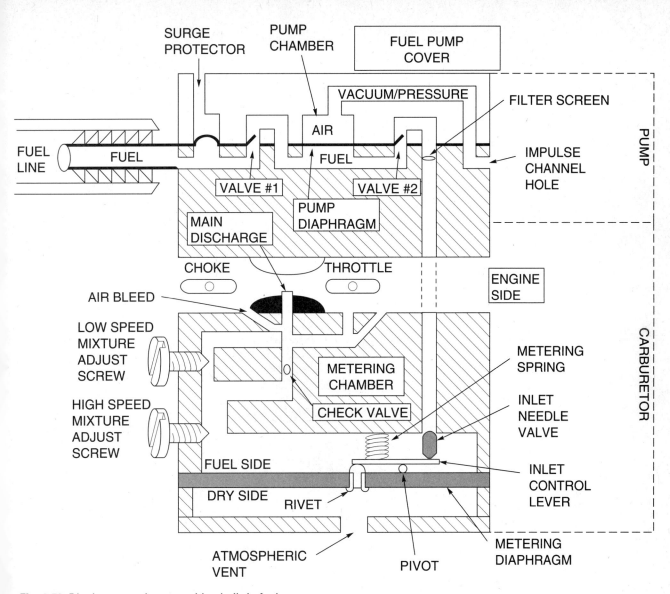

Fig. 4-79 Diaphragm carburetor with a built-in fuel pump.

crankcase. This hole may be plugged with foreign material or from improper gasket installation.

The surge protector is installed in many diaphragm carburetors. When the demand for fuel is low and the pressure in the fuel tank is great, the stress is relieved by the flexible part of the diaphragm expanding so that the excess pressure is relieved.

DIAPHRAGM CARBURETOR OPERATION

Fuel flows from the fuel pump to the inlet needle valve. This valve opens and closes according to the movement of the metering diaphragm. The dry side (lower) is exposed to atmospheric pressure through an atmospheric vent in the bottom cover. The fuel side is influenced by the degree of vacuum in the venturi, as shown in figure 4-83.

As fuel is elevated into the venturi through either the main discharge tube or the idle ports, atmos-

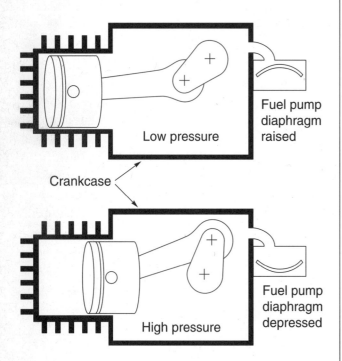

Fig. 4-80 Each revolution produces two changes in crankcase air pressure.

pheric pressure moves the diaphragm upward against the calibrated metering spring. This depresses the inlet control lever, allowing the inlet needle valve to open. Fuel can then flow from the pump side into the metering chamber and through the idle and high speed channels. The inlet needle may have a viton tip, which not only resists the effects of exotic fuels, but is more resistant to wear. When the discharge of fuel decreases or ceases from the idle or main ports, the incoming fuel pushes the diaphragm downward against the inlet lever and the flow of fuel is halted.

DIAPHRAGM ENGINE OPERATIONS

Starting a cold engine requires a rich fuel mixture. The choke shutter is in a closed position, as shown in figure 4-84, which exaggerates the vacuum in the venturi. A larger quantity of fuel is drawn in both the main discharge and the idle circuits.

When the engine is started, the choke shutter is opened to allow additional air to mix with fuel, as shown in figure 4-85. Fuel is drawn up the main discharge port, and the volume of fuel is controlled by the high speed adjusting needle. Some carburetors are assembled with a fixed high-speed jet. With a fully fixed jet, the jet hole size is calibrated to supply the correct amount of fuel required by the engine at high-speed operation. This orifice size is

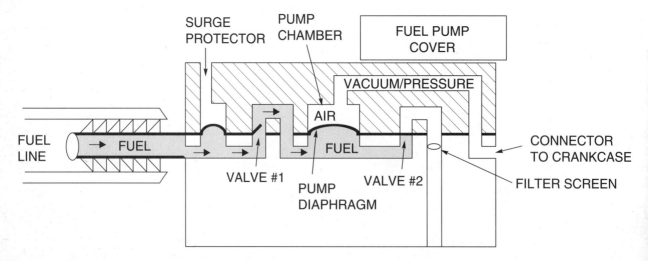

Fig. 4-81 Negative impulse.

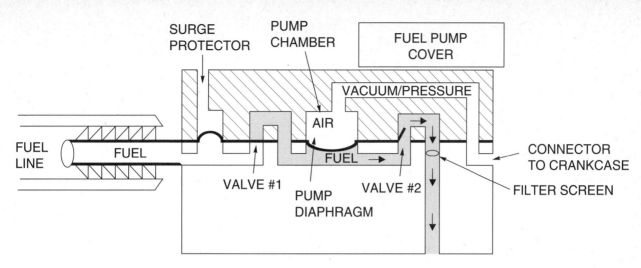

Fig. 4-82 Positive impulse.

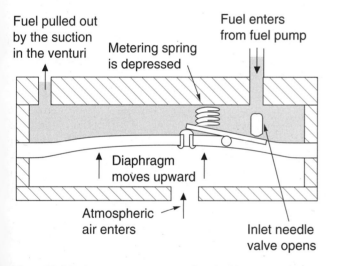

Fig. 4-83 Diaphragm moves upward as fuel is removed.

critical and should not be altered in any manner. Clean only with compressed air.

When the engine is idling, as shown in figure 4-86, the throttle plate is narrowly open. Engine suction is now permitted only through the low-speed discharge ports, and the volume is controlled by the low-speed mixture adjusting screw.

DIAPHRAGM CARBURETOR SERVICE TIPS

To ensure long life and top performance, a regular general overhaul is advisable. Before beginning disassembly, remove all excess dirt and saw dust from the carburetor. Do not use a cloth, as tiny lint particles are likely to adhere to the components and cause malfunction. Always keep a clean work surface.

Remove pump cover

Beneath the fuel pump cover is the pump gasket. Around the pump gasket there should be clean, well-defined imprints, indicating properly sealed-off areas of the pump surface, such as the fuel intake chamber, pulse chamber, and the inlet and outlet valve areas, shown in figure 4-87 (page 147). Any cross-leaking between these areas may cause starting and high-speed problems.

Inspect pump valves

Check the flaps of the valves for any sign of excessive wear, peeling, rupture, or distortion, as shown in figure 4-88. Make certain that they rest flat against the pump surface. If either one is curled, the diaphragm should be replaced.

Inspect inlet screen

The inlet screen is located in the chamber above the inlet needle on the fuel pump side of the carburetor, as shown in figure 4-89. This screen filters fuel to the metering chamber. The screen mesh is very small and difficult to identify. Clogging of the

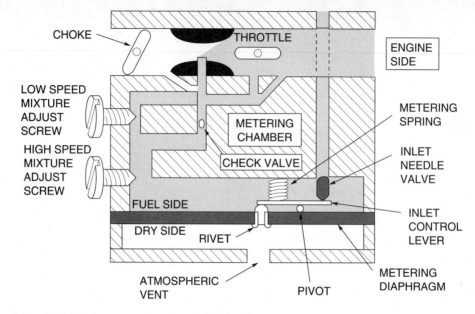

Fig. 4-84 Diaphragm carburetor choke circuit.

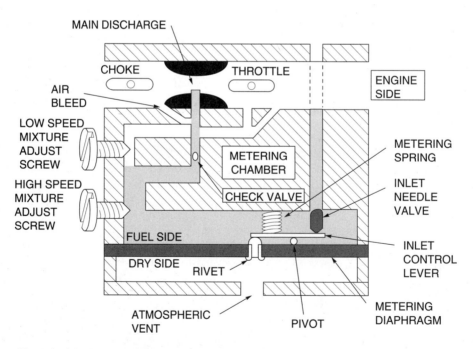

Fig. 4-85 Diaphragm carburetor high speed circuit.

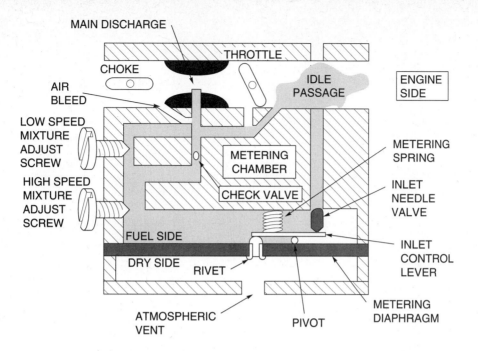

Fig. 4-86 Diaphragm carburetor low speed circuit.

screen will restrict fuel flow, affect acceleration, and impair high speed performance. Remove this delicate mesh carefully with a sharp object and clean it thoroughly.

Metering cover and diaphragm

The vent hole is exposed to the atmosphere, where it may become clogged with dirt. Check for tearing or peeling of the gasket in this area. The diaphragm must be flexible and show no signs of deterioration, as shown in figure 4-90.

Lever height

Lever height must be adjusted properly (check service manual), and the lever must move easily, as shown in figure 4-91. Disassemble the inlet seat components. Hold the inlet lever down with your index finger when removing the fulcrum pin retaining screw, as there is a tension spring directly underneath. Check the inlet lever for wear at the point of contact. Keeping the inlet needle/fulcrum spring in place during assembly can be very tricky.

Fig. 4-87 Well defined imprints indicates proper sealing.

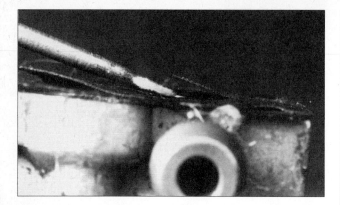

Fig. 4-88 Check valves for sealing.

Fig. 4-89 Remove and clean fuel screen.

Fig. 4-90 Atmospheric vent hole.

Fig. 4-91 Lever height must be adjusted.

A very small amount of clean, lithium-based grease can be placed in the well, where the spring seats. This will hold the spring erect and lessen the chance of it dropping out.

High- and low-speed needles

Remove the high- and low-speed mixture adjustment needles and check for rusting or damaged threads, as shown in figure 4-92. Roll them on a flat surface to inspect for a bend. A bent needle point can damage the precision machined orifice.

Remove welch plug

This can be accomplished by using a pointed punch or by drilling a hole in the center using a 1/8-inch drill. When a new plug is put in, expand it with a punch.

Rinsing and air cleaning

With all the pump and metering components and mixture needles removed, clean the parts in a solvent. After rinsing, blow thoroughly through all the channels. Do not use drills or any hard metal objects

Fig. 4-92 Check needle for damage.

to clean away obstructions. Soft tag wire may be used carefully, as shown in figure 4-93.

Pressure test

When the carburetor is fully assembled, pressure test the inlet seat to detect any leaks that may remain. Connect the pressure inflator to the fuel inlet fitting and apply five PSI, as shown in figure 4-94. If there is no leakage around the inlet needle, this should hold steady for about four seconds. Depress the metering diaphragm with a pointed instrument, such as a pencil, and repeat this test.

A blow-off pressure test can also be done. Continue to pump until the inlet needle is unseated. This will usually occur between fifteen to twenty-five PSI. The dial needle will be seen to drop and should stabilize above five PSI.

Final inspection

Check the throttle plate and choke shafts. They should open and close freely. When closed, the throttle plate should completely seal the throttle bore.

Fig. 4-93 Clean passages with air and soft tag wire.

DIAPHRAGM CARBURETOR PRE-ADJUST

There are only three basic carburetor adjustments: the Low-speed (L), High-speed (H) fuel mixture needles, and the screw for setting the idle speed. Make initial adjustments before starting the engine, as shown in figure 4-95.

Initial adjustments

1. *Gently close the high- and low-speed mixture needles by turning them clockwise until they are lightly seated. The points of the needles are delicate and rest in a metered orifice. They can be damaged by over-torquing.*

2. *Turn the screws outward about 1-1/2 turns. This is usually a rich (excess fuel) condition because most carburetors are designed for best fuel flow at about one turn open. Start tuning the engine from the rich side of the adjustment.*

3. *Turn the idle speed screw outward until it does not touch the idle stop. Then screw it back inward until it barely touches the throttle plate lever. At this point you can pre-adjust the screw by turning it in an additional 1-1/2 turns.*

Final adjustments

1. *Start the engine and flick the trigger a few times. If necessary, reset the idle speed so that the*

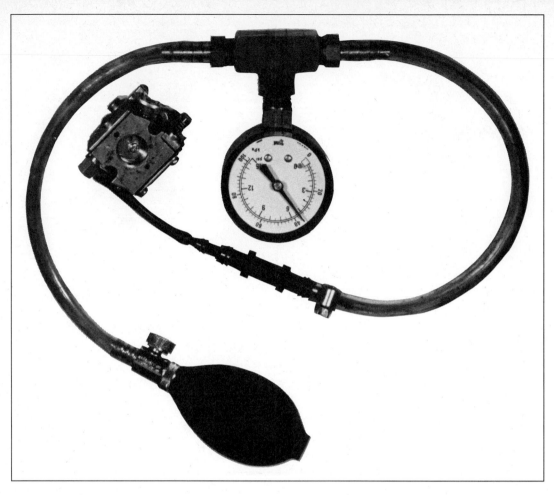

Fig. 4-94 Pressure test after assembly.

clutch is not engaging. Most small two-stroke-cycle engines idle between 2500 and 3000 RPM.

2. Turn the low-speed mixture needle inward about 1/8-turn from the initial setting. The engine speed will increase slightly because you are going from a rich condition to a "less rich," or "lean," condition. If the clutch starts to engage, back off slightly on the idle speed screw. Accelerate the engine to flush out excess fuel from the crankcase and repeat, alternating from low-speed needle to idle speed screw and back again. About the third time you turn the low-speed needle, engine speed will decrease slightly. Stop at this point. You are just on the lean edge of an best idle mixture. If you squeeze the trigger, the engine will probably stumble a

little during acceleration. "Richen" (turn counter-clockwise) the low-speed mixture needle just enough to get quick, smooth acceleration. Use the idle speed screw one last time to achieve correct idle speed.

3. The high-speed mixture needle is adjusted at wide open throttle plate with a load on the engine. Close the needle clockwise just far enough to hear the engine increase in speed and break into the fringe of a clean, or "bumblebee," sound. This adjustment is too lean to do work. Open the needle counter-clockwise far enough to hear again the rich sound of four-cycling as the speed decreases slightly. What you are doing here is tuning by ear, like a piano tuner. Remember, the fuel is the only source of

Fig. 4-95 High and Low mixture adjusting screws.

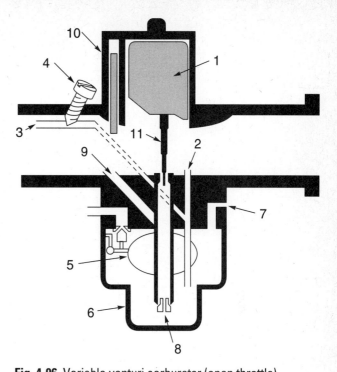

Fig. 4-96 Variable venturi carburetor (open throttle).

lubrication for a two-stroke engine, and that rich four-cycle sound is good insurance. If you set it too rich, you will lose power and invite faster carbon build-up. The engine should break into a clean sound when you put it to work. Take caution to avoid running the engine too long at wide open, no-load situation. You should be able to set the adjustment in about ten or fifteen seconds.

VARIABLE VENTURI CARBURETOR

This is a type of carburetor found on many motorcycles and some small air-cooled two-stroke cycle engines. Since the venturi size limits maximum power, a large venturi is optimal for maximum engine performance. When a large venturi is used for performance, the trade-off is poor performance at midrange and low engine speeds. One solution to the problem is the use of the variable venturi carburetor which varies the size of the venturi for

different loads so that the necessary suction in the venturi is adequate at all operating speeds.

THROTTLE SLIDE

The throttle slide, marked with #1 in figure 4-96 and figure 4-97, is raised or lowered to change the restriction in the venturi. This, in turn, controls the engine's speed from the idle to high-speed position. In the position, pictured in figure 4-96, the air flow is at a maximum, causing the fuel to flow from the main discharge tube (#11). While in the position, as pictured in figure 4-97, the air flow is limited to a small area above the idle passage (#2).

IDLE PASSAGE

The idle passage, marked with #2 in figures 4-96 and 4-97, is where the fuel is discharged when the throttle slide (#1) is in the down position. The idle passage operates from the idle position to approximately 1/8 throttle. The amount of fuel that is discharged is regulated by the idle air bleed (#3), low-speed mixture adjustment screw (#4), and the idle passage (#2).

IDLE AIR BLEED

The idle air bleed, marked with #3 in figures 4-96 and 4-97, provides for premixing of air and fuel in the idle passage (#2) before the fuel reaches the venturi.

LOW-SPEED MIXTURE ADJUSTMENT SCREW

The low-speed mixture adjustment screw, marked with #4 in figures 4-96 and 4-97, controls the amount of air mixed with the fuel in the idle passage. Sometimes referred to as an idle air adjustment, this screw, when it is turned inward, will reduce the idle circuit air to make it more rich in fuel.

BOWL FLOAT AND INLET NEEDLE

The bowl float and inlet needle, marked with #5 in figures 4-96 and 4-97, controls the level of the fuel reservoir to provide a constant distance from the fuel's surface to the venturi.

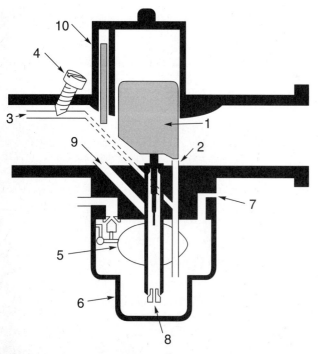

Fig. 4-97 Variable venturi carburetor (closed throttle).

FLOAT BOWL

The float bowl, marked with #6 in figures 4-96 and 4-97, holds the fuel reservoir and provides a place for any sediment to gather. A float bowl drain plug may be included to conveniently remove any deposit.

BOWL VENT

The float vent, marked with #7 in figures 4-96 and 4-97, permits atmospheric air to enter the area above the surface of the fuel in the float bowl. The difference in pressure from the atmosphere (high) to venturi (low) pushes the fuel through the main and idle circuits.

MAIN JET

The main jet, marked with #8 in figures 4-96 and 4-97, limits the amount of fuel that can exit the main discharge tube during WOT (wide open throttle). This preset orifice regulates the proper delivery of fuel and air to provide maximum power and economy at WOT. The air-fuel ratios can be changed by switching to another main jet with a different sized orifice.

HIGH-SPEED AIR BLEED

The high-speed air bleed, marked with #9 in figures 4-96 and 4-97, provides for premixing of air and fuel for better vaporization in the main discharge tube (#11) before the fuel reaches the venturi.

CHOKE

The choke, marked with #10 in figures 4-96 and 4-97, slides down to block the majority of air flow through the carburetor and provide increased vacuum in the venturi, which permits a rich-in-fuel mixture to enter the combustion chamber for easier cold engine starts.

MAIN DISCHARGE TUBE

The main discharge tube, marked with #11 in figure 4-96, permits a fuel mist to enter the venturi from three-quarter to full throttle opening. The midspeed needle (#12) does not limit the amount of fuel discharged at wide throttle openings.

MIDSPEED NEEDLE

The midspeed needle, marked with #12 in figure 4-96, controls the discharge of fuel from the main discharge tube (#11) from one-quarter to three-quarter throttle opening. The midspeed needle is connected to the center of the throttle slide (#1) and can be adjusted by either extending or retracting it from the throttle slide. This provides for an air-fuel adjustment at mid-range speeds.

CARBURETOR REBUILD STEPS

Difficulties with fuel systems usually originate from improper carburetor settings, or from dirt, gum, or varnish in the carburetor. The urgency of cleaning will depend upon application and operating conditions. To clean thoroughly, it will be necessary to completely disassemble the carburetor and fuel system. All parts should be cleaned in solvent. Blow out all passages with compressed air and replace all worn and damaged parts. Always use new gaskets.

1. *Clean engine*
   A clean engine is necessary for proper inspection of broken or missing parts.

2. *Drain fuel from tank*
   Empty the fuel from the tank in a clear glass container. Inspect for any contamination that may be present.

3. *Remove fuel tank*
   When convenient, remove the fuel tank for better access to the carburetor parts.

4. *Drain fuel from carburetor*
   Many carburetors have a drain valve at the bottom of the carburetor fuel reservoir that allows the fuel to be drained. If the drain is not present, loosen the adjusting packing nut or the bowl nut slightly so the fuel may drain.

5. *Disconnect throttle plate cables*
   Detach all cables that operate the speed of the engine.

6. *Remove air filter*

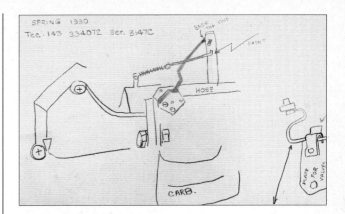

Fig. 4-98 Sketch linkage before removal.

Remove and inspect the air filter for damage or neglect.

7. *Sketch and remove linkage*
   The throttle is attached to the governor system by a wire that should be marked, or a sketch be made, so that it can be reassembled easily, as shown in figure 4-98.

8. *Remove carburetor*
   The carburetor can now be removed and put into a small tray so that all the parts can be kept together. Inspect the carburetor body for cracks, holes, and other wear or damage.

9. *Remove float bowl*
   Remove the bowl nut and float bowl and bowl gasket, as shown in figure 4-99.

10. *Disassemble the carburetor*
    Disassemble the carburetor parts, as shown in figure 4-100. Inspect the float for dents or holes and check the float hinge for wear. Remove the inlet valve seat, using a #5 crochet-hook to hook it and pull it out, as shown in figure 4-101. Order new parts for the assembly, as shown in figure 4-102.

11. *Remove rubber parts*
    Remove the idle mixture screw with the spring, O ring, and washer. Inspect the idle mixture needle for a bent needle point or a groove in the tip of the needle, as shown in figure 4-103. Replace the needle if it is bent or grooved. Inspect the throttle plate and choke shaft and plate assemblies for wear or excessive play. Remove all rubber or plastic parts except the fuel inlet.

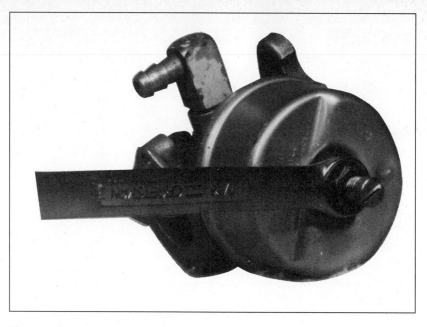

Fig. 4-99 Remove bowl nut.

12. *Remove throttle shaft (optional)*
    If worn, the throttle shaft should be replaced. Rotate the throttle plate shaft to the close position and remove the throttle plate screw. The throttle shaft seals should be replaced if present.

13. *Remove choke shaft (optional)*
    If worn, the choke shaft should be replaced. Grasp the choke valve and remove it from the

Fig. 4-100 Disassemble carburetor.

choke shaft. Remove the choke shaft and felt or foam washer.

14. *Remove the welch plugs*
    The welch plugs are used to cover certain passages that were drilled at the factory. The plugs may be removed, as shown in figure 4-104.

15. *Soak in cleaner (15 minutes)*
    Put the carburetor into a cleaning solution for fifteen to thirty minutes. If the carburetor cannot be cleaned in this time frame, the carburetor body should be replaced.

16. *Wash parts in warm water*
    Wash all the parts in warm water to remove the carburetor cleaning solution.

17. *Clean passages*
    Inspect for wear, damage, cracks, or plugged openings. Use compressed air to clear plugged openings. Be careful not to have the air pressure higher than 40 PSI, otherwise you may propel some dirt or other contamination into the casting of the carburetor body, which may prevent proper flow of fuel. Soft tag wire may be used to probe passages for obstructions.

Fig. 4-101 Remove inlet valve seat.

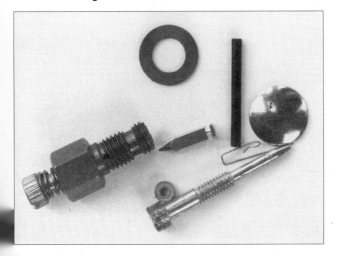

Fig. 4-102 Order repair kit.

18. Lay out parts in clean area

    Keep the assembly area clean from dirt and debris. All components should be cleaned and inspected before the assembly begins.

19. Install welch plugs

    Install welch plugs with a pin punch slightly smaller than the outside diameter of the plug, as shown in figure 4-105. After the plug is installed, seal the outside edge of the plug with fingernail polish, as shown in figure 4-106. Only the welch plugs located on the outside of the carburetor should be sealed with nail polish.

20. Install and adjust float assembly

    The blunt end of the #5 crochet hook can be used to install the new seat into place, as shown in figure 4-107. Install and adjust the float, as

Fig. 4-103 Inspect idle mixture needle.

Fig. 4-104 Remove welch plug.

Fig. 4-105 Install welch plug.

Fig. 4-106 Seal with nail polish.

shown in figure 4-108, according to the procedures described earlier in this chapter.

21. *Install remaining parts*
    If the throttle shaft or choke shaft has been removed, a new screw should be used when installing the throttle or choke plate. The new screws have a dry adhesive that prevent the screw from vibrating loose and being ingested into the engine. Install the remaining parts as needed.

22. *Pre-adjust idle mixture needle*

Fig. 4-107 Install new inlet valve seat.

Turn the idle mixture needle inward (clockwise) until the needle lightly seats. Then turn the needle outward (counter-clockwise) one complete turn.

23. *Pre-adjust high-speed mixture needle*
Turn the high-speed mixture needle inward (clockwise) until the needle lightly seats. Then turn the needle outward (counter-clockwise) 1-1/2 turns.

24. *Remount carburetor*
Use a new gasket for proper sealing.

25. *Install linkages*
Refer to the disassembly sketches.

26. *Install air cleaner*
Clean or replace the air cleaner element.

Fig. 4-108 Set float level.

27. *Connect speed control cable*
    After the speed control cables have been installed, insure that when the control is in the stop position that the grounding switch will work.

28. *Inspect fuel tank*
    Clean and inspect the fuel tank for debris. Clean the tank and reinstall on the engine.

29. *Connect fuel lines*
    Connect fuel lines and install a new fuel filter, if used.

30. *Add fresh fuel*
    Use gasoline that has not been in the storage can for more than three months.

31. *Check for leaks*
    Before starting the engine, check all areas of the fuel system for any leaks that could cause a hazardous condition.

32. *Start engine*
    Let the engine warm up for five minutes.

33. *Adjust carburetor*
    Adjust the carburetor as described earlier in the chapter.

SUMMARY

A carburetor is a metering device for mixing fuel and air. The correct mixture in the combustion chamber is essential for the engine to run properly. Liquid gasoline will not support rapid combustion without being changed to a mist or broken down or atomized into tiny drops. When liquid gasoline is introduced to the venturi, it is gasified by its collision with the rapidly moving air mass.

Briggs & Stratton manufactures three different basic types, designated as: Vacu-Jet, Pulsa-Jet, and Flo-Jet.

The Vacu-Jet can be identified by the single pick-up tube that extends into the fuel tank through the bottom of the carburetor.

The Pulsa-Jet can be identified by the two different sized fuel pipes that extend into the fuel tank. The long one transfers fuel from the large tank to the small fuel cup by means of a fuel pump, while the short tube transfers the fuel from the fuel cup to the venturi.

The Flo-Jet resembles the internal workings of the carburetors generally found on automobiles. It is a float-type carburetor.

Tecumseh produces both float-type carburetors and diaphragm carburetors. The newest addition to the carburetor family is the Vector, which has a simple construction and low production costs.

Questions

1. **A Briggs & Stratton Pulsa-Jet carburetor needs the air-fuel mixture adjustment. Mechanic A says there is only one adjusting screw. Mechanic B says there are two adjusting screws. Who is correct?**

    A. Only Mechanic A
    B. Only Mechanic B
    C. Both Mechanic A and B
    D. Neither Mechanic A or B

2. **The Tecumseh carburetor float level needs to be checked. Mechanic A says to level the float. Mechanic B says to use the Tecumseh adjusting tool. Who is correct?**

    A. Only Mechanic A
    B. Only Mechanic B
    C. Both Mechanic A and B
    D. Neither Mechanic A or B

3. **The gas/oil mixture used in a two-cycle engine has been properly mixed. The carburetor high-speed adjustment is adjusted too lean. Mechanic A says that may cause the engine to fail because of improper lubrication. Mechanic B says that may cause the engine to fail because it will overheat. Who is correct?**

A. Only Mechanic A
B. Only Mechanic B
C. Both Mechanic A and B
D. Neither Mechanic A or B

4. A Walbro carburetor is used on some chain saws. Mechanic A says it is likely to be a diaphragm carburetor. Mechanic B says that it likely that it has a fuel pump built into the carburetor.
Who is correct?

A. Only Mechanic A
B. Only Mechanic B
C. Both Mechanic A and B
D. Neither Mechanic A or B

5. The high-speed mixture is adjusted on a carburetor. The ratio of air and fuel must be made "richer." Mechanic A says to turn the screw clockwise. Mechanic B says to turn the screw counter-clockwise.
Who is correct?

A. Only Mechanic A
B. Only Mechanic B
C. Both Mechanic A and B
D. Neither Mechanic A or B

6. A chain saw engine will run in many positions. Mechanic A says this is because of its float carburetor. Mechanic B says this is because of its diaphragm carburetor.
Who is correct?

A. Only Mechanic A
B. Only Mechanic B
C. Both Mechanic A and B
D. Neither Mechanic A or B

7. Gasoline in a fuel tank has been contaminated with water. Mechanic A says it will stay on the surface and can be skimmed off. Mechanic B says that it will dissolve in the gas and eventually be burned.
Who is correct?

A. Only Mechanic A
B. Only Mechanic B
C. Both Mechanic A and B
D. Neither Mechanic A or B

8. An engine starts and runs briefly after a small quantity of fuel is put into the spark plug hole, but does not start when fuel is added to the carburetor's throat. Mechanic A says that the problem may be that the exhaust valve is not seating properly. Mechanic B says the problem may be the carburetor mounting gasket.
Who is correct?

A. Only Mechanic A
B. Only Mechanic B
C. Both Mechanic A and B
D. Neither Mechanic A or B

9. A lawnmower stops running after a short period of time. The fuel tank filler cap is removed and a nearly full tank of fuel is found. The cap is re-installed and the engine starts easily. The lawnmower operates for a short time and the problem continues. Mechanic A says that the vent in the fuel tank cap may be blocked. Mechanic B says that the fuel may have the wrong octane rating.
Who is correct?

A. Only Mechanic A
B. Only Mechanic B
C. Both Mechanic A and B
D. Neither Mechanic A or B

10. The machine starts and runs well under all conditions for some length of time, then it loses power and finally quits. If you leave it for a short time, it will start and run for awhile and then quit again. Mechanic A says to check for a dirty, partially plugged fuel filter or a fuel line routing that has humps and hollows where vapor lock may occur. Mechanic B says to check the valve gap clearance. Who is correct?

A. Only Mechanic A
B. Only Mechanic B
C. Both Mechanic A and B
D. Neither Mechanic A or B

11. The equipment operator tells you the unit won't run. There is no fuel getting to the engine. Mechanic A says to check for fuel in the fuel tank. Mechanic B says to remove the fuel line at the carburetor and check for fuel flow. Who is correct?

A. Only Mechanic A
B. Only Mechanic B
C. Both Mechanic A and B
D. Neither Mechanic A or B

12. The engine is low on power but uses a lot of fuel. Black smoke comes out of the muffler, It misfires most noticeably on rough ground and in corners. The spark plugs foul out after the machine runs for only a few hours. Mechanic A says the engine needs new rings and should be overhauled. Mechanic B says to check the choke plate operation and make sure it will open all the way during operation. Who is correct?

A. Only Mechanic A
B. Only Mechanic B
C. Both Mechanic A and B
D. Neither Mechanic A or B

13. The engine surges badly at a low idle, but runs better at high idle. It runs adequately under load, but it starts hard. Mechanic A says that, most likely, there is an air leak in the inlet manifold. Mechanic B says that the carburetor is adjusted too lean. Who is correct?

A. Only Mechanic A
B. Only Mechanic B
C. Both Mechanic A and B
D. Neither Mechanic A or B

14. The engine starts well and runs fine at low idle. It starts to surge some at high idle and surges badly when under load. Mechanic A says the carburetor is adjusted too rich. Mechanic B says that there is a leaking float bowl inlet jet. Who is correct?

A. Only Mechanic A
B. Only Mechanic B
C. Both Mechanic A and B
D. Neither Mechanic A or B

15. What is a passage not found in a Tecumseh dual system carburetor? Mechanic A says it is the internal atmospheric vent. Mechanic B says that it is the idle transfer passage. Who is correct?

    A. Only Mechanic A
    B. Only Mechanic B
    C. Both Mechanic A and B
    D. Neither Mechanic A or B

16. Tecumseh Series I carburetors with an idle system use which type of adjustment system? Mechanic A says only the solid fuel adjust system. Mechanic B says only the air adjust system. Who is correct?

    A. Only Mechanic A
    B. Only Mechanic B
    C. Both Mechanic A and B
    D. Neither Mechanic A or B

17. To properly clean a Tecumseh carburetor, Mechanic A says to remove all non-metallic parts except fuel inlet, remove welch plugs, soak for up to 30 minutes, and probe passages with small drills. Mechanic B says to remove all non-metallic parts except fuel inlet, remove welch plugs, soak for up to 30 minutes, and blow all passages with compressed air. Who is correct?

    A. Only Mechanic A
    B. Only Mechanic B
    C. Both Mechanic A and B
    D. Neither Mechanic A or B

18. Which of the following parts should be replaced after cleaning a carburetor? Fuel inlet, needle and seat, float, welch plugs, main nozzle. Mechanic A says the float, welch plugs, and main nozzle. Mechanic B says needle and seat, welch plugs and gaskets. Who is correct?

    A. Only Mechanic A
    B. Only Mechanic B
    C. Both Mechanic A and B
    D. Neither Mechanic A or B

19. What is the easiest way to distinguish a Series III Tecumseh carburetor from a Tecumseh Series IV? Mechanic A says by the number of shutter screws in the choke plate. Mechanic B says by the bosses on the side of the carburetor body. Who is correct?

    A. Only Mechanic A
    B. Only Mechanic B
    C. Both Mechanic A and B
    D. Neither Mechanic A or B

20. How can the welch plugs be sealed after installation? Mechanic A says that fingernail polish can be used because it will not dissolve in gasoline. Mechanic B says to spray the area with paint. Who is correct?

    A. Only Mechanic A
    B. Only Mechanic B
    C. Both Mechanic A and B
    D. Neither Mechanic A or B

CHAPTER 5

Ignition Systems

INTRODUCTION

Small air-cooled engines are internal spark combustion engines in which the burning of the fuel is initiated by a flow of electrons through an air gap in the spark plug. The spark plug's air gap creates a high resistance against the flow of electrons, so the voltage must be high enough for the electrons to leap the air gap and produce the heat energy needed to ignite the compressed mixture of fuel and air. Manipulating and controlling the electrons to accomplish this task will be the focus of this chapter.

ELECTRICAL CONCEPTS AND COMPONENTS

ELECTRICITY

Molecules and substances are combinations of 80 different natural elements composed of similar atoms. All matter is composed of atoms which contain the major parts called **protons, electrons,** and **neutrons,** as shown in figure 5-1. The proton and the neutron are about equal in their mass and account for the majority of the mass of the atom. The electron also has mass, but it is very small relative to the other atomic particles and accounts for little of the mass of the atom. The **proton** has a positive charge (+), while the **electron** is negative (−), and the **neutron** has no charge.

A neutral atom has an equal number of protons and electrons. The protons and neutrons are located at the core of the atom, sometimes called the nucleus. The electrons move around the nucleus at different levels. The electrons that are farthest from the nucleus can sometimes leave one atom and trav-el to another. An atom that can lose an electron easily is characterized as a "good conductor."

Electricity is a form of energy created by the movement of electrons from one atom to another and which can be used to produce light, heat, magnetism, and chemical changes. The flow of electrons can be initiated by friction, induction, and chemical phenomena.

VOLTAGE

The electrons move when an electrical pressure is applied, as shown in figure 5-2. This pressure is called either voltage or EMF (ElectroMotive Force). Electrical pressure to move the electrons is created

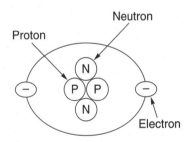

Fig. 5-1 The atom contains the major parts called protons, electrons and neutrons.

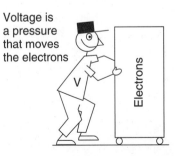

Voltage is a pressure that moves the electrons

Fig. 5-2 Voltage is a pressure that moves the electrons.

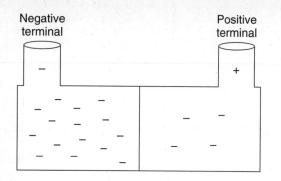

Fig. 5-3 Negative terminal contains excessive electrons.

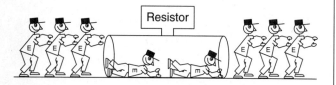

Fig. 5-4 Resistance is the force that opposes the flow of electrons.

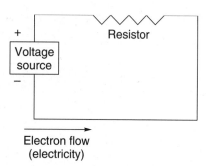

Fig. 5-5 The current in a circuit depends on the voltage and the resistor size.

when one material has a surplus of electrons and the other has a lack of electrons. The surplus of electrons will always attempt to move to where there is a lack of electrons to equalize the imbalance. A battery is a good example, where one of the terminals has an excess amount of electrons (negative) trying to get to the side that is lacking the electrons (positive), as shown in figure 5-3.

CURRENT

The electron's rate of movement or flow is called **current**. The more electrons that move through an area in a certain amount of time, the higher the cur-

rent will be. The unit of current is the **ampere**, or simply amp. The number of amps is dependent on the size of the conductor, the conductiveness of the material, and the pressure of electrons applied.

Resistance is the force that opposes the flow of electrons, as shown in figure 5-4. The unit of resistance measurement is the ohm (Ω). A high ohm resistor can slow the flow of electrons in a circuit. The amount of current in a circuit that has a resistor will depend on the resistance and the voltage applied, as shown in figure 5-5.

CONDUCTORS AND INSULATORS

Some materials hold all their electrons tightly and are classified as good insulators. Other materials allow at least one of their electrons to be extracted easily and are called good conductors. Most metals are good conductors which allow electrons to be passed between their atoms easily.

BATTERY

A battery, as shown in figure 5-6, is an electro-chemical device that can store electrical energy. One part of the battery has an excessive amount of electrons, called the negative (−) terminal, and another

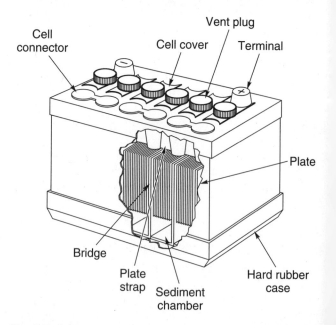

Fig. 5-6 A battery is an electro-chemical device that can store electrical energy.

part is lacking in electrons, called the positive (+) terminal. The terminals are isolated from each other internally and the imbalance, or electrons, is stored until a conductor is connected between the two poles.

COIL

The typical ignition coil's circuit has from 100 to 180 primary circuit windings of #20 copper wire, coated with varnish so that the windings will be insulated from each other. The secondary circuit has 10,000 windings of thin, varnished #38 copper wire, as shown in figure 5-7.

A core made of numerous soft-iron strips is placed in the center of the winding. The core laminations permit a more efficient coil. As the magnets move past the iron core, the magnetism is concentrated in one direction and then rapidly reversed. Because iron itself is an electrical conductor, large eddy currents would be generated in a solid piece of iron and act as an electro-magnet that oppose the change of direction of the magnetic field in the iron, thus slowing it down and producing damaging heat. By splitting the iron core into many thin laminations, the eddy currents are reduced. The slight amount of oxide between each lamination acts as enough insulation to prevent the eddy currents from traveling across between one lamination and the next.

The coil assembly is enclosed in an epoxy compound, so it becomes a single solid mass that is protected from moisture and vibration. The ratio of turns from the primary to the secondary determines the voltage "step-up." The coil magnifies the prima-

ry voltage of 150 volts by at least 100 times to give 15,000 volts in the secondary circuit.

CONDENSER/CAPACITOR

A condenser or capacitor is installed electrically in parallel across the points, as shown in figure 5-8. It is made from two long strips of electrical conductor foil, called plates, and is separated by insulating paper, as shown in figure 5-9. The foil strips and insulation are rolled together, then placed in a container and sealed. The capacitor acts like a small storage battery that provides an easy route for the electrons to accumulate rather than jump an air gap, which also allows exact ignition timing by diverting the current in the primary as soon as the points open. Without the capacitor, the voltage in the primary circuit would force the electrons across the point air gap as it opens, and the precise current sus-

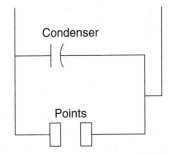

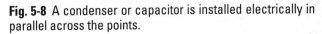

Fig. 5-8 A condenser or capacitor is installed electrically in parallel across the points.

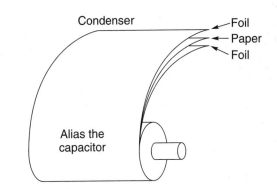

Fig. 5-9 The condenser is made from two long strips of foil and separated by paper.

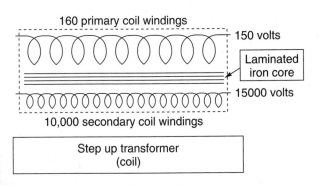

Fig. 5-7 Step-up transformer (coil).

pension necessary for proper engine timing would be difficult. When the coil's magnetic field starts to collapse as the points open, a current is induced in the primary circuit or primary windings. This induced voltage can be as high as 250 volts and is high enough to force electrons across the gap between the points. The electrons would form an arc and absorb electrical energy which would reduce the available voltage in the secondary coil.

When the points close, all the electrical energy which is stored in the capacitor is discharged to the circuit and expedites the subsequent current flow in the primary circuit.

ELECTRON FLOW THEORY

When a conductor is connected across the terminals of a battery, the electrons flow from the negative to the positive poles. This direction is electron flow theory, as shown in figure 5-10.

CURRENT FLOW THEORY

The conventional view of current flow is from the positive pole to the negative, as shown in figure 5-10. This concept has the electrical energy moving in the direction of the "holes" produced when the electrons move.

An analogy is the old-time bucket fire brigade, who moves water from the well to the fire. The fireman near the fire grabs the bucket filled with water from the person next to him. That person is left without a bucket and turns and grabs the water bucket from the person next to him. The lack of bucket then flows to the next, all the way to the well. The water (electrons) moves one way, while the lack of bucket (holes) moves the other. No matter how the movement is viewed, the water arrives at the fire.

Either electron flow or current flow theory is acceptable, but throughout this text we will explain the movement of electricity with the current flow theory.

MAGNETS (PHYSICS)

Magnetism is a characteristic of substances which attract or repel other substances. Few natural materials possess this quality; however, it is possible to magnetize certain materials, either temporarily or permanently. Iron and iron alloys are the most common materials that can be magnetized, as shown in figure 5-11. Soft-iron, unlike hardened steel, cannot be converted into a permanent magnet.

The strength of a magnet is concentrated at two poles, called north and south. If two magnets are moved near each other, the unlike poles attract each other and the like poles repel. The magnetic poles are surrounded by a magnetic field. A magnetic field consists of lines of magnetic force or flux, as shown

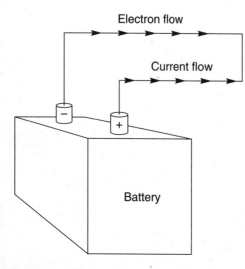

Fig. 5-10 Electron flow vs. current flow.

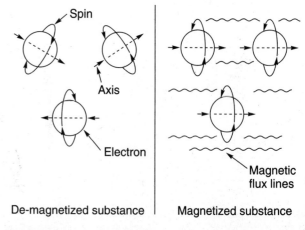

Fig. 5-11 Magnetized vs. De-magnetized substance.

in figure 5-12. Theoretically, each line of force departs from the north pole and reenters the magnet at the south pole.

A temporary magnet can be created by passing a current of electricity through a wire that is coiled around a piece of iron, as shown in figure 5-13. Permanent magnets are made from materials that will retain their magnetic poles through most adverse conditions. Alnico permanent magnets are made by a combination of aluminum, nickle, and cobalt. New ceramic magnets are available which develop very high magnetic strength in a short distance. When using the ceramic magnet vs. an Alnico magnet, the size of the magnet can be reduced by 50 percent without any loss in flux strength, as shown in figure 5-14.

INDUCTION

Induction is a process by which some things, having magnetic (magnet) or electrical (electrical current) properties, produce similar qualities in adjacent objects, usually without direct contact, as shown in figures 5-15 and 5-16. When a permanent magnet, having two poles, is moved in the proximity

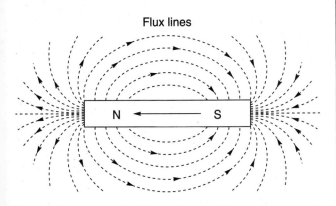

Fig. 5-12 A magnetic field consists of lines of magnetic force or flux.

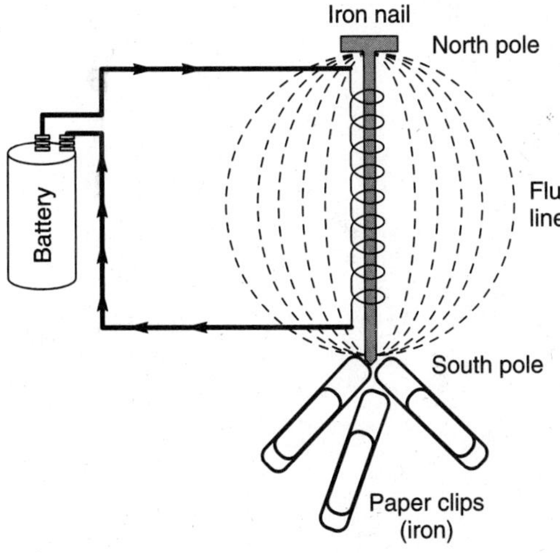

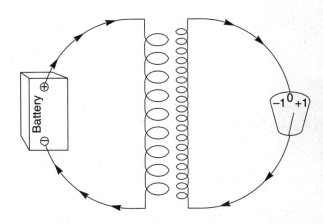

Fig. 5-14 High strength ceramic magnets in a flywheel.

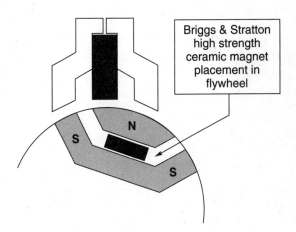

Fig. 5-13 A temporary magnet can be created by passing a current of electricity through a wire that is coiled around a piece of iron.

Fig. 5-15 Induction of electrical current from one circuit to another.

of a coil of wire, a current is generated which is induced to the adjacent circuit. A similar result occurs when a current flow is generated when a battery is hooked into the circuit. Current is induced only when the current in one coil is changing, such as when the magnetic flux increases or decreases in the windings or when the battery circuit fluctuates from no current to full current flow.

SPARK PLUG CONSTRUCTION

The spark plug consists of a center electrode surrounded by a ceramic insulator that is held by a steel shell, as shown in figure 5-17. The center electrode provides one side of the air gap that the spark must jump, while the remaining side is formed by the ground electrode that is attached to the shell. Even though most spark plugs look alike, they have differences that are not always apparent.

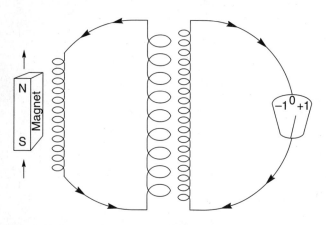

Fig. 5-16 Current is induced only when the lines of flux are moving.

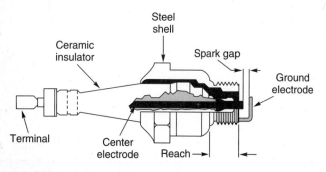

Fig. 5-17 Spark plug anatomy.

ELECTRODE TEMPERATURE

The electrode area should be kept between 700° to 1500° F. Less than 700° F will cause a build-up of carbon and oil on the spark plug electrode, which will eventually stop the engine, and above 1500° F, will cause the combustion mixture to ignite before the spark occurs. In order for the spark plug to maintain the ideal "self-cleaning" temperature, it is designed with a "heat range" to match the engine. A "cold" spark plug readily dissipates the spark plug's heat, while a "hot" spark plug conducts the heat less readily.

The spark plug that is recommended for an engine is designed for average engine use. An engine operated at a slow speed for an extended period of time may need to shift to a "hotter" spark plug, while one operated under heavy load and fast speed will require a "cooler" spark plug.

CENTER ELECTRODE MATERIAL

The electrode can vary in composition. The common steel center electrode can be substituted by either copper or platinum alloy to allow greater concentration and intensity of the spark, which promotes self-cleaning and longer life.

A copper core spark plug runs hotter at low speeds and cooler at fast speeds than does a normal spark plug. It is designed to better resist fouling and overheating.

A platinum alloy spark plug needs a lower voltage for a spark to occur and has a wider heat range than the standard spark plug.

SPARK PLUG REACH

The reach is the distance between the seat and the end of the thread, as shown in figure 5-18. The seat

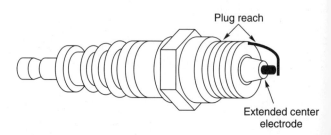

Fig. 5-18 The reach is the distance between the seat and the end of the thread.

is where the spark plug touches the cylinder head and forms a leakproof seal. If the spark plug reach is too short for the engine, the electrode will not be suitably situated in the combustion chamber for proper combustion. If it is too long, the electrodes may protrude into the combustion chamber far enough to strike the piston head or valve.

SPARK PLUG SIZE

Some spark plugs are reduced in the length and are called "shorty" plugs. The primary reason for the different sizes is a restricted area around the engine, which makes it not advisable to use a normal length spark plug. The spark plugs can be used in the same engine because the reach, heat range, and electrode design are similar.

SPARK PLUG SERVICE

The spark plug should be removed, replaced (or cleaned), and adjusted occasionally.

NOTE: DO NOT USE A SAND OR BEAD BLASTER TO CLEAN THE SPARK PLUG. SMALL PARTICLES MAY REMAIN IN THE SPARK PLUG AND ENTER THE ENGINE'S CYLINDER WHILE THE ENGINE IS RUNNING, CAUSING SEVERE DAMAGE. THE FOLLOWING ARE THE CORRECT SERVICING STEPS:

1. *Use a new spark plug if the insulation is broken or if the electrode is pitted or burned. When reusing an old spark plug, clean the spark plug with solvent, such as carburetor cleaner and a wire brush, and then blow out completely with compressed air.*

2. *Check and adjust the spark plug's air gap with a wire feeler gage, as shown in figure 5-19. A flat*

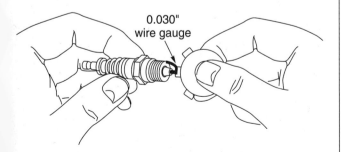

Fig. 5-19 Check and adjust the spark plug's air gap with a wire feeler gage.

gage will ride on the high points and give a false reading.

3. *Install the spark plug and tighten to 180 inch-pounds of torque (15 ft-lbs.).* **Note:** *if a torque wrench is not available, screw in the spark plug as far as possible by hand and use a spark plug wrench to turn it an additional 1/8- to 1/4-turn. When inserting the spark plug into the cylinder head, it is important to initially use your hands for at least two revolutions without any tools to avoid "cross-threading" the threads in the cylinder head.*

SPARK PLUG SEIZURE

Spark plug seizure in an aluminum cylinder head is generally caused by improper installation alignment, as well as by excessive torque. When a spark plug is hot, the threads become distorted from the heat of expansion. If the spark plug is removed while hot, it may damage the threads in the aluminum head. The spark plug threads return to the proper dimensions when the spark plug is cooled. Any spark plug that begins to bind during installation or removal is likely damaging the aluminum threads. It is a good practice to run a thread chaser through the threads in the cylinder head to restore thread cleanliness after you have removed the spark plug. The threads should be coated with an anti-seize compound before installation for easier future removal.

Overtorquing a steel-shelled spark plug into an aluminum alloy head can easily distort the softer aluminum threads. Generally, proper spark plug torque in an aluminum head engine is no greater than 15 ft.-lbs (180-inch pounds).

DAMAGED SPARK PLUG THREADS (CYLINDER HEAD)

The spark plug threads in an aluminum cylinder head can be damaged easily by the harder steel threads on the spark plug. If damage has occurred, the following are possible solutions:

Thread chasing

A tool (tap) that matches the threads of the spark plug may be used to clean the threads from any

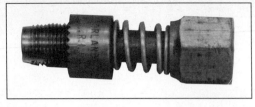

Fig. 5-20 A tool (tap) that matches the threads of the spark plug may be used to clean the threads from any deposits or distortions.

Fig. 5-21 Spark plug thread insert.

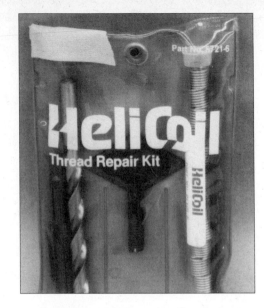

Fig. 5-22 Helicoil spark plug repair thread.

Helicoil thread

This is a threaded insert product used to restore damaged threads by drilling them out, retapping the new hole, and then screwing a springlike set of threads into the newly tapped hole, as shown in figure 5-22.

This technique is a popular method used to restore damaged threads. A special drill is used to cut the damaged threads, and a special tap is used to cut a new thread. A springlike set of threads included in the kit is then twisted in the newly tapped hole with a special tool. The new threads are stronger than the original because of the work hardening of the aluminum in the head and the use of a steel alloy for the new thread.

New cylinder head

The easiest method is to use another cylinder head in good condition, especially if a good "used" one is available. Less labor time is involved in this process.

TWO-STROKE-CYCLE ENGINE SPARK PLUG

Engine lubricating oil is burned with the fuel in the two-stroke-cycle engine. The tendency for the

deposits or distortions, as shown in figure 5-20. Care is necessary to start the tap so that it matches the original threads. Lubricate the tool with wheel bearing grease so that all the dirt and chips adhere to it rather than dropping in and contaminating the combustion chamber.

Thread insert

This process can be used for threads that cannot be cleaned or reconditioned. A kit is used that provides a drill to oversize the hole. A larger new thread is cut, and a metal insert is twisted in and fastened to the original cylinder head, as shown in figure 5-21.

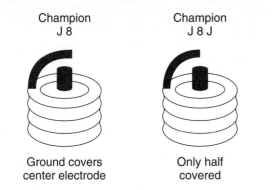

Champion
J 8

Champion
J 8 J

Ground covers
center electrode

Only half
covered

Fig. 5-23 One method to reduce plug fouling is to reduce the spark area in the ground electrode so that is extends only half the center electrode.

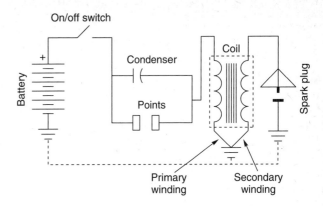

Fig. 5-24 The battery provides a low voltage (12V) source of direct current.

spark plug to contaminate and foul out in this environment demands that the electrode be designed so that the spark heat is more intense in order to burn away the deposits.

Modified ground electrode type

One method used is to reduce the spark area in the ground electrode so that it extends only half the center electrode, as shown in figure 5-23.

> **NOTE:** MANY TIMES THE TWO-STROKE-CYCLE TYPE OF SPARK PLUG CAN BE USED IN A FOUR-STROKE-CYCLE ENGINE FOR LONGER SPARK PLUG LIFE WHEN THE OIL CONSUMPTION IS EXCESSIVE BECAUSE OF INTERNAL WEAR.

Extended electrode type

Some spark plugs have a longer center electrode that extends beyond the steel shell, as shown in figure 5-18. This design retains more heat at lower engine speeds, which increases the heat range to burn away the deposits. At higher speeds, when a colder spark plug is needed, the extended tip is cooled by the flow of the incoming air-fuel mixture.

BATTERY IGNITION SYSTEM

INTRODUCTION

The dependence on a bulky battery makes this system the least used in a small air-cooled engine. The advantage of the system is that a consistent spark is produced at low starting speeds.

BATTERY IGNITION SYSTEM DESCRIPTION

The battery provides a low voltage (12V) source of direct current, as shown in figure 5-24. When the on/off (ignition) switch is closed, a voltage is supplied to the points and condenser from the battery. No current is flowing because the points are open and the condenser will not conduct the current. When the breaker points are closed, the voltage across the points decreases to almost zero and a current flows through the points. The current travels through the primary side of the coil and back to the negative side of the battery. The current through the primary windings creates a slow-building magnetic field in the coil, which induces a voltage in the secondary windings, but the voltage produced in the secondary is too low to allow the electrons to jump the spark plug's air gap. The electrons are allowed to flow through the primary windings long enough to create the maximum magnetic field possible. The points are then opened and, with the assistance of the condenser, the current in the primary circuit is suddenly stopped. The condenser/capacitor soaks up the electrons in the primary circuit as the points open to facilitate the sudden stoppage that creates a fast break down in the magnetic field around the coil. The rapidly collapsing field induces a large voltage (15,000–20,000 volts) into the secondary windings, which produces a sufficient voltage for the electrons to leap the spark plug's gap.

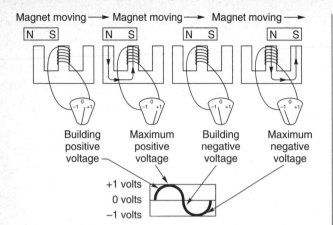

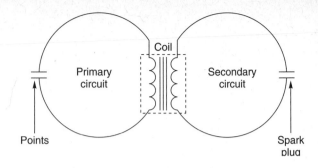

Fig. 5-25 As a magnet passes an iron core that has been wrapped with a wire, an alternating current is produced.

Fig. 5-26 Magneto contains a primary low voltage and a secondary high voltage circuit.

MAGNETO IGNITION SYSTEM

As a magnet passes an iron core that has been wrapped with a wire, an alternating current is produced, as shown in figure 5-25. A positive voltage is induced in the circuit as the magnets approach the iron core. When the magnets are moving away from the core, a negative voltage is induced. The adaptation of this basic alternating voltage system is the generating source of energy for a magneto system.

FEATURES

A magneto is a self-contained unit engineered to produce brief surges of electrical current. The voltage output can be high enough to cause current to leap across a spark plug gap and ignite the air-fuel mixture in the cylinder of an engine. Unlike the battery-coil ignition system, it is not dependent on an outside source of electricity. A magneto consists of two simple circuits: one called a primary circuit and the other the secondary circuit, as shown in figure 5-26. Both circuits have windings which surround the same iron core in the coil, and the magnets in the flywheel act on both circuits. Electrical voltage is induced in each coil by passing a magnetic field by them. Mechanical energy of the moving magnets is converted into electrical or electromagnetic energy. The coil acts as a "step-up" transformer to convert the low voltage in the primary circuit to a higher voltage in the secondary circuit.

SPARK PLUG VOLTAGE

The amount of voltage induced in the secondary circuit is determined by:

1. The strength of the magnetic field through the iron core. *The closer the magnets are to the iron core, the greater the magnetic field. The coil/armature air gap should be kept as small as possible without allowing the parts to touch during operation. Manufacturer's specifications are given for this gap.*

2. The speed at which the magnetic field reverses. *A fast-spinning flywheel creates a stronger reversing field. The breaker points must stop the primary circuit current flow quickly and abruptly for an efficient field reversal.*

3. The number of turns of wire *around the iron core in the primary and secondary circuit.*

4. The amount of voltage needed *to force electrons to jump across the spark plug gap. A smaller spark plug air gap will require a lower secondary voltage.*

BRIGGS & STRATTON MAGNETO

The theory of the system is similar to all magnetos, but with some physical engineering differences. Since the early 1960s, Briggs & Stratton has used a two-legged coil/armature and a three-pole magnetic field in the flywheel, as shown in figure 5-27 and 5-28. The magnets are embedded in the outside circumference of the flywheel, which allows greater magnet "fly-by" speed and high ignition output at low speeds.

Fig. 5-27 B&S 3-legged armature (left), B&S 2-legged armature (right).

The flywheel rotates and aligns the magnet with the legs of the armature/coil, as shown in figure 5-29. As the flux intensity changes, a voltage is induced in the primary and secondary windings. This relatively gradual change in flux field is not rapid enough to create the secondary voltage required to jump the spark plug gap in the circuit. Since the primary circuit is a closed loop (points closed), the voltage causes a current to flow. This current, in turn, creates a magnetic field (electro-magnet) around the windings of the coil, which is utilized to form a barricade to any further flux change. As the magnetic flux intensity changes through the iron core, a voltage is induced.

The magnets continue their rotation so that two different magnet poles line up with the coil/armature legs, as shown in figure 5-30. The flux attempts to reverse, but it is blocked by the magnetic field around the coil. When the magnets align directly below the legs, the points open and the current is stopped in the primary circuit, as shown in figure 5-31. The magnetic field and the resistance to

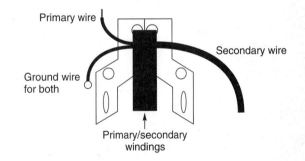

Fig. 5-28 Briggs & Stratton two-legged coil/armature.

the flux flow collapse suddenly. This abrupt change in flux, along with an instantaneously collapsing magnetic field, causes a high voltage in the primary and secondary windings. The voltage is sufficient in the secondary to cause a current flow through the air gap of the spark plug. The high resistance of the atmospheric air at the spark plug causes an intense amount of heat (spark), which can be used to start the fuel mixture on fire (ignite, cause combustion).

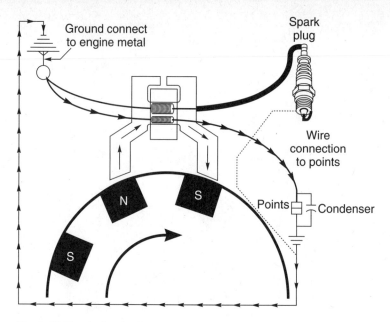

Fig. 5-29 North-South alignment of magnets.

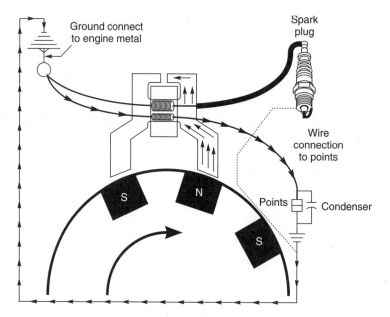

Fig. 5-30 South-North alignment of magnets.

TECUMSEH MAGNETO

As the flywheel rotates, the powerful magnets mounted in the flywheel pass the coil. As the magnet's poles pass the area of the center leg of the stator, a magnetic field is concentrated through the laminations from the magnet's north pole, as shown in figure 5-32. This causes the generation of a current in the coil's primary winding while the ignition points are closed.

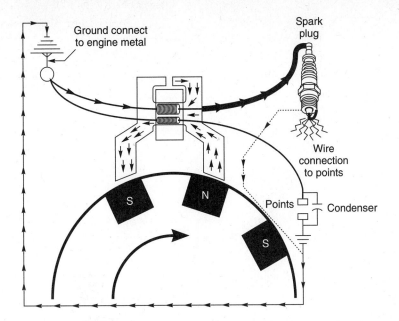

Fig. 5-31 Sudden surge of magnetic flux from north to south pole of the magnet as the points open.

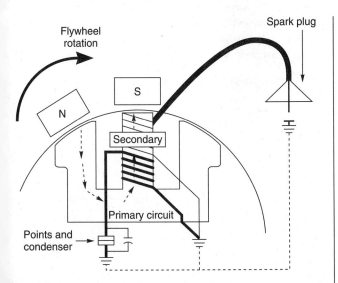

Fig. 5-32 South pole of magnet aligned over the center leg of the laminations.

Since the points are closed, a current flow induced in the primary coil creates a second series of magnetic lines around the core which oppose a decrease in field intensity even though the magnets are moving away. The flux movement is blocked by the opposing field in the primary circuit, as shown in figure 5-33. When the flux intensity reaches a maximum, the points open and, with the help of the condenser, abruptly halt the flow of current in the primary circuit. The collapsing magnetic field and the rush of magnetic flux aid each other to produce a high voltage which is "stepped-up" in the secondary circuit to produce a spark across the spark plug's gap, as shown in figure 5-34.

LOW TENSION IGNITION SYSTEM

This system uses a rapid surge of electrical current in the primary circuit to generate a spark at the spark plug. It differs from the conventional magneto and the battery ignition system, which depend on the rapid collapse of electrical current in the primary circuit. This system was used for a few years, mainly in motorcycles and snowmobiles, until the solid state ignition systems were introduced. The advantage of this system over the conventional magneto is that it allowed the ignition coil to be located

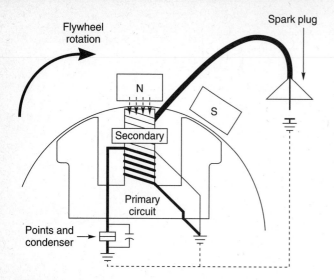

Fig. 5-33 North pole of magnet aligned over the center leg of the laminations.

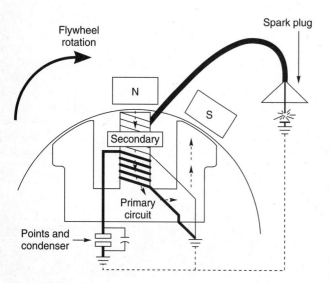

Fig. 5-34 Sudden surge of magnetic flux from north to south pole of the magnet as the points open.

in a cooler area with a resultant reduction of current through the points and a greater durability.

The system consists of a modified alternator composed of an *exciter coil,* as shown in figure 5-35, which produces an alternating current when the magnets move by it. When the points are closed, the

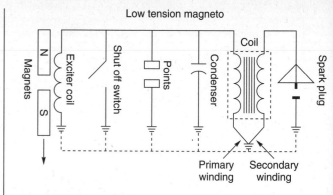

Fig. 5-35 Low tension magneto.

current produced by the alternator in a circuit by-passes the ignition coil. At precisely the correct moment, when the current is at the maximum, the points open and the current now "rushes" through the primary side of the coil, producing a sudden change in the magnetic field which is sufficient to produce a spark in the secondary gap. Since the points in this system only direct the flow of current to another area rather than stopping the flow, the possibility of arcing at the points is reduced and longer point life is achieved.

The stop, or shut-off, switch is connected in parallel and operates by shorting the system. The points are only necessary for generation of spark at low speeds. The alternating current produced by this system when the engine is quickly spinning is sufficient to provide the quick increase and collapse of the magnetic fields necessary for a spark to occur.

ELECTRONIC IGNITION SYSTEM

Most current engines incorporate electronic ignition. The points and condenser, which were a major cause of ignition failure, are replaced by a solid state switching system that:

1. is maintenance free.

2. has no moving parts.

3. is waterproof.

4. never needs adjusting.

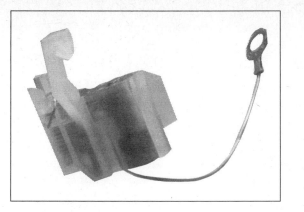

Fig. 5-36 Magnetron® unit.

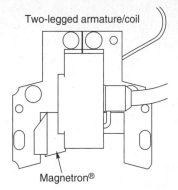

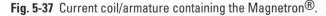

Fig. 5-37 Current coil/armature containing the Magnetron®.

5. *provides constant spark efficiency.*

6. *increases spark plug voltage.*

Since solid state ignition components can be damaged readily by heat, electronic units must be placed in a well-ventilated area. Engine manufacturers that already had their ignition parts located outside the flywheel, in the cooling air flow, were able to adapt faster to the new technology, and, in the case of the Briggs & Stratton Magnetron®[1] unit, it could be retrofitted on existing coil/armatures.

TRANSISTOR TYPE (MAGNETRON®)

Briggs & Stratton introduced the Magnetron®, shown in figure 5-36, ignition in the late 1970s, and most engines produced since 1982 include this solid state feature. The reliability of the unit has been extremely high (covered by a five-year warranty), the failure rate being one per 500,000 units sold. The Magnetron® revolutionized the industry by offering the use of an electronic module on most of the older point and condenser engines produced since 1962, provided they were equipped with a two-legged coil/armature, as shown in figure 5-37. The same type flywheel, coil/armature, and spark plug are used with the Magnetron® as with the points and condenser ignition systems. Prior to 1962, a large three-legged armature coil was used which did not permit retrofitting the Magnetron®.

---

[1]Magnetron is a registered trademark of the Briggs & Stratton Co.

The Magnetron® simply replaces the points and condenser found in the conventional ignition system with a package that contains a *trigger coil* and an *integrated circuit chip.* When the unit is installed, it performs the same function as the points and condenser, i.e., to interrupt the primary circuit at a precise moment for proper spark output. The transistor circuitry and the trigger coil are encapsulated in an epoxy-sealed nylon case. A description of the theory of operation is outlined at the end of the chapter.

Simple installation

The installation of the Magnetron® on an older engine is a simple process. Labor time saved and dependability of the system make it more cost effective than replacing the points and condenser. The wire coming from the points is cut as close as possible to the dust cap and spliced to the Magnetron®. The procedure is outlined at the end of the chapter.

A harder flywheel key, included in the kit, must be used to replace the original key when the flywheel is removed. In older ignition systems, the soft flywheel key protected the crankshaft and the flywheel from damage if the lawnmower blade hit a solid object and stopped suddenly. The force of the sudden impact would be absorbed by shearing the flywheel key, which then prevented the engine from starting. The Magnetron® is not dependent upon the points opening for the spark to be produced, so the engine will spark whenever the magnets pass the coil/armature. If the key is sheared slightly, the engine may suddenly "pull the rope back" when attempting to start the engine. The harder key is used to prevent the flywheel key from shearing easily, but it also allows more stress to be applied to more expensive parts of the engine, such as the crankshaft and flywheel.

A new type of Magnetron®, introduced in 1988, has seen only limited use. The new ignition system uses the same components found in a standard Magnetron® with the addition of a Silicone Controlled Rectifier (SCR). The SCR is used to open the electronic switch that fires the spark plug at about 15° before Top Dead Center or BTDC (retarded spark), at low speeds for easy starting. As the engine speeds up, the voltage induced in the trigger coil of the Magnetron® increases and, working with the SCR, advances the spark to 23° BTDC for better performance.

Magnetron® retrofit advantage

The efficient operation of the older point ignition system is dependent not only on the points and condenser, but on other mechanical factors, as well, which include the following:

1. *The plunger hole must not be worn, or else oil will enter the point and condenser chamber.*

2. *The crankshaft area that pushes the plunger up and down must be clean and in good condition.*

3. *The breaker point plunger must not be worn.*

4. *The crankshaft magneto bearing must not have more than 0.005-inch wear.*

The Magnetron® solid state ignition is not dependent on these conditions. Marginal deficiencies in these areas can be bypassed with the use of the Magnetron® system.

When not to use the Magnetron®

The Magnetron® module **will not fit** the following:

1. *Three-legged coil/armatures produced before 1962, which were larger-sized, and had more windings. Improved magnet technology allowed the change to a smaller, two-legged coil/armature.*

2. *Some cast-iron engines. Current cast-iron*

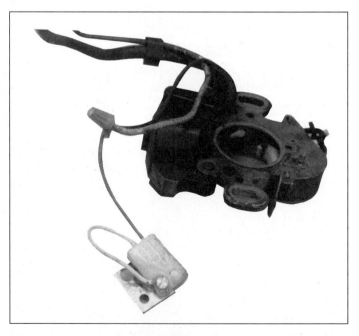

Fig. 5-38 Transistor type ignition used by Kohler and "third parties" for conversion from conventional to electronic ignition systems.

engines have the Magnetron® installed, but previous cast-iron engines with points have magnets (in the flywheel) whose polarity is not compatible with the Magnetron®. If necessary, the flywheel may be returned to the factory and the magnets can be re-polarized so that the solid state unit can be used.

3. Constant parasitic loads attached. The Magnetron® ignition requires a slightly higher starting speed than conventional points. The point ignition system is still preferred for situations where the engine load is attached when starting and rapid rotation is difficult (such as an air compressor).

TRANSISTOR TYPE ("POSTAGE STAMP")

The "postage-stamp" transistor unit attaches to the blower housing and is a system used in many solid state units produced by "third-parties" that are sold to convert other four-stroke-cycle and two-stroke-cycle engines from conventional point and condenser to solid state, as shown in figure 5-38. A description of the theory of operation is outlined at the end of the chapter.

CAPACITOR DISCHARGE TYPE

A capacitor discharge ignition system (CDI) is another type of solid state system, as shown in figure 5-39. As the magnets in the flywheel rotate past the charge coil, electrical energy is produced in the module, as shown in figure 5-40. This energy is transferred to a capacitor, where it is stored until it is needed to fire the spark plug. A description of the theory of operation is outlined at the end of the chapter.

The magnets continue rotating past the trigger coil, where a low voltage signal is produced and closes an electronic switch by a Silicone Controlled Rectifier (SCR). The energy which was stored in the capacitor is now transferred through the SCR switch to a transformer, where the voltage is increased from 200 volts to 25,000 volts.

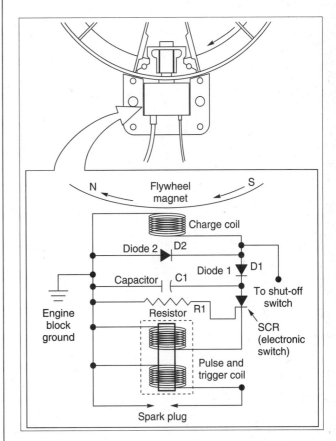

Fig. 5-40 Tecumseh CDI ignition system.

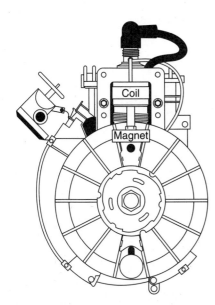

Fig. 5-39 Tecumseh CDI ignition system.

STOP SWITCH CIRCUIT (MAGNETO)

Stopping a running engine can be accomplished by removing the spark from the plug. This can be done by shorting the secondary circuit (spark plug) or the primary circuit (points or solid state).

NOTE: THE ENGINE SPEED SHOULD BE IN THE IDLE POSITION BEFORE SHUTTING OFF THE IGNITION SO THAT A LARGE AMOUNT OF UNBURNED FUEL IS NOT CYCLED THROUGH THE ENGINE INTO THE HOT MUFFLER, WHICH COULD CAUSE A BACKFIRE.

SECONDARY GROUNDING

In this method, the spark is given an easier alternate route to follow in the circuit instead of passing through the gap of the spark plug. Some engines have clips next to the spark plug that can be "snapped" against the spark plug to provide a path from the spark plug wire to the metal of the engine, as shown in figure 5-41. This can also be accom-

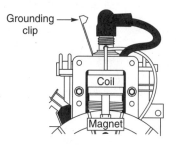

Fig. 5-41 Engine shut-off by grounding secondary circuit at the spark plug.

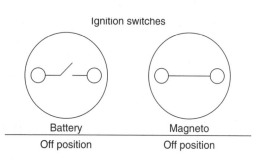

Fig. 5-42 Battery vs magneto shut-off switches.

plished by taking a metal object, such as a screwdriver blade, and, while touching the metal fins of the engine, move it so it touches the metal end of the spark plug wire.

NOTE: DO NOT STOP THE ENGINE BY PULLING OFF THE SPARK PLUG WIRE FROM THE SPARK PLUG UNLESS IN AN EMERGENCY. THIS CAUSES AN OPEN CIRCUIT, AND THE HIGH VOLTAGES PRODUCED WILL ATTEMPT TO FIND A GROUND TO COMPLETE THE CIRCUIT. MANY TIMES, THE SPARK WILL BE TRANSFERRED TO THE INTERNAL COMPONENTS OF THE IGNITION COIL OR TO OTHER SOLID STATE PARTS, WHICH WILL DAMAGE THEM.

PRIMARY GROUNDING

This type of stop switch is different in a battery ignition system than in a magneto system. In a battery ignition system, the primary circuit is interrupted by *opening the switch circuit,* causing the battery current not to enter the primary circuit. In the magneto system, the *switch circuit is closed,* causing the primary current to by-pass the points, as shown in figure 5-42. The two switches may look similar, but may not be interchanged.

The magneto system can be enabled simply by opening the grounding circuit either by cutting the wire or detaching the ground wire screw. Many times, the technician will attempt this to isolate the cause of a "no spark" condition. He uses this method to determine whether the fault is in the primary circuit or in the ignition switch, as shown in figure 5-43.

SAFETY INTERLOCKS

Many times, safety interlocks are added to the equipment to prevent the starting of the engine in an unsafe mode. The safety switches may be wired-in parallel, and any one of them that is in the closed position will not allow the engine to start, as shown in figure 5-44. Other types are wired-in series, where all switches must be open in order for the engine to start. When troubleshooting a "no spark" condition, it is necessary to disconnect the common wire at its junction block on the primary circuit (the point where all the wires from the safety switches converge on the engine).

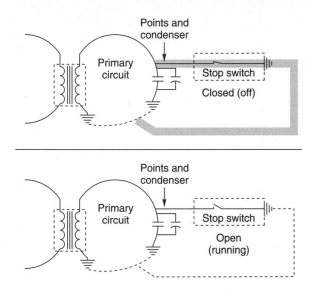

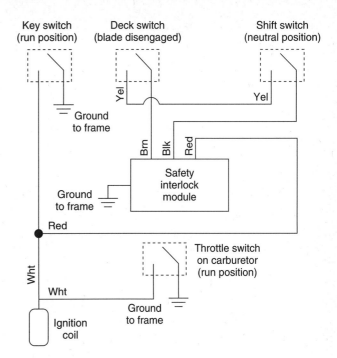

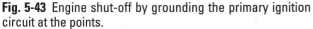

Fig. 5-43 Engine shut-off by grounding the primary ignition circuit at the points.

Fig. 5-45 Engine will start only if switch is open. Once it starts, the switch may be closed.

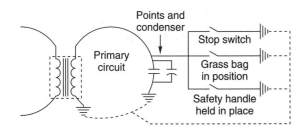

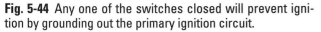

Fig. 5-44 Any one of the switches closed will prevent ignition by grounding out the primary ignition circuit.

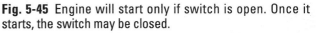

between the red and brown wire in the module, which prevents the ignition system from stopping if any of the switches connected to it are closed.

IGNITION SYSTEM MAINTENANCE (TUNE-UP)

Conventional ignition tune-up consists of replacing:

1. the points and condenser

2. the spark plug

3. setting the armature/coil air gap

4. inspecting the flywheel key

Solid state electronic ignition tune-up involves only:

1. setting the air gap

2. replacing the spark plug

It is important to note that an ignition tune-up alone will *not* restore lost engine power, as shown

There are some instances where a dangerous situation evolves into a safe situation. A tractor may be prohibited from starting by grounding the ignition unless it is in neutral position. Once the engine is started, the ignition circuit must not be grounded (stopping the engine) when transmission is put into a forward gear. This can be accomplished by the use of a safety interlock module, as shown in figure 5-45. The safety interlock module senses the voltage produced in the primary circuit. When the voltage is low or absent, the red wire internally connects to the brown wire to enable the ground circuit. If any one of the switches are closed, spark will not occur. When the engine starts running, the voltage from the primary circuit opens the connection

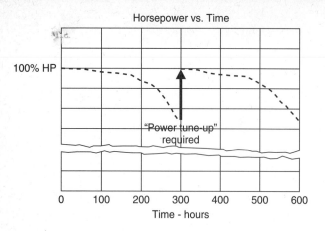

Horsepower vs. Time

Fig. 5-46 Power tune up is required every 100 hours of operation.

in figure 5-46. The power tune-up should also include removing the cylinder head and cleaning the cylinder, cylinder head, top of piston, and around the valves with a non-abrasive cleaning method. The valves should be inspected and refaced or reseated if required. The carburetor and governor should be adjusted properly and the cylinder cooling fins cleaned.

POINT REPLACEMENT

The breaker points must conduct the electrons easily when they are closed. To check the breaker point resistance, set an ohmmeter on the low scale and connect one lead of the meter to the stationary point bracket and the other lead to the movable point. A reading of 0.2 ohms or less should be indicated for good breaker points.

NOTE: BE SURE TO ZERO THE OHMMETER BEFORE EACH READING.

If the meter reading is above 0.2 ohms, clean the points with 600-grit sand paper. After lightly sanding the point surface, wipe the points clean of all contamination. The edge of clean paper money provides a good lint-free paper. Check the point resistance again. If good, the points can be reused. If resistance is still above 0.2 ohms, replace the points.

IGNITION TIMING

This is the time at which the spark occurs at the spark plug in relation to the position of the piston in

its travel in the cylinder bore. Correct ignition timing is necessary to utilize full combustion power to produce the maximum useful work from the fuel and lowest exhaust emissions. If the ignition timing is too advanced, the combustion pressures oppose the rising piston with a resulting loss of power, abnormal stresses to the engine parts, and detonation that causes a "ping" or "knock" in the engine. If the ignition timing is too retarded, the effect will be incomplete combustion, low power output, overheating, and increased exhaust emissions.

Briggs & Stratton ignition timing

The timing of the ignition is not adjustable, and it is extremely dependent on proper point gap setting. It is important that the following method be used to install the points and set the gap. The breaker point gap affects the ignition timing. If the breaker point gap is wider than the specification, the timing will be advanced. This may result in a "kick-back" when starting the engine. If the point gap is too close, timing will be late. This will result in a power loss. In either case, the maximum voltage will not reach the spark plug at the proper time.

When installing the points and condenser, turn the shaft until the keyway lines up with the plunger, as shown in figure 5-47. This will insure that the

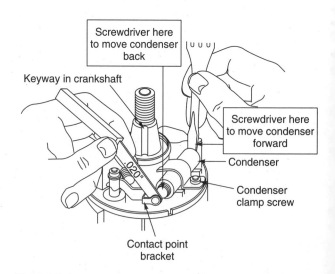
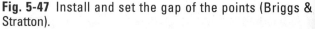

Fig. 5-47 Install and set the gap of the points (Briggs & Stratton).

plunger is off of the timing flat on the shaft. Use the plastic depressor tool or a special tool available to install the primary lead and the ground wire. This plastic tool is furnished with each new set of breaker points. Install the points and condenser clamp securely. Use a screwdriver to pry the condenser backward or forward until the proper point 0.020-inch gap is achieved and measured with a feeler gage. The condenser clamp will hold the condenser tight enough so it will not loosen during installation, yet will remain still loose enough to adjust for proper point gap. When installing the dust cover, make sure that it is a snug fit to prevent dust from entering. Put a sealer, such as Permatex #2, at the opening for the primary lead wire. It is very important to keep the dust and moisture away from the breaker point area.

Tecumseh ignition timing

The ignition timing is adjustable on a Tecumseh engine equipped with points and condenser.

1. *Set points. Turn the engine over until the rubbing block on the points is on the high side of the cam. Set points at 0.020-inch, as shown, in figure 5-48, using a small blade feeler gage.*

2. *Determine TDC. Check the specifications for proper BTDC position of the piston. Install the dial indicator and rotate the crankshaft until the piston is at TDC. This is the point where the needle switches direction on the dial of the gage. Loosen the dial face plate and rotate it so*

the needle is set at zero. Set the piston at the specified BTDC by rotating the crankshaft from TDC, counter-clockwise (from the magneto side) to the proper specification, as shown in figure 5-49.

3. *Adjust stator. Remove the leads from the point terminal and then reinstall and tighten the nut and washer. Attach a continuity device onto the point terminal and to a good ground on the engine. Loosen the two bolts holding down the stator and rotate the stator until the points just open or until the continuity light or ohmmeter indicates a break of the circuit. Tighten the stator bolts and check for proper timing by rotating the crankshaft counter-clockwise (from the magneto side) and then clockwise, watching for the light or continuity to interrupt. When the points open and the light goes out, observe the dial indicator for the proper specification, as shown in figure 5-50. A good setting will vary about 0.003-inch from the specification.*

4. *Clean points. Close and clean the points by sliding some lint-free paper back and forth between the contacts, as shown in figure 5-51. The corner of some clean paper currency works well. Manually open the points when removing the paper to eliminate any paper fibers from remaining between the contacts. The cam felt is permanently lubricated and no oil should be added to it. If the felt is damaged, remove it and coat the surface of the cam with some cam*

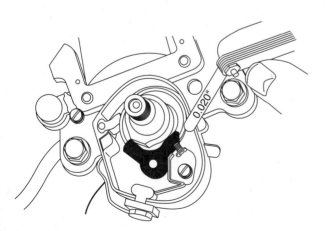

Fig. 5-48 Install and set the gap of the points (Tecumseh).

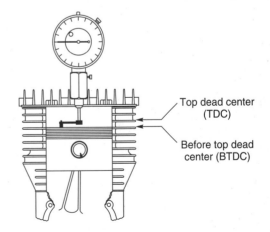

Top dead center (TDC)

Before top dead center (BTDC)

Fig. 5-49 Set the piston at the specified before top dead center.

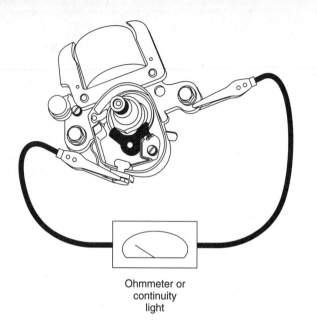

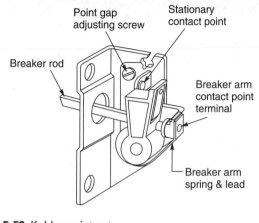

Fig. 5-52 Kohler point set.

Fig. 5-50 Rotate crankshaft and notice when the points open.

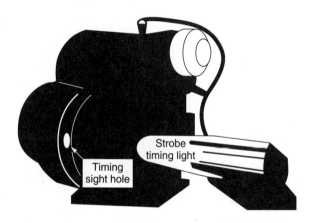

Fig. 5-53 Kohler sight hole for ignition timing.

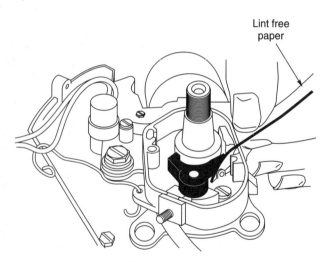

Fig. 5-51 Clean points before installing cover.

grease. This is a special grease that will not migrate to other parts when it is warm.

Kohler ignition timing

On most Kohler engines, best overall performance is obtained if the spark occurs about 20° of crankshaft rotation BTDC. The basic timing of the spark in relation to the engine position is established through proper adjustment of the point gap while the ignition cam is in the maximum lift position. The breaker point assembly is designed with one stationary contact and one movable contact, which is part of an adjustable plate, as shown in figure 5-52. By shifting this plate, the moment of point opening is changed in respect to movement of the ignition cam, thus causing the spark to occur earlier or later in the engine cycle, depending on which way the plate is shifted. Opening the gap advances the timing, and closing the gap retards the timing.

The initial timing setting is made by adjusting the breaker point gap to 0.020-inch, as checked with a feeler gage. This setting should allow the engine to be started.

The final timing adjustment is made with the use of an automotive timing light while the engine is

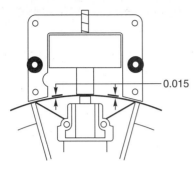

Fig. 5-54 Adjust air gap to manufacturer's specifications.

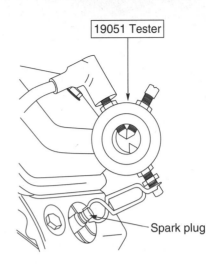

Fig. 5-55 Briggs & Stratton ignition tester (part #19051).

operating. All Kohler engines are equipped with a timing sight hole and timing marks on the flywheel. Connect the timing light per the manufacturer's instructions. Start the engine and shine the timing light into the sight hole. On magneto ignition engines, the sight hole will be in the bearing plate on the side opposite the carburetor, as shown in figure 5-53. On battery ignition engines, it will be in the blower housing on the carburetor side. Adjust the breaker to bring the spark mark (S) into the center of the sight hole. For twin cylinder engines, refer to the appropriate service manual for sight hole location and timing procedure.

ARMATURE AIR GAP SETTING
External mounted coil

Adjust the air gap to the manufacturer's specifications, as shown in figure 5-54. The armature air gap is important in that it affects the flywheel speed needed to deliver sufficient voltage to the spark plug. If the air gap is too great, it is necessary to crank the engine over much faster in order to produce a spark. If the air gap is too small, a spark will be produced at quite a low cranking speed, but the magnets may rub against the legs of the armature and destroy the unit.

Internal mounted coil

The internal air gap cannot be altered, but it is possible to check it. Normally, the clearance should range from 0.007-inch to 0.017-inch for proper operation. If the gap is too large, the lamination assembly must be replaced. Checking the air gap can be

accomplished by using some plastic tape to cover the ends of the laminations. Check the tape for thickness, and then apply one or two layers. Replace the flywheel to the crankshaft and torque the flywheel nut. Turn the flywheel around about twice and return it to its original position. If the flywheel magnets have cut through the tape, there is too little air gap, and the high spots on the stationary laminations can be dressed down with a file. Care should be taken so that each of the three legs of the lamination be equidistant from the rotating flywheel magnets. In case the magneto is not operating properly and one or two strips of tape does not touch the flywheel, then apply another strip.

IGNITION PROBLEMS

The follow sequence should be followed to troubleshoot the ignition system. The steps proceed from the easiest, most probable causes of failure, to the most difficult:

CHECK FOR SPARK

Determine whether the spark is present or adequate.

Test with spark plug in

With the spark plug in the engine, install a spark tester between the spark plug wire and the spark

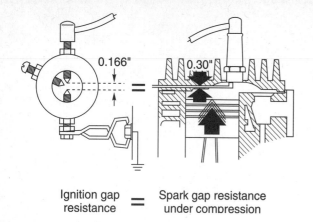

Ignition gap resistance = Spark gap resistance under compression

Fig. 5-56 Large gap simulates the normal spark plug gap under the pressures of compression.

plug. The Briggs & Stratton ignition tester (part #19051—old tool, or part #19368—new tool), shown in figure 5-55, has a large gap adjusted to 0.166-inch. This large gap is sufficient to test the points and condenser system as well as the solid state system. The extra point on the tester compensates for the ionization of the air that occurs at the tester point area when a series of sparks jump the gap, as shown in figure 5-56. It keeps the spark jumping straight and the tester accurate. The opening is covered by a transparent plastic on both sides to increase the accuracy. Attempt to start the engine and observe if a spark is present. A spark across the tester indicates that the ignition system can produce adequate voltage necessary to start the engine. This means that the spark plug is o.k.

NOTE: IN SOLID STATE IGNITIONS LIKE THE MAGNETRON®, THE SPARK MAY BE GOOD, BUT THE ENGINE TIMING MAY BE INCORRECT BECAUSE OF A SHEARED FLYWHEEL KEY. REMOVE THE SPARK PLUG AND OBSERVE THE ELECTRODE AREA. AN ENGINE WHICH HAS A WET SPARK PLUG (WITH FUEL) AND GOOD SPARK, BUT WHICH MAY NOT START, MAY INDICATE A DAMAGED KEY AND INCORRECT TIMING. A DRY SPARK PLUG INDICATES THAT A CHECK OF COMPRESSION AND CARBURETION SHOULD BE MADE.

Test with spark plug out

If a spark is not noticed with the spark plug in, remove the spark plug for easier engine rotation and install the tester between the spark plug wire and some metal of the engine to gain a good ground. Attempt to start the engine, and if a spark is noticed at the tester, the spark plug is defective and should be replaced. If there still is no spark, then proceed to the next step.

ISOLATE IGNITION

A lack of spark at the tester's gap may be caused by either the ignition system or by an equipment-related problem. Reduce the number of possible causes by isolating the engine from the equipment. To do this, locate the terminal where the armature/coil's primary wire connects to the equipment wiring and separate them. Repeat the ignition test. *No spark* at the tester indicates an ignition system problem, and *spark* across the tester's gap indicates an equipment problem.

CHECK ARMATURE AIR GAP

Check the manufacturer's service manual for the appropriate coil/armature air gap dimension. The highest ignition voltage will be obtained with air gap adjusted to the minimum standard. For example, a typical 3.5 HP Briggs & Stratton engine will require an air gap from 0.006-inch to 0.010-inch. To obtain the maximum ignition output, the gap should be kept near the 0.006-inch specification.

Some coil/armature air gaps can not be checked or adjusted. Many of the older Tecumseh and Kohler models place the coil/armature underneath the flywheel, which makes access to adjustment gap impossible. A "rule of thumb" is that the magnets should not rub against the legs of the coil/armature. Look for marks that indicate the magnets touching the legs. Electrical tape can be temporarily stretched and laid on the coil's legs before the flywheel is in place. Install the flywheel and rotate it so the magnets move by the legs. There should be noticeable marks on the electrical tape when the flywheel is removed. After the adjustments are made, test again for spark. If no spark is noticed, then proceed to the next step.

INSPECT THE FLYWHEEL KEY

The only purpose of the flywheel key is to align the flywheel and magnets in the correct relationship with the crankshaft for proper spark production.

Many mistake its function as a driver of the fly-wheel. The tapered shaft surfaces, when the flywheel nut is properly torqued, provide the connection by which the rotation of the crankshaft is transferred to the flywheel. It is possible to place the flywheel on the crankshaft in the proper location and torque it down. If the engine does not stop suddenly, the engine will run properly.

The keys are made of materials of different hardness according to the manufacturer. The softer the key, the more protection you have against expensive repairs when the key is sheared. A vertical shaft engine used on a rotary lawnmower is prone to hitting objects with the blade that will stop the engine suddenly. The sudden stop of the crankshaft will allow the momentum of the flywheel to continue to rotate. If the key is soft, then it will shear without damaging other parts. A harder key will resist shearing and possibly cause a cracked flywheel or a damaged crankshaft.

The trend had been to use a softer key until the advent of the solid state ignition. In the point and condenser ignition system, when the flywheel key is sheared slightly, the ignition voltage output would fall below the requirements of a good spark, as shown in figure 5-57. Many times, an engine with a sheared key will have enough voltage to cause a spark when the spark plug is removed from the cylinder head and laid against a ground (the metal of the engine is a good one), but the engine will not start when the spark plug is inserted into the engine. This is caused by the higher voltage needed to jump the air gap where the air molecules are condensed in the compression stroke of the engine. The engine would not start until the key was replaced. In the solid state ignition system, the harder key is used because the spark will occur even if the key is sheared. The ignition timing will be changed and will cause an advance in the engine timing, which can make the engine hard to start and give a "kick-back" that will possibly destroy an electric starter or hurt your fingers as the starter rope is suddenly pulled back by the engine.

If the key is found defective, it should be replaced and the ignition tested again. If there still is no spark, proceed to the next step.

CHECK POINTS OR SOLID STATE
Points and condenser

If the breaker points are burned, pitted, or oily, they should be replaced. If the area inside the breaker point dust cover is dirty or oily, check to find out where the dirt or oil is coming from.

1. REPLACEMENT

A new set of points and condenser should be installed and adjusted to the manufacturer's specification at this point. Clean the point surfaces by inserting the corner of a clean dollar bill into the points while they are in the closed position and pull it out slowly. Check for any smear left on the corner of the dollar and repeat the process until all residue is removed.

2. SPARK TIMING

Check the spark timing specifications and adjust, if possible. Briggs & Stratton spark timing is obtained by proper gap adjustment of the points during installation. Tecumseh spark timing is accomplished by rotating the stator so the points open at the proper time (See section on Tecumseh timing). Kohler has timing marks on the flywheel that must align with a marker as the points open. The point gap is adjusted for the proper timing.

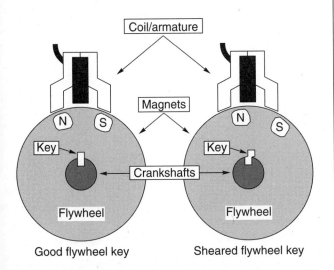

Fig. 5-57 The proper alignment of the magnets to the coil's legs is disturbed when the points open on an engine that has a sheared flywheel key. This results in a reduction of spark plug voltage.

3. PLUNGER/CAM FOLLOWER

The points are opened and closed by either a plunger or a cam follower. These are moving parts and should be inspected for wear during the tune-up. If there is excessive wear in these areas, it is recommended that the system be upgraded to solid state ignition, if possible.

4. MAGNETO BUSHING

If this bushing is worn excessively, proper point movement will be affected when the engine is operating. The timing that was set when the engine was not running will change because of the oscillations of the crankshaft in the worn hole. The amount of wear on the crankshaft journal and bearing can also affect the armature air gap. As the shaft moves up and down between the bearing, the gap will be decreased or increased. A quick check for play can be made by grasping the flywheel and pushing up and down while checking the armature air gap clearance in both positions.

Solid state

If the solid state unit is a integral part of the coil/armature, then the unit should be replaced. These units are not repairable and, at this point, it is logical and proper to replace parts.

NOTE: BEFORE REPLACEMENT, MAKE SURE THAT THE ARMATURE/COIL WAS NOT INSTALLED BACKWARDS.

If the solid state unit can be removed from the coil/armature, disconnect it and proceed to the next step. It will be checked in a later step.

CHECK MAGNETS

The magnets rarely fail, but many magnets are now being glued onto the flywheel. Sometimes the magnets can break off during operation. The magnets may not be glued on again, so the flywheel must be replaced. The magnetism can be checked by laying the flywheel on a wooden surface and dangling a screwdriver about one inch above it. The magnets should attract the screwdriver. Even though the magnets are permanent, they can lose their magnetism by a hard drop of the flywheel or by storing them stacked up on each other.

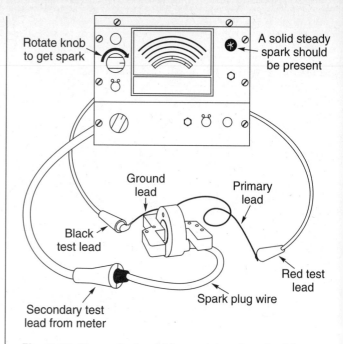

Fig. 5-58 The coil should be tester using the Merc-o-tronic™.

COIL TESTING

The coil should be tested using the Merc-o-tronic™ or Graham-Lee tester™, if one is available, as shown in figure 5-58. If one is not available, then the armature/coil should be replaced with a new or good used one. Before replacing the coil, inspect for broken or loose wires. The cost of the tester compared to the cost of a new armature/coil prohibits many businesses from acquiring a coil tester.

REPLACE SOLID STATE

If there is still no spark after the above steps have been followed, the solid state module must be replaced.

HOW THE MAGNETRON® WORKS

A transistor is frequently compared to a faucet. It is a controlling device that can allow electrons to pass through it like a conductor or stop the flow of electrons, like an insulator. The faucet handle can electrically regulate the state of the transistor.

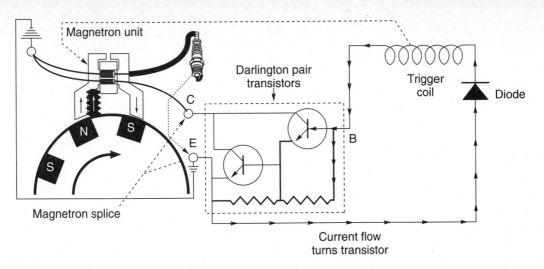

Fig. 5-59 Magnetron schematic. Magnetron turns on the primary circuit in the coil/armature.

The integrated chip sealed inside the Magnetron® contains many small electrical components, but the main part is the Darlington NPN transistor pair that works as a power amplifier, as shown in figure 5-59. Voltage created in the trigger coil circuit affects the base (B) terminal of the IC chip. When a voltage is present at the transistor pair base (B) terminal, the unit allows a continuity to be formed in the primary circuit between the collector (C) and the emitter (E), which replicates closed points. When the voltage is removed from the base (B), continuity discontinues between the collector (C) and emitter (E) and simulates the points opening.

A diode in the circuit converts the alternating voltage (current—AC) induced in the trigger coil to a one-direction voltage (current—DC). While the flywheel rotates, the north and south poles of the magnet line up with the legs of the coil/armature. This alignment induces a current in the Magnetron® trigger coil in the forward direction (diode gate opens). This current stimulates the base (B) terminal on the transistor and allows the coil's primary circuit to close. Electrical current is allowed to flow in the primary windings, as described in the magneto discussion.

The magnet continues its rotation, and opposite magnetic poles now align with the coil/armature legs. The voltage induced in the Magnetron® trigger coil reverses, but the current flow to the base (B) of the transistor is blocked by the diode and opens the primary circuit, as shown in figure 5-60. Continuity is broken in the coil's primary circuit just as if the points had opened to create a spark at the spark plug.

The newer Magnetron® is produced as a molded integrated part of the coil/armature, and the total package is replaced if defective.

HOW THE TRANSISTOR STYLE SOLID STATE IGNITION WORKS

Kohler makes an electronic retrofit kit that replaces the breaker points and condenser in the "K" series magneto engine (part #25-757-10). This unit may not be used to replace the points and condenser used on a battery ignition. The transistor unit attaches to the blower housing and is a "postage stamp" style system. The design is used in many solid state units produced by "third-parties" that are sold to convert other four-stroke-cycle and two-stroke-cycle engines.

The design of the unit is dependent upon the use of a magneto to generate a voltage that the unit can sense. When the voltage produced is high enough, the spark plug will fire the engine. Figure 5-61 shows the two transistors in the unit (Tr 1, Tr 2), which are normally closed to any current through them. As the magnets move by the primary coil (L 1), a voltage is applied to the circuit. The resistance of R 2 is so high that very little current gets by to turn on transistor 2 (Tr 2). The value of R 1 is low enough for a current to flow through it to induce a

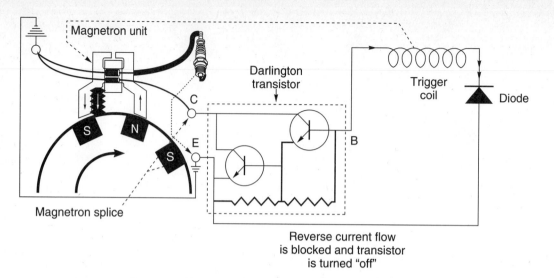

Fig. 5-60 Magnetron® schematic: Magnetron® turns off the primary circuit in the coil/armature causing a spark in the secondary circuit.

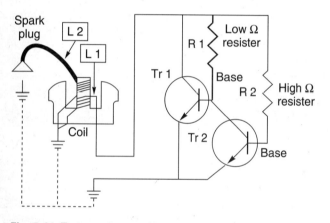

Fig. 5-61 Tr 1 conducts primary current to simulate points closed.

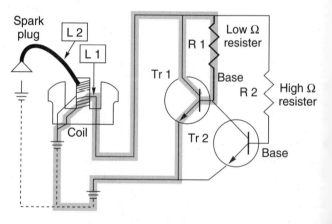

Fig. 5-62 Transistors open primary circuit simulating points opening.

voltage on the base of transistor 1. The voltage is sufficient to "turn-on" the transistor, which closes the primary circuit like the points closing in the conventional system.

At the proper timing of the ignition, the voltage reaches a level that activates transistor 2 is, as shown in figure 5-62. When transistor 2 becomes a conductor, the voltage drop between resistor 1 (R 1) and transistor 2 (Tr 2) is almost zero, which deactivates transistor 1 and opens or disconnects the primary circuit, as shown in figure 5-63. The sudden current shut-off through Tr 1 creates a high enough

voltage induced in the secondary windings (L 2) for proper ignition.

HOW THE CAPACITOR DISCHARGE TYPE WORKS (TECUMSEH PRODUCTS)

The charge coil is able to produce very high primary voltages (300–400 volts) that can be utilized to produce extremely high secondary voltages (25,000–45,000 volts). The charge coil induces an

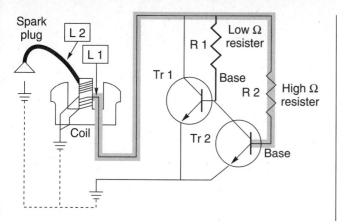

Fig. 5-63 Transistors open primary circuit.

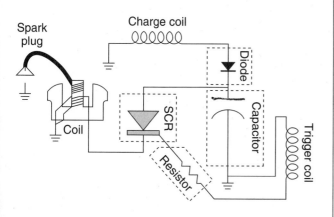

Fig. 5-64 The capacitor imitates a battery by storing a large amount of potential electrical energy.

alternating voltage when the flywheel magnets move near it. The diode connected in series functions as a "half-wave rectifier" and allows current to flow into the large capacitor but not out. This direct current charges the capacitor when the SCR (Silicon Controlled Rectifier) switch is in the "off" state. The capacitor imitates a battery by storing a large amount of potential electrical energy, as shown in figure 5-64.

The trigger coil is "excited" as the flywheel magnets move by it, and the voltage generated in the trigger coil causes the SCR gate to open when it is high enough to pass the resistor. When the SCR gate opens, the stored electrical energy in the capacitor moves quickly through the SCR and also through the primary windings of the coil. The sudden change in the magnetic field around the core of the coil pro-

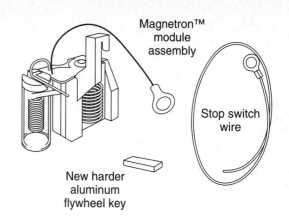

Fig. 5-65 Magnetron® retrofit kit (#394970).

duces a large voltage which is "stepped-up" to a higher voltage in the secondary circuit, causing a spark to occur.

Ignition advance is common to all CDI systems. It is controlled by the speed of the magnets and the voltage in the trigger coil. The spark will occur sooner as the engine speed increases, causing up to a six-degree spark advance necessary for good performance.

MAGNETRON® INSTALLATION

The Magnetron® can be quickly and efficiently installed using common tools and supplies.

MATERIALS

Materials needed are:

- Solder iron and 60/40 rosin core (electrical) solder. Do not use acid core solder, which could corrode the coupling and cause failure.

- Sand paper

- Electrical tape

- Small fingernail scissors or knife

- Magnetron® unit and new flywheel key (#394970), shown in figure 5-65.

- Plug for plunger hole (optional) (#231143), as shown in figure 5-66.

- Pliers

#231143
B&S
plunger
hole plug

Fig. 5-66 Option plug for plunger hole.

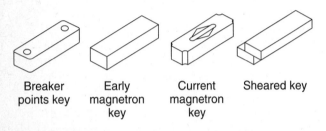

| Breaker points key | Early magnetron key | Current magnetron key | Sheared key |

Fig. 5-67 A harder flywheel key must be used when installing the Magnetron®.

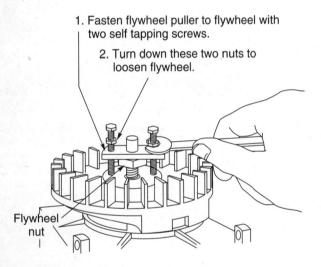

1. Fasten flywheel puller to flywheel with two self tapping screws.

2. Turn down these two nuts to loosen flywheel.

Flywheel nut

Fig. 5-68 Removal of Briggs & Stratton flywheel with appropriate tool.

PROCEDURE
Remove flywheel

The flywheel key must be replaced with a new harder key, as shown in figure 5-67. Access to the point cover is accomplished by removing the flywheel. Unthread the crankshaft end nut and use a

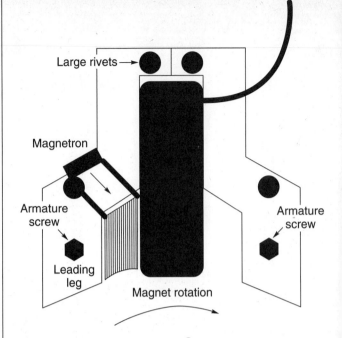

Large rivets →

Magnetron

Armature screw

Leading leg

Magnet rotation

Armature screw

Fig. 5-69 Position of Magnetron® on armature.

puller to remove the flywheel, as shown in figure 5-68.

Cut primary wire

Note that the surface of the coil/armature before removal shows four large rivet heads and the two screws holding the armature to the engine, as shown in figure 5-69. Unbolt and remove the coil assembly from the engine. The Magnetron® ignition system module will eventually be inserted between the leading armature leg (or as the observer faces the armature, the leg to his left) and the armature coil. After cutting the primary and stop-switch wires where they meet the point cover, as shown in figure 5-70, the armature is turned over because most of the installation will take place from the "backside." It is not necessary to remove the cover or the points, but if they are removed, the plunger hole must be plugged with the optional plunger hole plug.

Install Magnetron® to coil

The module is pressed on the leading leg of the coil/armature using force, as shown in figure 5-71.

190   Schuster / Small Engine Technology

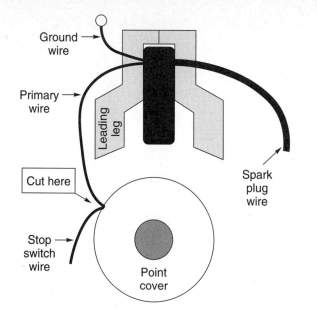

Fig. 5-70 Location of area to cut wires for installation of Briggs & Stratton's Magnetron®.

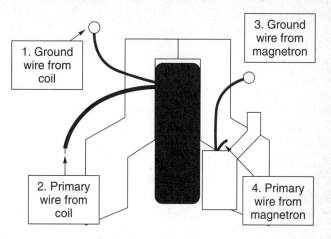

Fig. 5-72 Magnetron® and coil/armature wire identification.

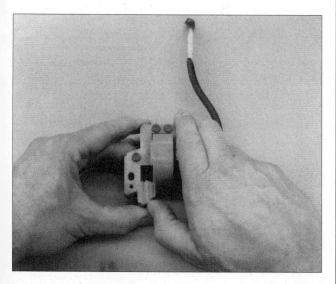

Fig. 5-71 The module is pressed on the leading leg of the coil/armature using force.

There is an interference fit, so considerable force is required to slide it in. The leading leg is the leg that makes the first contact with the magnets on the flywheel as the flywheel travels in a clockwise direction when the flywheel is facing you. Place the coil in its normal position on the engine with the big

heads of the rivets facing upward. The side that the magnets first come into contact is where the module should be placed on the coil/armature. Slide the module all the way on until it snaps into place.

Trim primary wire

Turn the coil/armature over so that the small rivets are facing upward, and observe the four wires that are exposed on the coil with the Magnetron® installed, as shown in figure 5-72. There is (1) the ground wire from the coil (which usually has eyelet attached), (2) the primary wire from the coil (bare end), (3) the ground wire from the Magnetron® (has eyelet), and (4) the primary wire from the Magnetron® (bare end). Before the wires can be connected, the primary wire from the coil must be trimmed and sanded. Take the coil/armature primary wire and bend it over to the primary wire of the Magnetron®. Estimate how much wire is necessary to reach it and cut off the remaining length. Trim about 3/4-inch of insulation from the end of the wire. The insulation is loose around the wire, and it is best to trim it away with a small pair of scissors, as shown in figure 5-73. A wire insulation stripper has a tendency to nick the wire and cause a weak spot which may separate with vibration. The red-colored shellac on the wire is also an insulator that must be removed with a piece of sand paper or by scrapping.

Solder primary wires

Connect the sanded wire and primary wire of the Magnetron®, either under the clip or just twist and solder them, as shown in figure 5-74. The old condenser can be used to depress the spring clip, if it is used. If the new stop switch wire is used on the engine, strip the end of the wire and add it under the spring clip with the primary wires. Twist the wires together with a needle nose pliers. Solder the twisted wires with fine gage solder and then trim off any excess wires.

Solder ground wires

Twist the two ground wires together, degrease, and solder them. Make sure that one of the wire eyelets is able to reach the top mounting bolt hole on the leading leg of the coil/armature.

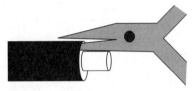

Cut insulation

Fig. 5-73 Cut away insulation rather than use a wire stripper tool.

Eliminate vibration problems

Put electrical tape over the connections to prevent the wires from breaking due to vibration fatigue. For extra insurance, the area can be coated with Permatex #2, but this is not usually necessary.

Optional step

Even though it is not necessary to remove the nonfunctional points, condenser, plunger, and cover, many prefer to remove them. If they are removed, the plunger hole must be sealed by the tapered plug that can be purchased separately.

The plug (part #231143) is driven into the hole with a flat punch and a hammer, and the sides of the taper may be sealed with red (#271) or blue (#242) Loctite®[2].

NOTE: DO NOT ALLOW THE LOCTITE® TO DRIP DOWN TO THE CRANKSHAFT LOCATED AT THE BOTTOM OF THE PLUNGER HOLE. THE LOCTITE® MAY CAUSE THE CRANKSHAFT TO SEIZE IN THE MAGNETO BUSHING.

---

[2]Loctite is a registered trademark of the Permatex Co.

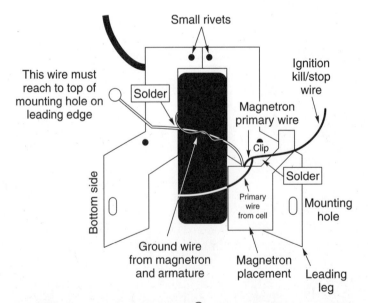

Fig. 5-74 Connect the Magnetron® wires with the wires of the coil/armature.

Replace coil/armature

Place one of the screws through the ground wire and attach the coil/armature.

Install new key

A new, harder flywheel key must be installed with the Magnetron®. In a point ignition system, the spark ceases when the key is sheared, while in the solid state system the key may shear, but the engine will still spark.

Install the flywheel and torque it to specification.

Adjust coil/armature air gap

Check the specifications for the air gap dimension. Use a feeler gage or a piece of old microfiche to set the armature air gap, as shown in figure 5-75. Loosen the coil/armature mounting bolts to pull the armature as far away from the magnets as possible and then tighten one of the bolts. Rotate the flywheel so that the gage can be between the coil/armature and the magnets. While leaving the gage in place, loosen the bolt and let the magnets pull the armature against the gage. Tighten the bolts to **35 in-lbs.** and remove the gage by rotating the flywheel.

Connect stop-wire

Remove the eyelet from the new stop-wire if it is not needed. Connect the wire to the stop-switch.

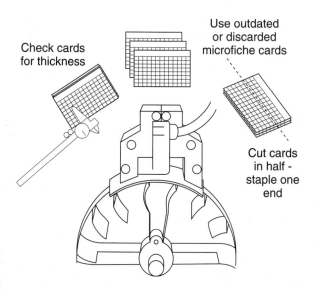

Check cards for thickness

Use outdated or discarded microfiche cards

Cut cards in half - staple one end

Fig. 5-75 Use a feeler gage or a piece of old microfiche to set the armature air gap.

SUMMARY

Electricity is a form of energy created by the movement of electrons from one atom to another and which can be used to produce light, heat, magnetism, and chemical changes. Electrical pressure, called voltage, which moves the electrons, is created when one material has a surplus of electrons and the other has a lack. The electron's rate of movement is called **current**. The typical ignition coil's circuit has from 100 to 180 primary circuit windings of #20 copper wire coated with varnish so that the windings will be insulated from each other.

The spark plug consists of a center electrode surrounded by a ceramic insulator that is held by a steel shell where the center electrode provides one side of the air gap that the spark must jump, while the remaining side is formed by the ground electrode that is attached to the shell.

The battery ignition system is powered by a low voltage (12V) source of direct current and produces a spark by suppling energy to a ignition coil and to the points and condenser.

A magneto ignition system is a self-contained unit engineered to produce brief surges of electrical current across a spark plug gap and ignite the fuel mixture in the cylinder of an engine. Unlike the battery-coil ignition system, it is not dependent upon an outside source of electricity.

Most current engines incorporate an electronic ignition. The points and condenser, which were a major cause of ignition failure, are replaced by a solid state switching system that:

1. *is maintenance free.*

2. *has no moving parts.*

3. *is waterproof.*

4. *never needs adjusting.*

5. *provides constant spark efficiency.*

6. *increases spark plug voltage.*

Many times, safety interlocks are added to the equipment to deter the starting of the engine in an unsafe mode. The engine may not start if the operator leaves the safe mowing area or if the grass bag is not attached properly.

Questions

1. **An engine is found to have an oil fouled spark plug. Further tests show that the cylinder and rings are worn, causing excess oil to enter the combustion chamber, but the customer does not want to spend the money for an overhaul. Technician A says to install a colder spark plug to help prevent spark plug fouling. Technician B says to install a hotter spark plug to help prevent spark plug fouling. Who is correct?**

   A. Only Technician A
   B. Only Technician B
   C. Both Technician A and B
   D. Neither Technician A or B

2. **The solid state ignition is better than the conventional point system. Technician A agrees with this statement because it can produce higher voltage in the secondary circuit. Technician B agrees with this statement because it is sealed from contamination. Who is correct?**

   A. Only Technician A
   B. Only Technician B
   C. Both Technician A and B
   D. Neither Technician A or B

3. **The Briggs & Stratton Magnetron® is installed on a 3.5 horsepower engine. Technician A says that it will spark despite a broken key. Technician B says that it will not spark when the flywheel key is sheared. Who is correct?**

   A. Only Technician A
   B. Only Technician B
   C. Both Technician A and B
   D. Neither Technician A or B

4. **A Tecumseh 3.5 horsepower engine is found to spark every time the piston reaches the top of the cylinder. Technician A says that this is normal. Technician B says that the points should be replaced. Who is correct?**

   A. Only Technician A
   B. Only Technician B
   C. Both Technician A and B
   D. Neither Technician A or B

5. **An electric circuit design contains a battery, resistor, and a continuity switch. Technician A says that the resistor will block the flow of electrons when the switch is closed. Technician B says that the resistor will slow the flow of electrons depending on the rating of the resistor. Who is correct?**

   A. Only Technician A
   B. Only Technician B
   C. Both Technician A and B
   D. Neither Technician A or B

6. **The following statement is made about an ignition system: Technician A says that the greater the speed of the magnets, the greater the voltage induced in the coil. Technician B says that the secondary windings produce a higher voltage than the primary windings. Who is correct?**

A. Only Technician A
B. Only Technician B
C. Both Technician A and B
D. Neither Technician A or B

7. **A small horsepower Briggs & Stratton engine's spark plug is firing too early. Technician A says that this could be caused by too small a spark plug gap. Technician B says that this could be caused by an improper point gap. Who is correct?**

A. Only Technician A
B. Only Technician B
C. Both Technician A and B
D. Neither Technician A or B

8. **A spark plug is examined and found to be light tan in color. Technician A says this indicates that the engine was running "rich." Technician B says it indicates that the engine was "burning oil." Who is correct?**

A. Only Technician A
B. Only Technician B
C. Both Technician A and B
D. Neither Technician A or B

9. **A lawnmower engine is stopped suddenly when it hits a steel pipe. It will not restart. Technician A says that it could have a sheared flywheel key. Technician B says that it could have a bent crankshaft. Who is correct?**

A. Only Technician A
B. Only Technician B
C. Both Technician A and B
D. Neither Technician A or B

10. **A cooler spark plug than recommended is used in a four-stroke-cycle engine. Technician A says that the engine will run cooler and have longer "life." Technician B says that the engine will foul a spark plug faster. Who is correct?**

A. Only Technician A
B. Only Technician B
C. Both Technician A and B
D. Neither Technician A or B

11. **The armature air gap should be measured with the dial caliper. Technician A agrees with this statement. Technician B disagrees with this statement. Who is correct?**

A. Only Technician A
B. Only Technician B
C. Both Technician A and B
D. Neither Technician A or B

12. **Points and condenser are used in the primary circuit to initiate a spark. Technician A says that a spark occurs when the points close. Technician B says that a spark occurs when the points open. Who is correct?**

A. Only Technician A
B. Only Technician B
C. Both Technician A and B
D. Neither Technician A or B

13. **When the flywheel key is sheared on a breaker point controlled ignition system, the engine will not produce a good spark. Technician A states that this is because the voltage has been reduced in the secondary circuit. Technician B states**

that it is caused by the incorrect alignment of the magnets with the legs of the armature.
Who is correct?

A. Only Technician A
B. Only Technician B
C. Both Technician A and B
D. Neither Technician A or B

14. Technician A says that the ignition coil is part of the primary circuit. Technician B says that the ignition coil is part of the secondary circuit.
Who is correct?

A. Only Technician A
B. Only Technician B
C. Both Technician A and B
D. Neither Technician A or B

15. A spark plug, used in an engine that operates primarily at slow speeds, continually fouls with excessive deposits. Technician A says to use a spark plug that retains more heat (hotter spark plug). Technician B says to use a spark plug that loses its heat more rapidly (colder spark plug).
Who is correct?

A. Only Technician A
B. Only Technician B
C. Both Technician A and B
D. Neither Technician A or B

16. Technician A says that the condenser (capacitor) in a breaker point ignition system reduces arcing at the points. Technician B says that the condenser

(capacitor) in a breaker point ignition system aids in the collapse of the magnetic field in the coil.
Who is correct?

A. Only Technician A
B. Only Technician B
C. Both Technician A and B
D. Neither Technician A or B

17. Electricity is a type of energy. Technician A says that the energy is transferred when the proton moves from one atom to another. Technician B says that the energy is transferred when the electron moves from one atom to another.
Who is correct?

A. Only Technician A
B. Only Technician B
C. Both Technician A and B
D. Neither Technician A or B

18. The amount of voltage generated at the spark plug depends on a variety of factors. Technician A says that two are:
   1. Type of ignition system
   2. Speed of the flywheel magnets
Technician B says that two are:
   1. The size of the armature air gap
   2. The numbers of turns of wire in the coil
Who is correct?

A. Only Technician A
B. Only Technician B
C. Both Technician A and B
D. Neither Technician A or B

19. To avoid damage to the spark plug threads in an aluminum cylinder head, certain precautions should be observed. Technician A says to:
    1. Tighten the spark plug with the proper torque
    2. Coat the spark plug threads with an anti-seize compound

    Technician B says to:
    1. Only remove the spark plug when the engine is at operating temperature
    2. Sand blast the threads to clean any carbon off before installing

    Who is correct?

    A. Only Technician A
    B. Only Technician B
    C. Both Technician A and B
    D. Neither Technician A or B

20. Technician A says that the Magnetron® can be retrofitted on all older Briggs & Stratton point ignition systems. Technician B says that a harder flywheel key must be used when switched from point ignition to the Magnetron®.

    Who is correct?

    A. Only Technician A
    B. Only Technician B
    C. Both Technician A and B
    D. Neither Technician A or B

CHAPTER 6

Governor Systems

INTRODUCTION

Governor systems, included on almost all air-cooled engines, regulate the carburetor's throttle opening to achieve the engine speed desired by the operator. If a high engine speed is desired but the engine load is light, such as when a lawnmower is operated out of the grass, the governor system will close the throttle to adjust for the power needed. When the mower is moved into grass and the engine load is greater, the throttle will be opened to increase the engine power without a change in engine speed.

Many automobiles have a governor system installed as an accessory, which is called "cruise control." When the driver turns on the cruise control and sets the vehicle speed at 55 MPH, it is the same as setting the output speed of the engine. The driver may now remove his or her foot from the accelerator pedal, and the governor, or cruise control, will regulate the power. If the automobile must climb a hill, the throttle is automatically opened and adjusted for more power. If the car is moving down a slope, the throttle closes to reduce the power or force.

The assembly and service of the governor system can be frustrating without having an understanding of the basic theory. It can be time consuming to try to determine how the linkages are connected, but even though the governor systems vary in physical attributes between engines, they all follow a simple fundamental theory. Understanding the basic principles will allow the technician to make correct and efficient repairs.

PURPOSES

The purposes of the governor system are to:

1. *Protect the engine*

2. *Provide operating convenience*

3. *Ensure blade speed safety*

GOVERNOR PROTECTS THE ENGINE

The maximum power an engine can produce is called the horsepower rating. Air-cooled engines develop their maximum horsepower at a specified RPM (revolutions per minute), as shown in figure 6-1. If the Briggs & Stratton 3.5 horsepower engine is operated at a speed slower than 3600 RPM, less than the maximum horsepower is developed. If 3600 RPM speed is surpassed, engine power weakens and internal wear increases.

When engine speed is not regulated or is operated at RPM above that is necessary for maximum horsepower, the possibility of internal metal engine parts breaking increases. For example, one possible cause for a broken connecting rod is overspeeding the engine. The rod will break away from the crankshaft, and the metal pieces, hurled inside the engine, can cause damage to many other components and may even exit the crankcase wall.

The governor system limits the engine RPM in order to protect against premature wear and damage to internal components. The governed engine speed should be adjusted at or below the RPM necessary for maximum horsepower.

Some operators think that by disconnecting the governor system, the engine will produce more power. Rarely is this true, as the engine may operate faster but power will be decreased. For example, an engine that develops its maximum horsepower at 3600 RPM may be used on a mini-bike. The operator is not satisfied with the top speed of the equipment, so the governor is bypassed or adjusted to increase the engine speed. The mini-bike will now increase its top speed, but engine power and engine life will be decreased. A better way to modify the equipment for more speed is to change the ratio between the engine driving pulley or sprocket and the driving wheel.

GOVERNOR OPERATING CONVENIENCE

The governor system allows the operator to choose the speed of the engine's output shaft and then lets the governor automatically regulate the carburetor throttle to provide the necessary power to maintain the set speed.

A lawnmower that is started on a cement pad and set in the fast position by the operator does not have to be readjusted when it enters the heavy grass. The governor system automatically adjusts the power for the load encountered.

A good example of not only the convenience of a governor system, but also the necessity, is an electric generator powered by an air-cooled engine. The generator components must be rotated at a specified speed to maintain the required 120 volts. The governor system maintains the critical speed by automatically adjusting the engine power for a light load, such as when a 100-watt bulb is lighted by the generator, or when a more demanding load of a 1/2 horsepower motor utilizes 1200 watts of power.

GOVERNOR CONTROLS SAFE BLADE SPEED

In order to prevent it from breaking, a lawnmower blade must not spin too fast. Safety specifications for power lawnmowers require that the blade tip speed travel at less than 19,000 feet per minute. When the blade is directly connected to the output shaft on a rotary lawnmower, the blade tip speed is a function of engine RPM. A slower blade tip speed requires a lower engine speed.

If the maximum blade tip speed is exceeded, the metal components of the blade may break off and send pieces flying out from under the mower deck. This unsafe condition must be avoided by setting the governor system to limit the blade speed. A chart is provided to convert the length of a blade to the maximum blade tip speed, as shown in figure 6-2.

The recommended speeds listed *are approximately 200 RPM below the maximum* and are calculated from the fact that if an engine is operated at 3600 RPM (60 revolutions per second, or one revolution per 1/60th of a second), and a 20-inch blade is used, the distance which the blade tip travels in one revolution is 63 inches. This converts to 18,900 feet per minute (215 MPH), which is **below** the maximum safe blade tip speed of 19,000 feet per minute. If a 24-inch blade is used at 3600 RPM, the distance which the blade tip must travel in one revolution is 75 inches. This converts to 22,500 feet per minute (255 MPH), which is **above** the maximum allowed

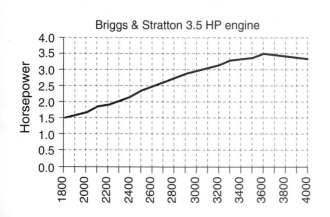

Fig. 6-1 If the engine is operated at a speed slower or faster than 3600 RPMs, less than the maximum horsepower is developed.

Blade length	Recommended maximum rotational R.P.M.
18"	3800
19"	3600
20"	3400
21"	3250
22"	3100
23"	2950
24"	2800
25"	2700

Fig. 6-2 If the maximum blade tip speed is exceeded, the blade may break apart and send pieces flying.

blade tip speed. Since speed is a function of the distance covered in a period of time, and because the time of one revolution (1/60th of a second) is the same, the longer the blade is, the faster the blade tip speed will be, as shown in figure 6-3.

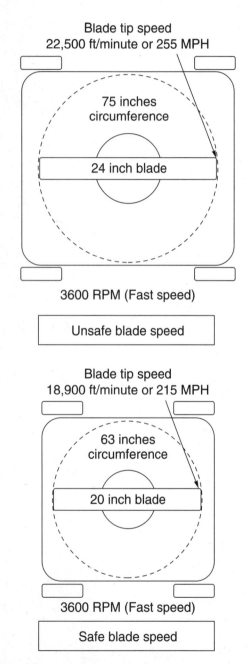

Fig. 6-3 Safe vs. unsafe blade tip speed.

The chart shows the top limit speed for safety. If the speed listed is above the ideal RPM for maximum power, the governor should be set lower than the listed chart velocity. If the chart indicates that the top limit speed is lower than the ideal engine power speed, then the top engine speed must be limited. When the engine velocity is limited below the maximum horsepower, a slight decrease in performance in heavy grass may be noted; but, under normal cutting conditions, the required horsepower is much less than the horsepower produced by the engine.

The blade speed can be calculated quickly using a calculator and inserting the blade length and engine speed into the formula:

Blade speed (feet/minute) =
engine speed x blade length x 0.262

eg., 3400 RPM x 18 inch blade x 0.262 = 16,034 feet/minute. This speed is safe because it is less than the 19,000 feet/minute maximum allowed.

To check the top governed speed on a rotary lawnmower with a cutter blade connected directly to the crankshaft:

1. *Measure the length of the cutter blade (diagonally across the cutting corners).*

2. *Set the speed control in the fast position. Adjust the carburetor mixture to obtain the maximum engine RPM.*

3. *Check the engine RPM with a tachometer.*

Top governed speed of the engine should not exceed the speed shown on the chart.

GOVERNOR TYPES

All governor systems must have a method for detecting the engine's speed. How the engine's velocity is sensed determines whether it is a pneumatic or mechanical system.

PNEUMATIC OR AIR VANE GOVERNOR

The pneumatic governor senses the engine speed by the force of air blowing on a flap attached to the

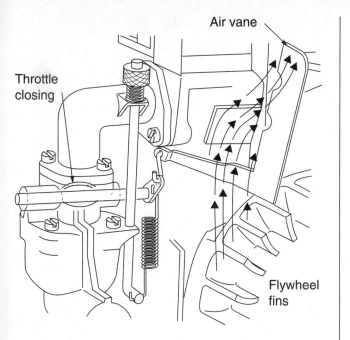

Fig. 6-4 The pneumatic (air vane) governor senses the engine speed by the force of air blowing on the air vane from the flywheel fins.

engine, as shown in figure 6-4. The flywheel contains fins that impels a stream of air as it spins. The air is directed by the surrounding sheet metal so that the force is applied to a flap or vane. The faster the engine spins, the greater the pressure exerted.

The pneumatic type is less expensive to manufacture and is acceptable for most applications where a minor fluctuation of set speed is allowable when a load is applied.

Governor sag is the term used to denote the temporary reduction of engine speed when more effort is required (increased load). When a lawnmower is moved from the sidewalk into heavy grass, the engine speed may decrease somewhat (governor sag) before the change of throttle position and engine power restores the set RPM. On equipment where governor sag must be kept to a minimum, a mechanical governor system is used.

MECHANICAL GOVERNOR

The mechanical governor is situated inside the engine block or underneath the flywheel. The

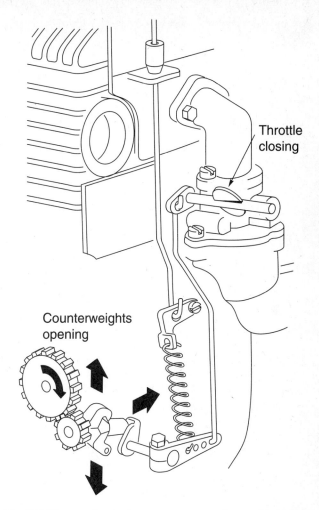

Fig. 6-5 The mechanical governor sensor the engine speed by a mechanism that is geared to the crankshaft or camshaft. The faster the sensor rotates, the greater the pressure applied.

engine speed is sensed by a mechanism that is geared to the crankshaft or camshaft, as shown in figure 6-5. The faster the mechanical speed sensor rotates, the greater the pressure created. This governor has more parts than the pneumatic governor and is more costly to produce, but it is more responsive to changing loads and retains the desired speed with minimal governor sag. An electric generator that depends upon the engine speed for the proper voltage uses a mechanical governor to limit governor sag and corresponding voltage drop when a load is applied.

GOVERNOR OPERATION

The governor works in a closed loop, which means that the system monitors any speed changes and adjusts the engine's power to maintain the desired engine RPM. If the engine speed drops below the set velocity, the governor system opens the throttle to increase the power. If the RPM rises above the set velocity, the governor closes the throttle and the engine loses power. The efficiency of the governor system will depend on how quickly it can sense and react to a change in speed and also whether the carburetor is adjusted properly.

OPPOSING FORCES

The governor mechanism can be compared to a "tug of war." The force in one direction is the engine power. The force in the opposite direction is the pull of a spring, as shown in figure 6-6. The "flag" in the middle of the rope is connected directly to the throttle so that when the flag moves in one direction, the throttle opens; when it moves the other way, the throttle closes.

Opposing forces control throttle

When the engine is not operating and the control lever is moved to the fast speed position, the force of the governor spring will pull the throttle wide open because there are no opposing forces, as shown in figure 6-7. When the engine is running, the air pressure from the pneumatic speed sensor will close the throttle if there is no opposing spring tension, as shown in figure 6-8.

Self-regulating closed loop

When the engine is started, the speed of the engine increases and applies a force opposite to the tug of the spring. The faster the engine speed, the greater the force. Eventually, the speed of the engine will overcome the tension of the spring and begin to close the throttle. But when the throttle closes, the engine speed and the pull against the spring decreases, which allows the spring to open up the throttle again. Now that the throttle is open, the engine speed increases, which allows the force of the speed sensor to increase and the throttle again begins to close. This cycle continues until a balanced position is found where both the spring force

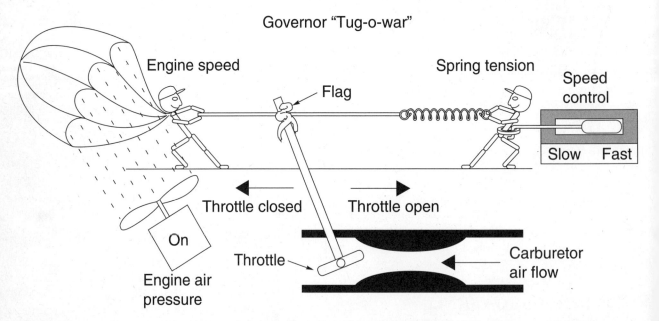

Fig. 6-6 The governor system can be compared to a "tug-of-war" between the force of the engine speed and the governor spring.

Governor "Tug-o-war"

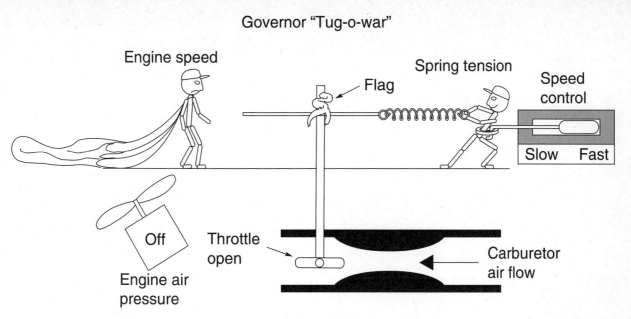

Fig. 6-7 When the engine is not operating and the control lever is moved to the fast speed position, the force of the spring will pull the throttle wide open.

Governor "Tug-o-war"

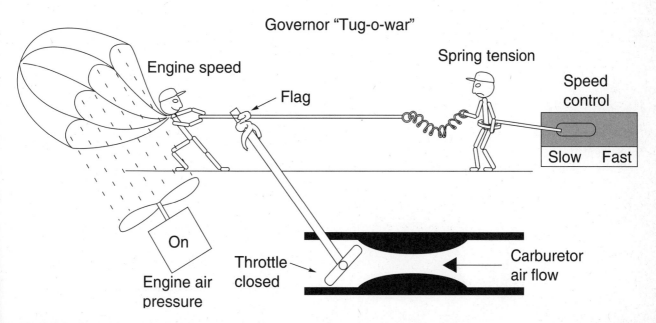

Fig. 6-8 When the engine is operating with the speed control in the slow position, the pressure from the speed sensor will close the throttle.

and the engine speed force neutralize each other at a certain throttle opening. This closed loop continues at such a quick pace that the operator should not hear any fluctuations.

BALANCED FORCES CONTROL ENGINE SPEED

When either side of the balanced forces are disturbed, a new balanced throttle position will occur. If the operator moves the control lever so that it increases the spring tension, the balanced throttle position will supply the engine with the power necessary for the increased speed. If the engine load increases, a new balanced throttle position will result in order to keep the engine speed constant.

GOVERNOR CONSTRUCTION

The physical characteristics of the governor system used on the small air-cooled engine may differ, but the construction principles are common to all. The speed sensor may be either mechanical or pneumatic. The linkage hook-up always includes a solid link from the speed sensor to the carburetor throttle. The engine speed attempts to close the throttle, while the spring tension tends to open it.

SPEED SENSOR

The speed sensor will exert a pressure that varies with the engine speed. The greater the engine speed, the greater the pressure.

Mechanical

The speed sensor for the mechanical governor is composed of a set of weights attached to an axis which rotates with the crankshaft of the engine. The weights are moveable and are thrown outward, by centrifugal force, as the speed increases. When the weights move away from their axis, a center pin is pushed upward to activate a lever which transmits the change outside the engine, as shown in figure 6-9. The faster the spin of the base, the more pressure transmitted by the center pin.

Air vane (Pneumatic)

The air vane senses the speed of the engine by the flow of air directed at it. The fins of the flywheel

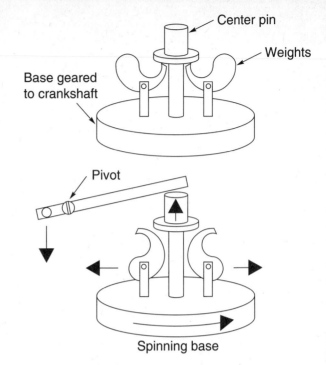

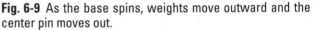

Fig. 6-9 As the base spins, weights move outward and the center pin moves out.

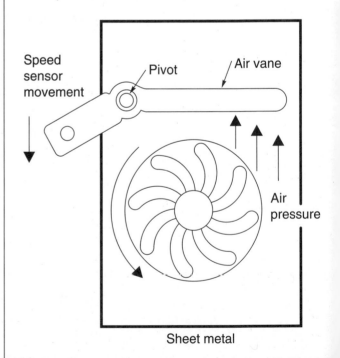

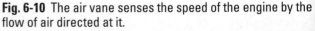

Fig. 6-10 The air vane senses the speed of the engine by the flow of air directed at it.

create the air pressure, which increases as the engine speed increases, as shown in figure 6-10.

CONNECTIONS AND SETUP

The easiest way to assemble the spring and governor link is to make a detailed sketch or photograph of the relationships before disassembling. The engine service manual provides illustrations of the connections.

By understanding only three principles of governor make-up, the technician can efficiently assemble and inspect the operation of any governor system. The following principles apply to all governors:

1. *A **rigid link** always connects the governor speed **sensor lever** to the throttle.*

2. *The **engine speed** sensor will always attempt to **close** the carburetor throttle while the engine is operating.*

3. *The **governor spring**, when stretched, will at all times attempt to **open** the carburetor throttle.*

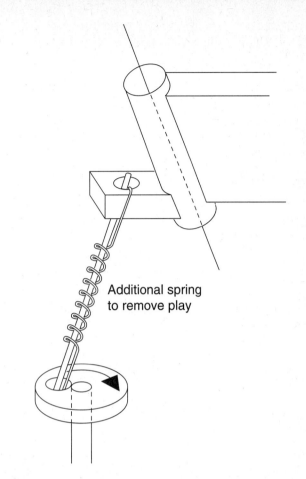

Additional spring to remove play

Fig. 6-12 An optional spring may wrap around the rigid link to remove any "play" at the connection points.

Rigid governor link

Generally, there must be a solid or rigid link between the governor speed sensor and the carburetor throttle, as shown in figure 6-11. Any movement of the speed sensor lever must immediately move the throttle. The governor link must be free to pivot at the connection point, and no other engine parts or sheet metal may touch the connection.

On some applications, a spring is wrapped around the rigid governor link, with each end of the spring connected to the holes in the sensor and the throttle. The spring tension "snugs" the throttle to the rigid link at one end and the sensor at the other, so that no play exists when movement of the sensor occurs. This setup makes governor control response quicker and more dependable, as shown in figure 6-12.

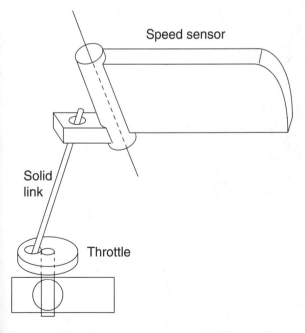

Speed sensor

Solid link

Throttle

Fig. 6-11 A solid link connects the speed sensor to the carburetor throttle.

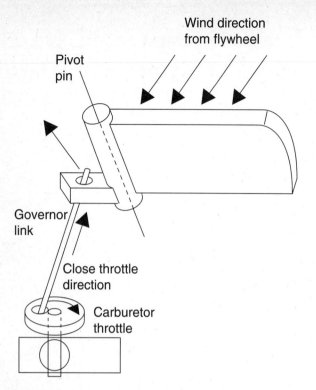

Fig. 6-13 The speed of the engine will always attempt to close the carburetor's throttle.

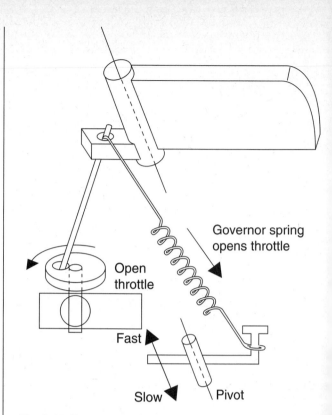

Fig. 6-14 Governor spring tension will attempt to open the carburetor's throttle.

Engine speed closes throttle

The force of the engine speed sensor lever, while the engine is operating, will always attempt to shut the throttle. Without any opposing forces, the engine will close the throttle when started and run in the idle position continually, as shown in figure 6-13.

Spring tension opens the throttle

The governor spring is the type that, if you grasp both ends of it and pull it apart, it will attempt to return to its original length. With no spring or no spring tension, the engine speed sensor movement will close the throttle. When the spring is stretched by the operator or the control knob, the increased tension will always attempt to open the throttle, as shown in figure 6-14.

When the engine is not operating, even a slight spring tension will open the throttle all the way because there is no opposing force from the engine

speed sensor. As soon as the engine is started, the engine speed builds and the "tug-of-war" begins between the engine sensor and the spring. The operator controls the tension of the spring to increase or decrease the engine speed.

GOVERNOR ADJUSTMENT

Preliminary governor adjustment is necessary for the mechanical governor, and the procedures are listed in most service manuals. These adjustments remove any possible play between the components which allows for a quicker adjustment to different loads. The basic rule of adjustment is that while the throttle is held in the wide-open position, all the "play" or air gaps are removed from the system, as shown in figure 6-15. An easy way to remember the procedure is that after the linkage is connected, rotate the carburetor throttle plate from the idle to the wide open position. At this point, the direction

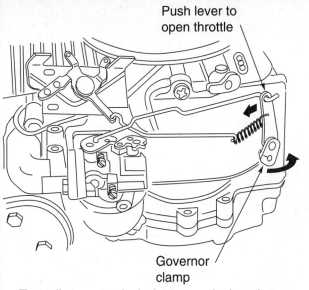

Push lever to
open throttle

Governor
clamp

Turn clip counterclockwise on vertical engines
(clockwise on horizontal engines)

Fig. 6-15 Preliminary governor adjustment is necessary for the mechanical governor, as shown on this Tecumseh engine.

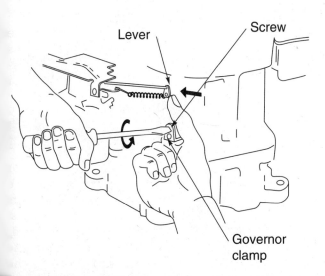

Lever

Screw

Governor
clamp

Fig. 6-16 A general rule is to note the rotation of the governor clamp when the throttle is moved from idle to wide open. While holding the throttle wide open, rotate the screw on the governor clamp the same direction to its limit and then tighten the governor clamp.

in which the governor arm rotates is the same direction you would rotate the governor shaft. Hold both the governor arm and governor shaft in this position and tighten the governor nut, as shown in figure 6-16. Any loose areas that cannot be removed by the initial procedures can sometimes be removed by using larger components, such as when a governor link with a thicker wire is used to remove the "play" created by worn holes in the speed sensor or carburetor throttle plate. In some applications, this inherent play is taken out by an additional spring wrapped around the wire and connected to the holes at both ends.

TOP SPEED ADJUSTMENT

It is important that you have access to a tachometer when adjusting the stretch of the governor spring so that safe operating speeds are not exceeded. An inexpensive but very accurate tachometer, sold by Briggs & Stratton or Tecumseh, is called a vibratach, and works by sensing engine vibrations (B&S part #19200; Tecumseh part #670156). The outside ring of the tool is rotated to lengthen or shorten the wire. When the sympathetic vibration is reached between the engine and the length of the wire, the wire will vibrate smoothly and in a wide arc, as shown in figure 6-17, and the engine speed can be read from the top window.

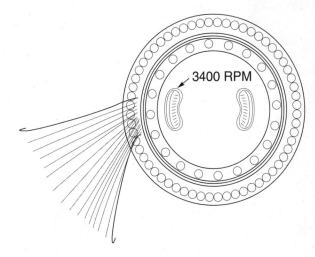

3400 RPM

Fig. 6-17 The vibra-tach is an inexpensive method for measuring engine speed.

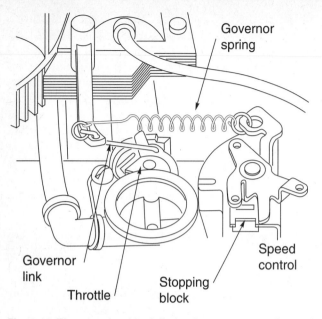

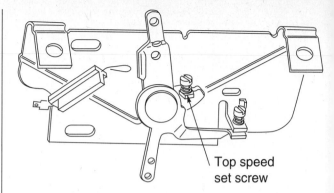

Fig. 6-19 The top speed set screw can adjust the top governed engine speed on this Tecumseh engine.

Fig. 6-18 The stopping block limits the top governed speed of the engine on the Briggs & Stratton engine.

When the engine speed control is put in the fast position, the spring is stretched, but the amount of stretch is limited. The maximum spring tension is controlled in a variety of ways. One method is to have a stopping block, as shown in figure 6-18. If a higher engine speed is needed for the application, a new spring with greater tension (see parts book) can be purchased, or some of the spring links can be removed and the spring end bent to conform to the original shape. This operation should be done only with an accurate tachometer for testing the new engine speed.

Some applications limit the top speed by a screw that can be adjusted either in or out, which will change the top engine speed, as shown in figure 6-19.

Armed with the theory of instruction and a good tachometer, you can modify the system. You can alter the spring to increase or reduce engine speed. Shortening a spring, by taking a loop from the spring, will increase the top speed. Reducing the tension of a spring by overstretching it beyond its limit of elasticity will decrease the top speed. **Warning:** Do not change or modify the system to exceed factory specifications.

GOVERNOR PROBLEMS

The following areas are possible governor problems along with their probable solutions:

HUNTING OR SURGING

The most common problem encountered with the governor system is when it "hunts" or surges. The problem is identified by engine pulsations or fluctuations varying between wide open and idle positions of the throttle, creating a wide range of engine speeds. The following procedures should be performed in the order listed until the problem has been corrected:

1. *Adjust the carburetor.*

2. *Adjust governor linkage.*

3. *Remove binding links.*

4. *Rebuild carburetor.*

Adjust the carburetor

Many times, either the idle or high speed circuit has not been adjusted properly or some of the passageways are blocked by dirt or foreign material. Since the governor is a balance of forces that depend on all circuits working properly, any one circuit that is at fault will cause the surging or hunting. The idle and high speed mixture screws must be correctly adjusted to minimize the problem.

Adjust governor linkage

The governor linkage adjustment must also be set so that there is no play in the system. Any play will delay the effects of the sensor and the resulting change in the throttle position.

Remove binding links

The governor linkage and spring must move freely when the speed control lever is at idle. The parts cannot rub against other portions of the engine, for any friction brings in a force that may assist one side or the other in the tug-of-war, and the system will not be correctly balanced. Often when an engine is rebuilt, it is also painted. The paint fills the holes and causes a binding at the pivot points. This condition must be corrected.

Rebuild carburetor

If the engine pulsations still persist, the carburetor should be disassembled and the low and high speed passages cleaned. A new carburetor rebuild kit, if available, should be used to replace the old, damaged, or worn parts.

OBSTRUCTED AIR FLOW

The pneumatic (air vane) governor may be affected by inappropriate air flow across the engine fins. Any debris that is caught in the fins or shroud may produce a restriction to the air flow and cause a false sensing of engine speed. The spring will open up the throttle and the velocity will increase, which may cause the engine to run at an inefficient, faster speed and lead to increased deterioration.

SUMMARY

Without an understanding of the basic theory, the assembly and service of the governor system can be frustrating. It can be time consuming trying to determine how the linkages are connected, but even though the governor systems vary in physical attributes between engines, they all follow a simple, fundamental theory.

The purposes of the governor system are to protect the engine, provide operating convenience, and ensure blade speed safety. All governor systems

must have a method for detecting the engine's speed. How the engine's velocity is sensed determines whether it is a pneumatic or mechanical system.

The pneumatic governor senses the engine speed by the force of air blowing on a flap attached to the engine. The mechanical governor senses the engine speed by a mechanism that is geared to the crankshaft or camshaft.

The governor works in a closed loop, which means that the system monitors any speed changes and adjusts the engine's power to maintain the desired engine RPM.

By understanding only three principles of governor make-up, the technician can efficiently assemble and inspect the operation of any governor system. The following principles apply to all governors: a rigid link always connects the governor speed sensor lever to the throttle; the engine speed sensor will always attempt to close the carburetor throttle while the engine is operating; and the governor spring, when stretched, will at all times attempt to open the carburetor throttle.

Preliminary governor adjustment is necessary for the mechanical governor in order to remove any possible play between the components, which allows for a quicker adjustment to different loads.

Questions

1. The following statements are made about governor systems: Technician A says that they are not necessary for engine operation, but if they are used, they will protect the engine from possible failure due to overspeeding. Technician B says that most air-cooled engines are equipped with a governor to increase the horsepower potential of the engine. Who is correct?

   A. Only Technician A
   B. Only Technician B
   C. Both Technician A and B
   D. Neither Technician A or B

2. A governor will limit the speed of the engine. An engine is operating at 3600 RPM with the speed control in the fast position; the limited top speed is 3600 RPM. Technician A says that the throttle position will remain at the same position when the mower is moved from the sidewalk into the heavy grass. Technician B says that the governor system will increase the power to the engine when the mower is moved from the sidewalk into the heavy grass. Who is correct?

   A. Only Technician A
   B. Only Technician B
   C. Both Technician A and B
   D. Neither Technician A or B

3. Good governor control is necessary on an engine used to power an electric generator because the voltage must not vary. The generator is supplying electrical energy to two electric motors and then two additional motors are turned on. Technician A says that the additional motors will cause the governor to speed up the engine in order to maintain the desired voltage. Technician B says that the governor will increase the throttle opening to account for the additional load. Who is correct?

   A. Only Technician A
   B. Only Technician B
   C. Both Technician A and B
   D. Neither Technician A or B

4. A safe blade speed is necessary for safe lawnmower operation. Technician A says that the maximum speed that the blade tip may spin is 19,000 feet/minute. Technician B says that blade speeds above the 19,000 feet/minute are allowed on lawnmower blades shorter than 18 inches. Who is correct?

   A. Only Technician A
   B. Only Technician B
   C. Both Technician A and B
   D. Neither Technician A or B

5. The speed of an engine is above recommendations. Technician A says one way to decrease the speed is to change the governor spring. Technician B says that the stretch of the governor spring could be limited to decrease the speed. Who is correct?

   A. Only Technician A
   B. Only Technician B
   C. Both Technician A and B
   D. Neither Technician A or B

6. The engine speed sensor must provide input to the governor system. Technician A says the pneumatic type sensor will attempt to close the throttle as the engine speeds up. Technician B says that the mechanical type sensor will close the throttle as the engine speeds up. Who is correct?

   A. Only Technician A
   B. Only Technician B
   C. Both Technician A and B
   D. Neither Technician A or B

7. Governor operation is the balance of opposing forces. Technician A says that the force in one direction is the speed of the engine, while the force in the other direction is the pull on the governor spring. Technician B says that the pull on the spring attempts to close the throttle plate, while the engine speed attempts to open the throttle plate.
Who is correct?

   A. Only Technician A
   B. Only Technician B
   C. Both Technician A and B
   D. Neither Technician A or B

8. An engine is hunting or surging when the speed control is placed in the fast position. Technician A says that this could be caused by an incorrectly adjusted carburetor. Technician B says that this could be caused by a sticking throttle shaft.
Who is correct?

   A. Only Technician A
   B. Only Technician B
   C. Both Technician A and B
   D. Neither Technician A or B

9. Three principles apply to all governors. One is that the engine speed sensor will attempt to close the throttle plate. Technician A says the other two are that a rigid link connects the speed control to the throttle and that the spring pressure will attempt to open the throttle plate. Technician B says that the other two are that the rigid link will attach the throttle plate to the choke and the spring pressure will attempt to close the throttle plate.
Who is correct?

   A. Only Technician A
   B. Only Technician B
   C. Both Technician A and B
   D. Neither Technician A or B

10. Preliminary governor adjustment is necessary on a mechanical governor. Technician A says that it is necessary to minimize governor sag. Technician B says that it is necessary to remove all the air gaps between governor components.
Who is correct?

   A. Only Technician A
   B. Only Technician B
   C. Both Technician A and B
   D. Neither Technician A or B

Engine Disassembly

INTRODUCTION

Many small air-cooled engines—those which are not economically feasible for a professional shop to repair—quickly find their way into a small engine repair class or into the hands of the hobbyist. Many of these engines can be restored to excellent operating condition with minimal expense and, when the labor charge is removed from the total cost of the repair, it then becomes economically feasible to recondition the engine with rewarding results.

TYPE OF ENGINE

The small four-stroke-cycle Briggs & Stratton engine is a very common engine and provides an excellent educational instrument for acquiring proper knowledge and procedures at minimal risk and cost. The focus of the tear-down in this chapter will be a small, vertical, air-cooled 3.5 horsepower Briggs & Stratton engine. A vertical or horizontal engine can be identified by the position of the crankshaft relative to the earth's horizon. A horizontal engine will have the crankshaft ends protruding from the engine parallel to the horizon, as shown in figure 7-1, while the vertical engine will have the ends at right angles to the horizon, as shown in figure 7-2. A typical application of the vertical engine would be in a rotary lawnmower, while a typical horizontal engine would be used to power a conveyor belt, roto-tiller, etc.

PRELIMINARIES

It is important that the mechanic follow proper disassembly procedure to guarantee effective and acceptable results.

EVALUATION

A preliminary evaluation encompasses valuable knowledge, which gained from questioning the customer as well as perceiving all relevant clues during the disassembly. This knowledge will assist the mechanic to present the most viable alternatives to the customer.

SCENARIO

An engine is brought to the mechanic for repair by the operator with the complaint that the engine lacks power and uses too much oil. The mechanic indicates that he will evaluate the engine and advise the customer of the alternatives.

Measurements and data are gathered about the wear, and the cost to rebuild the engine is compared to the cost of a new engine or a shortblock. The condition of the machinery driven by the engine also enters into the evaluation and, from the data collected, the mechanic will advise the customer regarding the choices and the mechanic's recommendation.

GOOD PROCEDURES

Correct, effective procedures should be used on an engine that is economically feasible to overhaul. The tear-down should be discontinued when additional data indicates that the engine should not be repaired. If the repair cost exceeds 50 percent of the replacement value it is good to recommend a new unit or replacement engine. Most customers appreciate being told that the mechanic intends to spend no more than one hour of labor evaluating the engine before making a recommendation. Even if the decision is not to repair the engine, the customer feels that the money is well spent.

TEAR-DOWN STEPS

The following tear-down steps provide two functions: One is to present an efficient order and explanation of the steps involved, and the other is to acquire information about the engine failure. Certain failures are very evident, but others will be detected during the disassembly process. Some tear-down inspection steps may be bypassed when the failure problems are evident.

ORGANIZATION

Before the disassembly begins, it is suggested that a piece of cardboard or small plastic bags be available to group and organize the disassembled parts, as shown in figure 7-3 and figure-7-4. It is very easy for parts to be lost if they are loosely tossed into the bottom of a container. The assembly is improved when you do not have to search and guess the part's position. Replacing the fasteners to their proper positions enhances a "professional looking" job.

Cardboard

Bolts and other parts can be placed in a piece of cardboard by piercing it with a sharp object and pushing the bolts through the cardboard, or by fastening them with some wire. The parts identification and many sketches regarding the different linkages can be recorded on the cardboard.

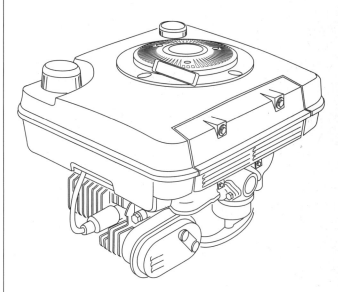

Fig. 7-2 Vertical air-cooled engine.

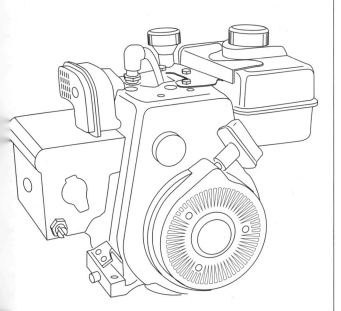

Fig. 7-1 Horizontal air-cooled engine.

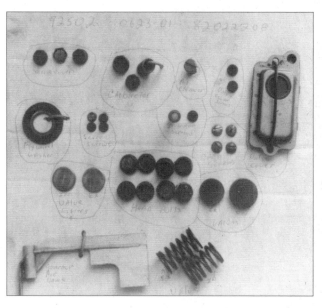

Fig. 7-3 Organization of parts using cardboard.

Fig. 7-4 Organization of parts using plastic bags.

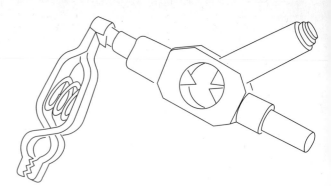

Fig. 7-5 Briggs & Stratton spark tester (part #19368).

Plastic bags

Plastic bags are used so that the fasteners and parts can be grouped in different pouches, with the additional advantage that notes can be made and placed into the bags. "Zip-locked" bags offer greater loss protection.

REMOVE ALL GAS FROM ENGINE

The fuel should be inspected and evaluated before it is removed from the tank.

Inspect fuel quantity

The lack of fuel will always cause a starting problem and, as trivial as it may seem, has caused many customers to bring their machinery in for repair.

Inspect fuel quality

Very old gasoline will have an odor of varnish. A volatility test of the gas can be made by placing the tip of a finger into the fuel, removing it quickly, and waving it around. Fresh fuel will vaporize quickly and cause a feeling of coolness on the finger.

Check for water in the tank

Any engine that has been left outside in a rain may have water deposits in the fuel tank. Since water is more dense than gasoline, it will settle on the bottom of the tank. Water in the fuel system will cause engine failure as a result of damaged or plugged carburetor passages.

Dispose of the fuel

The fuel, emptied from the fuel tank, should be disposed in accordance with local regulations.

IGNITION OUTPUT TEST
Install tester

A spark tester can be bought or made, as shown in figure 7-5. This tool, because of its large air gap (0.166-inches), can simulate the resistance in the spark plug during engine compression.

TEST WITH SPARK PLUG IN ENGINE An adequate ignition system produces enough voltage for the spark to simultaneously jump the gap in the tester and the spark plug. Install a spark tester between the spark plug wire and the spark plug while the spark plug is in the engine, as shown in figure 7-6. Activate the starter and if a spark is observed. It can then be assumed that the ignition system has sufficient voltage to ignite the mixture of fuel and air during the compression stroke.

TEST WITHOUT SPARK PLUG IN ENGINE If no spark is noticed in the previous test, bypass the spark plug and use the tester alone, as shown in figure 7-7. Remove the spark plug for easier engine rotation and ground the tester to the metal of the engine. Connect the spark plug wire to the terminal of the tester. Rotate the engine with the starting mechanism and observe if the spark jumps the gap. If a spark is observed, then sufficient voltage is available to start the engine, but the spark plug may be defective.

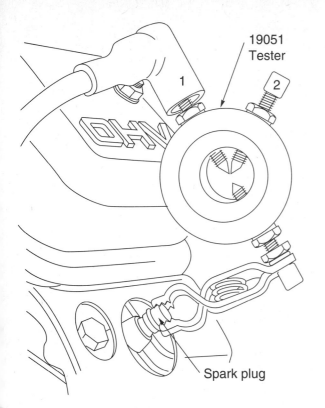

19051
Tester

Spark plug

Fig. 7-6 Test ignition output with spark tester connected between the spark plug and spark plug wire.

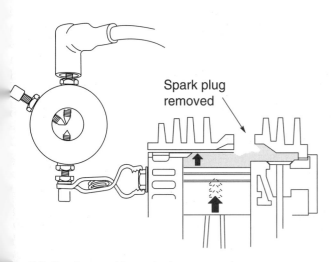

Spark plug removed

Fig. 7-7 Spark test with spark plug removed.

NO SPARK CONDITION If there is no spark with the tester, make sure that the engine switch is not in the "shut off" position. If it is, turn engine on and recheck. A no-spark condition must be corrected during the assembly. See the troubleshooting chapter for further testing of the ignition system.

PERFORM COMPRESSION TEST

The engine must have a minimum compression pressure in order to start. The lack of compression can be due to *leaking valves, bad rings, worn cylinder,* or a *leaking head gasket.* The following are ways to test compression (use the method that is most convenient and effective):

Activate starter

The simplest method is to just pull the starter rope or activate the electric starter. While the engine is rotating, a resistance to the rotation will be felt during the compression stroke. If no resistance is felt, further tests should be made. This test is the most undependable because all four-stroke-cycle engines have compression releases that reduce the compression pressures during the starting cycle. When an engine starts, the compression release either is disconnected mechanically or has little consequence with increased engine spinning.

Use compression gage

The best instrument for making a qualitative evaluation is the compression pressure gage, as shown in figure 7-8. Experience has shown that normal compression pressures vary from a 60 PSI (pounds per square inch) to a maximum of 90 PSI. Any reading less than 60 PSI may not have enough compression to start the engine. Any reading over 90 indicates a high compression that is typically caused by large carbon deposits in the cylinder or incorrect valve clearance, preventing the operation of the compression release.

Remove the spark plug and insert the compression gage into the spark plug hole. The carburetor's choke and throttle must be in the wide open position for the maximum intake of air to obtain the maximum compression gage reading. Activate the starter for at least six compression strokes and note the maximum compression reading.

Fig. 7-8 A compression pressure gage inserted in the spark plug opening.

Fig. 7-9 Cylinder leakage test equipment.

If there is low compression, the **wet test** will isolate the loss of compression to either *leaking piston rings* or *leaking valves.*

The wet test is accomplished by removing the spark plug, adding a teaspoon of engine oil into the cylinder through the spark plug hole, and then installing the compression gage into the spark plug hole of the engine. Activate the starter mechanism on the engine and note the reading on the compression gage. If the compression pressure increases from the earlier test, bad rings are indicated. The oil now creates a temporary seal around the rings, and the air is not allowed to escape during the compression stroke. If the pressure differs only slightly from the original reading, the valves can be the cause of the low compression.

Bounce test

Another method to test the compression is by performing the bounce test, which involves spinning the flywheel opposite to the normal rotation, with the spark plug installed, in order to bypass the compression release components used for easy starting. Remove the blower housing to expose the flywheel and disconnect the spark plug wire. Also remove any screens which may be attached to the flywheel that could injure your fingers during the test. Turn the flywheel counter-clockwise until you feel the resistance from the cylinder compression, then turn the flywheel clockwise one-quarter of a turn. At this point, give the flywheel a hard spin counter-clockwise, and the flywheel should bounce back if the compression is acceptable. If the flywheel spins through the compression stroke, the engine has low compression.

Cylinder leakage test

A cylinder leakage test may be performed if the equipment is available, as shown in figure 7-9. The test will indicate the condition of the cylinder and valve train. The cylinder is filled with air through the spark plug opening, and the engine is rotated until the piston is at the top of the travel on the compression stroke.

The leakage test will display the percentage of air lost when the piston is at the top part of the compression stroke. A cylinder leakage of 25 percent or less is acceptable.

The test will also disclose the source of the leak

While the cylinder is pressurized and the piston is on the top travel of the compression stroke, listen for air escaping from the carburetor, which will indicate a faulty intake valve.

Air leaking from the muffler will identify that a problem exists in the area of the exhaust valve, while an air leak from the carburetor venturi will identify that a problem exists in the area of the intake valve.

If air is detected exiting the crankcase breather, malfunctioning piston rings may not be sealing properly.

Any air leaking from the cylinder head indicates that a problem exists in the area of the cylinder head gasket or improper cylinder head torque.

INSPECT SPARK PLUG

Note the color and condition of the electrode. On a multi-cylinder engine, the spark plugs are compared to detect unusual firing conditions. The electrode should be a light tan in color without any excessive deposits. If it is oily, there could be an oil control problem. If the plug is black and sooty, there could be a carburetor adjustment problem.

DRAIN CRANKCASE OIL

Inspect and make note of the quantity and quality of the oil. A sticky or tar-like oil that is black in color could be caused by overheating of the engine oil. This may occur when:

1. oil is not changed regularly or frequently

2. proper oil level is not maintained

3. cooling system is not maintained

4. wrong oil viscosity is used

5. oil additives were used

6. engine runs lean

Dirty or gritty oil is an indication of dirt entering the engine, possibly from the oil fill area that was not cleaned before and after the oil plug or dipstick was removed or replaced. Oil level that is low may indicate internal wear from the lack of lubrication. Dispose of the oil in the proper container.

IDENTIFY MISSING OR BROKEN PARTS

The engines are assembled at the factory before they are painted. Any missing or changed engine parts will be readily identified by the different colors or shadows. Begin a parts listing at this point so that the assembly is not delayed by needed components.

WASH ENGINE AND EQUIPMENT

Before the engine is removed from the equipment, it should be degreased or pressure cleaned, if possible. Do not worry about water harming the engine, because all the disassembled parts will be cleaned during the rebuild. Be careful not to damage or destroy any specification decals that might be used for model identification.

REMOVE AIR CLEANER

Inspect the air cleaner for proper maintenance. An improperly maintained air cleaner will allow dirt to bypass the filter and enter the engine. When this occurs, the rate of wear on the moving components will increase by ten times or more, significantly reducing engine life. Inspect the carburetor for a loose throttle shaft that can be caused by dirt ingestion. Dirt bypassing the air cleaner system is drawn into the combustion chamber through the valve area and may cause worn valves and guides.

REMOVE BLOWER HOUSING

Dismantle as much of the sheet metal as possible. Note the manufacturer's identification numbers and record them. These numbers can be decoded for information concerning certain characteristics of the engine, as shown in figure 7-10. Note the cylinder fins. Check to see if any debris, wedged between the fins, is blocking the flow of cooling air. The cooling system is comprised of the air intake screen, blower housing, flywheel, cylinder block, and head. These parts critically control the cooling ability of the engine, and if blocked by grass clippings or other debris, the internal temperature of the engine will

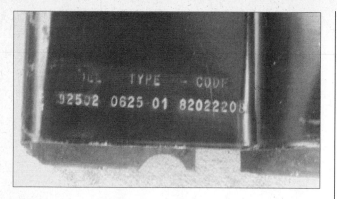

Fig. 7-10 Manufacturer's identification numbers.

increase, causing the engine to overheat. Poor engine cooling can cause a blown head gasket or high oil consumption.

REMOVE CYLINDER HEAD AND GASKET

Do not scrape the piston top clean. The amount of carbon on the piston's surface will give a clue to the engine's condition, as shown in figure 7-11. The larger the clean area, the greater the wear. A piston's top surface area that is half-clean indicates that the cylinder and rings are not sealing properly. A piston with good sealing between the rings and cylinder wall will show various deposits with only a small clean area.

An engine that has run for a moderate period of time will have adherent deposits. Wipe off any loose material. Notice that there is one area unsoiled by any deposits. All engines push oil from the crankcase, beyond the rings, and into the combustion chamber. As the rings wear, more oil escapes, proportional to the wear of the engine. The oil that is commonly used in modern air-cooled engines is a detergent type and will clean the piston top. Thus, the more oil that reaches the combustion chamber by being forced beyond defective rings, the greater the clean surface area of the piston top. Do not mistake the clean piston of a new engine with engine wear.

INSPECT CYLINDER WALL

Move the piston down in the cylinder and inspect the walls of the cylinder. Observe any irregularities

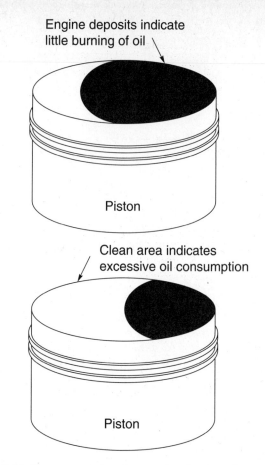

Fig. 7-11 The amount of carbon on the piston surface will indicate engine condition.

or grooves. An obvious ridge at the top of the ring travel usually means that the cylinder needs work. The original crosshatched grooves inside the cylinder walls are 0.0015-inch (one and one-half thousandths of an inch) deep. If they are absent in spots, the cylinder may be worn.

Notice where any long grooves originate and terminate. Grooves that begin from the bottom of the cylinder and extend partially up the cylinder walls indicate that abrasive dirt was present in the crankcase oil. Grooves that begin at the top indicate a defective or missing air cleaner. Look for galling. This is where the cylinder retains so much heat that it melts and smears the aluminum parts of the piston on the cylinder surface. A cylinder that is severely galled or grooved will have to be discarded or oversized.

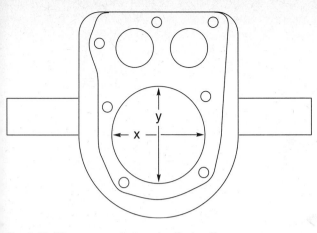

Fig. 7-12 Measure and record cylinder diameters.

MEASURE CYLINDER BORE

Measure the diameter of the cylinder bore with a telescoping gage and a dial caliper or micrometer. Initially, a quick inspection can be made by taking two measurements about one inch from the cylinder top; this is normally the area of maximum wear. Normal wear will cause an oval shaped bore where measurement Y (right angle to the crankshaft) will indicate greater wear than measurement X (parallel to the crankshaft), as shown in figure 7-12. Compare the greatest measurement to the manufacturer's specifications. The distance at point Y will wear faster from the increased friction of the piston being pushed and pulled by the rotating crankshaft.

Telescoping gage

The telescoping gage, as shown in figure 7-13, is used to measure hole diameters. It is a "T"-shaped measuring instrument that has a knurled nut at the end of the gage handle. The arms of the tool can be compressed with your fingers until their width is slightly smaller than the inside diameter to be measured. The knurled nut can be tightened to lock the arms in position. Insert the gage into the cylinder and loosen the knurled nut. A light spring inside the arms will push the adjustable end outward, and the gage will adapt to fit the cylinder. Make sure the telescoping end is at right angles to the cylinder wall, then tighten the nut and withdraw the gage. An easy method to obtain an accurate bore dimen-

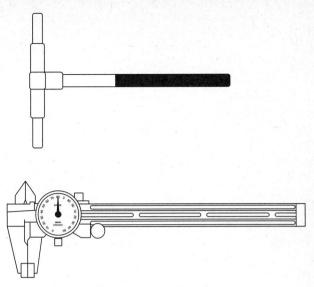

Fig. 7-13 Telescoping gage and dial caliper.

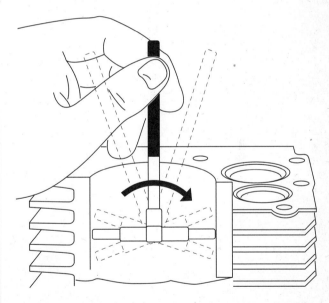

Fig. 7-14 Rock telescoping gage to get best measurement.

sion is to put the gage in the bore and purposely tighten the knurled nut while the gage is slightly cocked. With the knurled nut on the "T"-handle tightened, move the gage to the perpendicular position and continue the movement past the ideal position. The gage's arms will compress to fit the dimensions of the bore, as shown in figure 7-14.

Remove the gage carefully and measure the distance between the tips of the gage's arms.

Dial caliper

A dial caliper is accurate to measurements of 0.001-inch and is the tool of preference because of its cost effectiveness and ease of use. One tool can be purchased to precisely measure sizes from 0 to 6 inches, as shown in figure 7-13.

A micrometer can be used for the measurement, but a separate micrometer must be purchased for ranges 0–1 inch, 1–2 inches, 2–3 inches, etc., as shown in figure 7-15.

After some practice, the combination of the telescoping gage and dial caliper can provide accurate measurements, but initially there is much room for human error. Mistakes in measurement can be caused by improper feel of the telescoping gage, inappropriate placement of the telescoping gage in the dial caliper, or inaccurately reading the gage scale.

Bore gage

A bore gage is an easy, accurate measuring tool that is an alternative method to measure the cylin-der bore wear, as shown in figure 7-16 and figure 7-17. The gage is preset to the original bore specification in a special set-up, and then the gage is inserted into the cylinder to examine the deviations from the standard bore. Once the gage is set-up, it

Fig. 7-16 Engine bore gage.

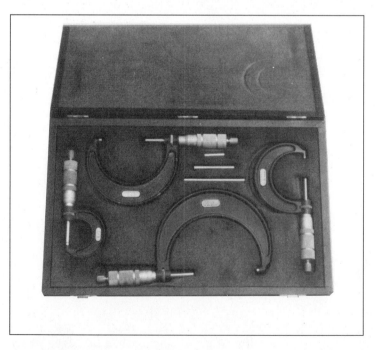

Fig. 7-15 Typical set of micrometers.

Fig. 7-17 Engine bore gage.

quickly and accurately allows for a direct readout of wear and out-of-round.

Record and interpret bore measurements

The cylinder bore (Briggs & Stratton) specification chart indicates that if one of the measurements is 0.003-inch or more wear than the standard bore, as shown in figure 7-18, the cylinder bore should be

---

Cylinder resizing

Resize if .003 or more wear or .0015 out of round on C.I. bore cylinders, .0025 out of round on aluminum bore cylinders.

Resize to .010, .020, or .030 over standard bore.

Fig. 7-18 Briggs & Stratton cylinder resizing specifications.

machined to increase the piston chamber to accept a larger piston and ring size (normally in increments of 0.010-inch, 0.020-inch, or 0.030-inch).

An alternative to machining the cylinder, if the wear is at least 0.003-inch but not more than 0.005-inch, is the use of chrome rings (from the manufacturer if available). The chrome metal is hard enough to wear down the cylinder wall, so that a good seal soon develops between the piston and cylinder.

Calculate the maximum out-of-round, which is the difference between the maximum measurement perpendicular (measurement Y) to the crankshaft compared to the maximum measurement parallel (measurement X) to the crankshaft. According to the specification chart, a difference that is 0.0015-inch or greater on a cast-iron engine, or 0.0025-inch on an aluminum engine, warrants machining the cylinder to a larger size.

The initial two measurements quickly indicate the condition of the cylinder, but in order to finally verify the condition, the cylinder should be measured at other depths, as shown in figure 7-19. Normally, the farther down the measurement is taken, the less wear will be observed unless abrasive dirt has contaminated the engine oil. The dirt moves upward in the bore, and the greatest amount of wear could be at the bottom of the ring travel.

If the manufacturer's specifications are not available, the piston can be removed, and the cylinder should be measured at its lowest area. The measurement observed in this area will reflect the original engine bore.

REMOVE ENGINE FROM EQUIPMENT

Remove blades, pulleys, clutches, etc., and dismount the engine. Proper use of "pullers" is necessary to remove pulleys and adapters from the crankshaft, as shown in figure 7-20. These parts are exposed to moist conditions and can be bonded tightly or "rusted" to the shaft. If the "puller" is not adequate to remove the adapter, then the use of heat, by either of the following two methods, should loosen the bond:

- One method mandates that the adapter or pulley be heated quickly to allow for the expansion

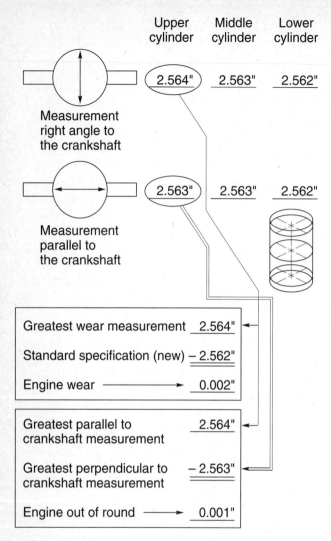

	Upper cylinder	Middle cylinder	Lower cylinder
Measurement right angle to the crankshaft	2.564"	2.563"	2.562"
Measurement parallel to the crankshaft	2.563"	2.563"	2.562"

Greatest wear measurement	2.564"
Standard specification (new)	− 2.562"
Engine wear	0.002"

Greatest parallel to crankshaft measurement	2.564"
Greatest perpendicular to crankshaft measurement	− 2.563"
Engine out of round	0.001"

Fig. 7-19 Engine measurement worksheet.

Fig. 7-20 Puller set-up used to remove crankshaft adapters.

of the metal parts in order to break the bonding. The adapter or pulley must be removed quickly with the puller before the crankshaft also heats up and expands.

• Another method involves heating the pulley and crankshaft simultaneously. Because the metals differ in structure, the expansion and contraction rates are usually different. The parts and the crankshaft are heated to the first signs of red-hot. The heat source is then taken away, and the metal is allowed to cool in the air until the red glow disappears. Water is used to finish the

cooling process quickly. The seal between the metal surfaces should now have been broken.

If the above procedure is not successful, then the adapter may have to be cut off. Be careful not to damage the crankshaft.

CHECK THE ARMATURE AIR GAP (OPTIONAL)

The gap between the magnets of the flywheel and the legs of the external armature must be adjusted properly to obtain the greatest voltage in the ignition system. A gap that is too wide may cause hard starting. Look up the armature air gap specification in the appropriate service manual for the engine you are working on and compare it to the actual mea-

Fig. 7-21 Use proper tool to hold the flywheel while the nut or clutch is removed.

surement. The gap is measured with a feeler gage or a special tool supplied by the manufacturer.

REMOVE FLYWHEEL

Use proper tools and pullers, as shown in figure 7-21 and figure 7-22, to remove the flywheel. Avoid the use of a knockoff tool, if possible, because of the damage that could occur to internal bearing surfaces. Examine the flywheel key for any damage. A broken key will inhibit the ignition system from working properly.

Refer to the manufacturer's service manual for the recommended flywheel removal tool.

REMOVE IGNITION COMPONENTS

Examine the metal surfaces of the flywheel for any cracks and/or abrasions. Any abrasions found may indicate component clearances that were not sufficient. Check to see if all the fins are present and verify that the flywheel keyway is intact. The keyway may crack when there is an abrupt stop of the crankshaft or improper torque to the flywheel nut.

The cause of the crack can be identified by its location. Set the flywheel on a table with the fins

Fig. 7-22 Use proper tool to remove flywheel to prevent damage to components.

facing upward. If the break is to the right of the keyway, the cause of the failure is improper flywheel torque because, as the crankshaft rotates in a clockwise direction, the crankshaft will accelerate faster than the flywheel, causing damage when the torque is below specification, as shown in figure 7-23.

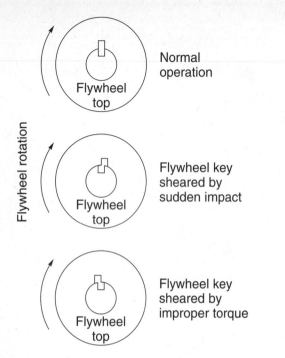

Fig. 7-23 Identify the cause of a sheared flywheel key

If the flywheel crack is to the left, the crack was caused by a sudden stop of the crankshaft. When the blade of a lawnmower hits a large, hard object the crankshaft stops quicker than the flywheel.

Different manufacturers use keys of varying hardness. The soft keys shear easily, and little damage will occur to the flywheel as a result. The tendency to use the softer key continued for many years. With the advent of the solid state ignitions, harder keys are being used. When the key is sheared in the point type of ignition, the engine will not start because of the weak spark, while in solid state ignitions, the spark will occur with a sheared key.

A slightly sheared key might change the engine's timing so that firing would be advanced enough to cause a "kickback" when starting the engine, accounting for many broken starters and sore fingers.

REMOVE CARBURETOR AND GOVERNOR LINKAGE

Be sure to make a clear drawing, video, or picture of the governor and carburetor linkage before dismantling them. A picture here is worth 1000 words

and two hours of assembly labor. Drain the gasoline from the carburetor before removal.

REMOVE VALVE COVER

On many engines, the valve cover contains the breather. The breather is a one-way valve that maintains crankcase vacuum. Air is allowed to escape from the crankcase of the engine when the piston is moving downward. Outside air is not permitted to enter when the piston moves upward and a vacuum is created. The air in the crankcase contains an oil mist which must be removed from the air as it leaves the engine through the breather valve and conserved.

There is a filter in the breather that collects the oil and allows it to drain back to the crankcase. Without a crankcase vacuum, the volume of air exiting the engine would saturate the filter with oil. This excess captured oil would not drain back to the engine, so it would exit through the breather tube. This tube could lead to the carburetor or it could just hang on the outside. If the tube leads to the carburetor, the oil will enter the intake air. This could cause the engine to "burn oil," which might be improperly interpreted as bad rings. If the tube exits to the atmosphere, the oil will spray over the exterior of the engine. Inspect the breather valve. The disc in the cover should not be distorted or pushed.

CHECK VALVE CLEARANCE (OPTIONAL)

Rotate the crankshaft, in the normal direction, until the piston is at the top of the compression stroke (valves closed), as shown in figure 7-24. Continue the crankshaft movement so that the piston is moving down (1/8- to 1/4-inch) on the power stroke between. At this position the valves will be seated. Check the gap between the lifters and the valve stem with a feeler gage. Check and compare the reading with the specification chart or service manual for the proper valve gap.

REMOVE VALVES

Remove the valves and valve springs using the proper tool, as shown in figure 7-25. Making sure that the exhaust valve and intake valve are stored together with their spring because some springs have different tension values and different lengths.

Normally, the strongest spring belongs to the exhaust valve. Examine the face of the valve and the seat in the block to determine if they were functioning properly. A valve that has not been sealing will be determined by the presence of combustion deposits on the valve face and seat abutting area.

CLEAN PTO CRANKSHAFT END

Remove all rust and metal burrs from PTO side of the crankshaft, as shown in figure 7-26. Often, the crankshaft is the same diameter as the bushing in the sump cover. The bearing surface will be damaged if this plate is resisting movement, and a hammer is used as a persuader. It is important to clean off the protruding crankshaft with sandpaper. When everything is clean, then the sump cover can be removed easily.

REMOVE CRANKCASE COVER

Place the engine on the table with the cylinder head downward, as shown in figure 7-27. This will

Fig. 7-24 Check valve clearance (optional step).

Fig. 7-26 Remove all rust and metal burrs from crankshaft.

Fig. 7-25 Use a valve removing tool.

Fig. 7-27 Place engine with cylinder head down when removing crankcase cover.

maintain the valve lifters in place until they are marked. Do not permit the valve lifters to drop out. Remove the crankcase cover by initially tapping it lightly to break the gasket bond. Usually the cover can be removed easily by using only one's hands. If the cover binds on the shaft, do not force it off with a hammer, as damage to the bearing surfaces may occur. When binding takes place, move the cover back to the starting position and search for the cause of the obstacle.

Possible obstructions include foreign particles in the oil seal area, protruding keyway surfaces, or a bent crankshaft. Remove the barriers by cleaning the oil seal region, filing any protruding keyway metal or cutting off the crankshaft.

Take out any oil pumps or slingers. On some engines with an auxiliary PTO shaft extending from the crankcase, the last cover bolt is hidden behind the shaft. The shaft must be removed first before the cover is removed. Other engines have the sump cover holding a bearing. The bearing must be unlocked before the cover may be removed.

IDENTIFY TIMING MARKS

Rotate the crankshaft until the timing marks on the crankshaft gear and the camshaft gear align, as shown in figure 7-28. If you cannot see timing marks, then make some with a metal scribe or punch on any of the mating gear teeth.

REMOVE CAMSHAFT

Remove the camshaft and any slinger or oil pump that is attached to it. Inspect the camshaft lobes and valve lifters. These components are specially hardened to withstand the forces of a running engine. They will, however, start to wear under a low-oil or no-oil condition.

REMOVE THE VALVE LIFTERS

The exhaust lifter can be found directly beneath the threaded exhaust port. Mark it with a piece of tape or paint, as shown in figure 7-29. A grinder may be used to put a mark on an area as long as it will not hinder the operation of the part.

REMOVE PISTON ASSEMBLY AND CONNECTING ROD

Prior to removing the rod bolts, you should make some punch marks on the mating metal surfaces, as shown in figure 7-30. It is a good practice to mark any metal parts before disassembly. The marks will show which pieces go together and their relative positions. Push up on the piston rod to remove the piston assembly. If there is resistance, remove the ridge at the top of the cylinder with a ridge reamer or sand paper. The piston and rod may be easily damaged during removal.

Inspect the piston and connecting rod. Blackening of the piston indicates excessive heat in the bore area. Combustion blow-by causes oil to burn on the

Fig. 7-28 Rotate crankshaft until timing marks align.

Fig. 7-29 Mark the exhaust lifter with tape or paint.

from the crankshaft by using muratic acid (hydrochloric). **Caution:** use only in a well-ventilated area.

Slight score marks caused by abrasive wear can be cleaned with crocus cloth soaked in oil.

Measure the three crankshaft journals with a dial caliper or a micrometer. Compare the reading to the specification given by the manufacturer, as shown in figure 7-31. It is important to determine whether the number from the manufacturer is a "new" crankshaft dimension, such as the specifications from

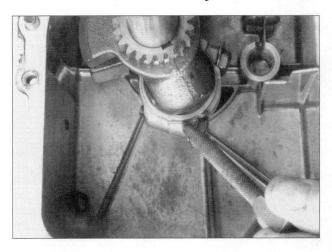

Fig. 7-30 Make punch marks on the mating metal surfaces.

piston skirt and increases oil consumption. The crankpin journal of the connecting rod can fail from a lack of lubrication, which will be identified by an aluminum transfer from the rod onto the crankpin.

REMOVE CRANKSHAFT

Remove any oil pumps or oil slingers. Observe any thrust washers that may be present.

INSPECT AND MEASURE CRANKSHAFT

Inspect all the rod journals and bushing areas for signs of poor lubrication or abrasive wear. Any heavy and continuous aluminum transfer will indicate a typical low-oil, no-oil, and/or overspeed failure. The aluminum may be chemically removed

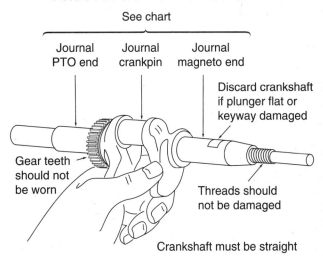

Fig. 7-31 Compare the micrometer reading to the specification.

Crankpin journal dia.	.8610 .8615
Crankshaft mag. main brg. dia.	.9985 .9990
Crankshaft P.T.O. main brg. dia.	.8735 .8740

Fig. 7-32 Is the manufacturer's number a new crankshaft dimension?

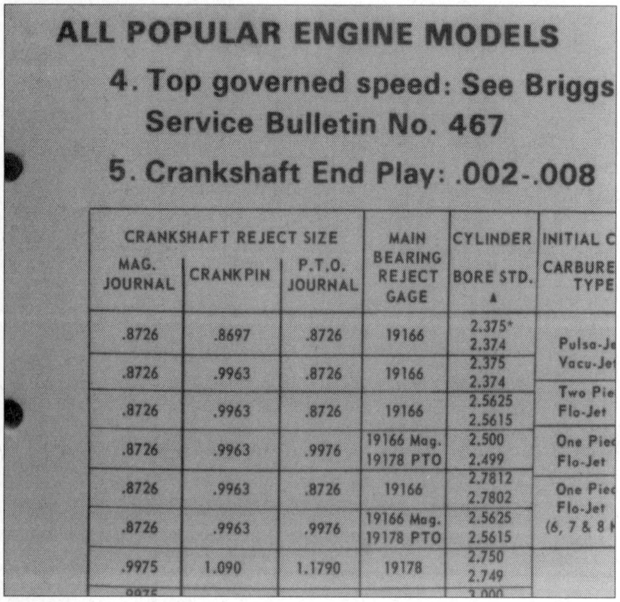

ALL POPULAR ENGINE MODELS

4. Top governed speed: See Briggs Service Bulletin No. 467

5. Crankshaft End Play: .002-.008

CRANKSHAFT REJECT SIZE			MAIN BEARING REJECT GAGE	CYLINDER BORE STD.	INITIAL C
MAG. JOURNAL	CRANKPIN	P.T.O. JOURNAL			CARBURE TYPE
.8726	.8697	.8726	19166	2.375* 2.374	Pulsa-Je Vacu-Je
.8726	.9963	.8726	19166	2.375 2.374	Two Pie Flo-Jet
.8726	.9963	.8726	19166	2.5625 2.5615	One Pie Flo-Jet
.8726	.9963	.9976	19166 Mag. 19178 PTO	2.500 2.499	One Pie Flo-Jet
.8726	.9963	.8726	19166	2.7812 2.7802	One Pie Flo-Jet (6, 7 & 8
.8726	.9963	.9976	19166 Mag. 19178 PTO	2.5625 2.5615	
.9975	1.090	1.1790	19178	2.750 2.749	
.9975				3.000	

Fig. 7-33 . . . a reject crankshaft specification?

Tecumseh and Kohler engines, as shown in figure 7-32, or a reject specification, like the one furnished by Briggs & Stratton, as shown in figure 7-33. Any reading you obtain that is greater than the reject specification signifies a journal that is "in specification." If the size is less than the reject value, the crankshaft must be replaced.

Any measurement should not vary more than 0.002-inch from the "new" crankshaft dimension. If all the crankshaft journals are "in spec" and the bearing surfaces are acceptable, the crankshaft may be used again.

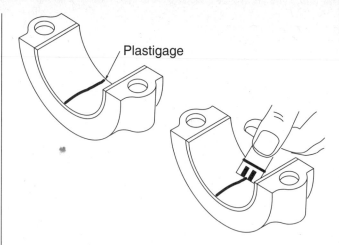

Fig. 7-34 Plastigage is a way to check running clearances.

Inspect the PTO side of the crankshaft with a straight edge to reveal a bent crankshaft. A bent crankshaft must be replaced because it is risky to straighten it and disrupt the metal characteristics—a situation which may cause further failure and danger.

PLASTIGAGE CRANKSHAFT CRANKPIN JOURNAL AND ROD

Plastigage provides a fast and accurate way to check running clearances, as shown in figure 7-34. It is a soft plastic that will flatten out to a measurable width when subjected to pressure; the measurement of the flattened plastic will be translated into a clearance between the crankpin journal and the connecting rod. The most common type is green plastigage, which is used to measure clearances from 0.001-inch to 0.003-inch.

Place the crankshaft into a holding device and lay a piece of plastigage on the crankpin journal of the crankshaft. Carefully install the connecting rod and rod cap without smearing the plastigage and torque the rod bolts to the proper tension. Carefully remove the rod assembly and check the width of the flattened plastic with the marks on the plastigage wrapper. The less the clearance between the crankpin journal and the connecting rod, the greater the flattening or width of the plastic.

Fig. 7-35 Measure the ring gap.

Ring gap reject sizes		
Model	Comp. rings	Oil ring
Aluminum cylinder models	0.035"	0.045"
Cast iron cylinder models	0.030"	0.035"

Fig. 7-36 The specifications will show either an acceptable range or a number that is the reject size.

Fig. 7-37 Remove the oil seals with a screwdriver.

REMOVE OLD PISTON RINGS

Take the old set of rings off the piston and install each ring individually into the cylinder. Center the ring by pushing it down with the piston. Move the ring to the area of maximum wear. Measure the ring gap, as shown in figure 7-35. This measurement can be compared to the specifications. The specifications either display an acceptable range or a single number that is the reject size, as shown in figure 7-36. Test each ring. If the cylinder is within tolerances, then an excessive ring gap would verify the diagnosis of "bad rings." A trick that is used by some mechanics to check a cylinder bore is to have a tested new ring put into the cylinder. If the ring gap is out of specifications, then the cylinder must have excessive wear. This is only a quick check and should be followed with further measurements.

REMOVE OIL SEALS

The oil seals may be removed by prying them with a screwdriver, as shown in figure 7-37.

INSPECT CRANKCASE BEARING SURFACES

Inspect the bearing surfaces for signs of wear or scoring. Briggs & Stratton provides a reject tool, with the part number listed in the main specification chart, that can be used to check if main bushing wear is excessive, as shown in figure 7-38. If the tool can be inserted into the hole, the opening is too large and must be reworked. A new metal lining or bushing may be inserted with the assistance of a special tool kit or the crankcase cover may be replaced.

If the "new" specification is provided, a telescoping gage and a dial caliper can be used to obtain the present measurement. The engine measurement should not be more than 0.002-inch from the given specification. Without a reject gage, a telescoping gage with a micrometer or dial caliper may be used for measuring. Subtract the measurement of the crankshaft from the measurement of the crankcase journal.

The magneto journal on many small engines must

Fig. 7-38 Is bearing wear excessive?

have the proper clearance. The points are opened and closed by the lobe or an indent on this area of the crankshaft. Excessive clearance will affect the engine's timing. One benefit of refitting an engine with an electronic ignition is that the timing is not affected by a worn magneto journal.

REMOVE MUFFLER

The muffler should be removed so that the complete valve chamber may be cleaned. Penetrating oil can be put on the bolts or threads. After a period of time, the muffler may be removed. The use of heat may assist the removal. Heat the threads with a torch and then let them cool. The expansion and contraction of the two metals will break the bond. The mufflers that twist in may be removed by a small oil filter wrench or a flywheel strap wrench, as shown in figure 7-39. The use of a vice grip or a pipe wrench may destroy the old muffler.

ENGINE REPAIR VS. REPLACEMENT

Consider the age and remaining life of the equipment powered by the engine. An aluminum cylinder engine will last about 300 hours, while the industri-

Fig. 7-39 Remove the twist in muffler with a small oil filter wrench or a flywheel strap.

al engine with a cast-iron cylinder and heavy duty valves will last an average of 2000 hours.

After the main components are measured and analyzed, the labor and parts cost of the following options should be calculated:

1. *Replacing the engine*
2. *Replacing engine block and main components (the shortblock)*
3. *Rebuilding the engine*
4. *Upgrading the current engine with an industrial shortblock*
5. *Replacing the whole machine*

Overhauling is the most labor-intensive choice and requires a good degree of mechanical skill, the correct equipment, and a sizable time investment. A shortblock contains all new internal engine parts (bearings, pistons, cylinder block, rods, etc.), but reuses many parts from your old engine.

At one end of the spectrum, repair of a small 3.5 HP aluminum engine used on a mower deck in fair condition may have the following prices calculated:

1. *Replacing the engine*......................................$140
2. *Installing the shortblock* $110
3. *Rebuilding the engine* $115
4. *Upgrading to industrial* $200
5. *Replacing the whole machine* $130

On the other end of the spectrum, a 16 HP cast-iron cylinder engine on a good lawn tractor may have the following prices calculated:

1. *Replacing the engine*......................................$900
2. *Installing the shortblock* $600
3. *Rebuilding the engine* $400
4. *Upgrading to industrial*..................................NA
5. *Replacing the whole machine* $2200

What would be your choice? What is best for the customer? The alternatives may not always be clear. Choosing the best repair alternative is up to the customer, but the mechanic's suggestion should be offered. Fixing a worn engine is more than a mechanical process comprised of working with the customer concerning the repair alternative, but also customer education regarding preventative maintenance to acquire the maximum machinery life.

SUMMARY

The small four-stroke-cycle Briggs & Stratton engine is a very common engine and provides an excellent educational instrument to acquire proper knowledge and procedures at minimal risk and cost.

It is important that the mechanic follow proper disassembly procedure to guarantee effective and acceptable results. After the main components are measured and analyzed, the labor and parts cost of the various options should be calculated to decide if the rebuild is possible.

Questions

1. **Technician A says that as an engine is disassembled, "clues" should be observed in order to theorize what causes an engine failure. Technician B says that the engine should be disassembled completely and then the parts should be analyzed for possible failure. Who is correct?**

   A. Only Technician A
   B. Only Technician B
   C. Both Technician A and B
   D. Neither Technician A or B

2. **Inspect the fuel during the disassembly. Technician A says that if the gas smells ok, it can be assumed to be in good shape. Technician B says that the volatility and odor should be observed to determine the condition of the fuel. Who is correct?**

   A. Only Technician A
   B. Only Technician B
   C. Both Technician A and B
   D. Neither Technician A or B

3. The compression of an engine should be analyzed. Technician A says the best way to do this is to pull the starter rope and feel for the resistance. Technician B says that the cylinder leakage is the most comprehensive test available. Who is correct?

   A. Only Technician A
   B. Only Technician B
   C. Both Technician A and B
   D. Neither Technician A or B

4. The "wet" test can be used to determine the cause of low engine compression. Technician A says that if the compression rises after the oil is introduced into the cylinder, the cause of the low compression is bad valve seating. Technician B says that the cause of the problem is the oil seals around the crankshaft. Who is correct?

   A. Only Technician A
   B. Only Technician B
   C. Both Technician A and B
   D. Neither Technician A or B

5. A Briggs & Stratton 3.5 horsepower engine is found to have a cylinder wear of 0.003-inch and an out-of-round of 0.001-inch. Technician A says that chrome rings could be used during the rebuild, if available. Technician B says that standard rings can be ordered and installed to rejuvenate the engine life. Who is correct?

   A. Only Technician A
   B. Only Technician B
   C. Both Technician A and B
   D. Neither Technician A or B

6. When the flywheel is removed, a tool should be used. Technician A says that the knockoff tool is the best choice. Technician B says that the best method is to loosen the flywheel nut and tap it firmly with a hammer while prying under the flywheel. Who is correct?

   A. Only Technician A
   B. Only Technician B
   C. Both Technician A and B
   D. Neither Technician A or B

7. The space between the valve lifter and the bottom of the valve stem can be measured with a feeler gage. Technician A says that this is an optional step. Technician B says that this measurement must always be taken before the valves are removed. Who is correct?

   A. Only Technician A
   B. Only Technician B
   C. Both Technician A and B
   D. Neither Technician A or B

8. The crankcase cover must be removed during the disassembly. Technician A says that the crankshaft must be sanded and filed so that the cover is removed easily. Technician B says that a hammer can be hit on the crankshaft while the sump cover is held to remove it if it is stuck. Who is correct?

   A. Only Technician A
   B. Only Technician B
   C. Both Technician A and B
   D. Neither Technician A or B

9. Technician A says that it is always more cost efficient to shortblock an engine rather than rebuild the components. Technician B says that the choice between the different types of engine repair should not be decided by the customer, but rather by the technician. Who is correct?

A. Only Technician A
B. Only Technician B
C. Both Technician A and B
D. Neither Technician A or B

10. Technician A says that tear-down steps should always be followed so that consistency and efficiency can be achieved. Technician B says that the tear-down should be done by one technician and another technician should do the assembly. Who is correct?

A. Only Technician A
B. Only Technician B
C. Both Technician A and B
D. Neither Technician A or B

CHAPTER 8

Engine Assembly

PREPARATION

The engine assembly chapter is earmarked for the small Briggs & Stratton air-cooled engine, but many of the steps are appropriate for other air-cooled engine brands. The small 3.5 horsepower air-cooled engine is rarely rebuilt for a customer because the cost of the labor and parts are more than the cost of a shortblock or new engine, but for the hobbyist and student it provides a good value. Many engines are discarded because of the high labor cost of repairs, but they can be rebuilt and restored to operation for a minimal parts cost.

The manufacturer's repair manual, as shown in figure 8-1 and figure 8-2, must be acquired for each engine rebuilt so that necessary specifications and adjustment procedures are followed.

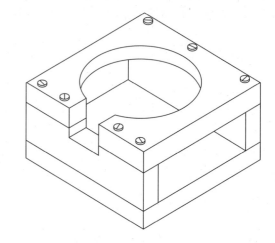

Fig. 8-3 Wooden engine stand.

Fig. 8-1 Manufacturer's repair manual—Briggs & Stratton.

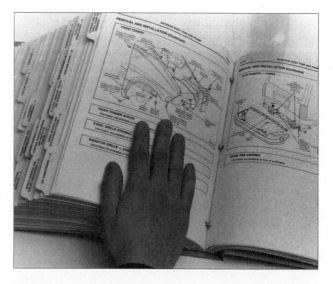

Fig. 8-2 A typical service manual (from Molinaro/Counterman's *Guide to Parts and Service Management,* copyright © 1989 by Delmar Publishers Inc. Used with permission.)

234 Small Engine Technology

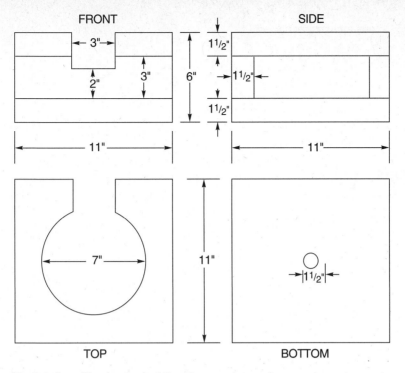

FRONT

SIDE

TOP

BOTTOM

Fig. 8-4 Specification for building the wooden engine stand.

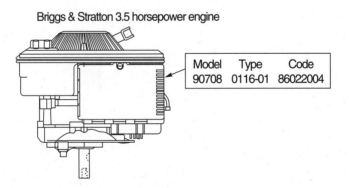

Briggs & Stratton 3.5 horsepower engine

Model	Type	Code
90708	0116-01	86022004

Fig. 8-5 Briggs & Stratton model #90708 engine.

After the engine has been disassembled, the parts are cleaned and inspected for any broken or missing pieces. Measurements are made on the engine parts before the assembly begins. Evaluation of engine parts in the disassembly steps will indicate if the engine is profitable to be rebuilt. A wooden engine stand can be easily made, which will allow easy access to all the parts, as shown in figure 8-3 and in figure 8-4.

IDENTIFY MODEL NUMBER

The theoretical Briggs & Stratton engine that will be rebuilt in this chapter has the model number 90708, as shown in figure 8-5, and can be decoded by using the Briggs & Stratton service manual. Since this number is found mainly on the sheet metal cover of the engine, as shown in figure 8-6, it is important to ascertain that the numbers belong to the correct engine.

Fig. 8-6 The sheet metal cover of the ending showing the model number.

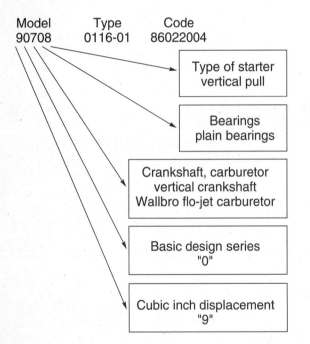

Fig. 8-7 The model number identifies certain engine characteristics.

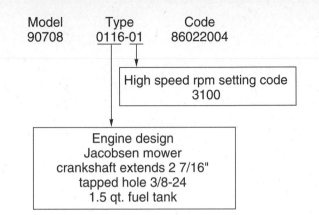

Fig. 8-8 *Type* numbers designate different variations in the engine.

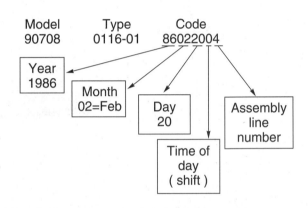

Fig. 8-9 The *code* number is a date code.

The model number is meaningful, as it identifies certain characteristics about the engine, as shown in figure 8-7. Decoding the model number may identify that the number is correct. The cover, where the numbers are found, can be interchanged easily with another engine. Often when parts are ordered, disappointment follows when the wrong parts are

received. Notice the paint color of the engine parts! The engines are sprayed with the same color at the end of the assembly line after all the components have been assembled. An engine that has different colored parts should signal the possibility of an incorrect model number, type, and code.

The "type"-numbers designate the different variations from the basic model number that the engine possesses, depending on the equipment the engine is designed to power, as shown in figure 8-8. A manufacturer may need a 3.5 HP engine with a crankshaft that extends only one inch from the engine to fit the lawnmower deck application, while another may need the same engine with a crankshaft that extends over two inches. The top engine speed of 3100 RPM for the example engine is less than

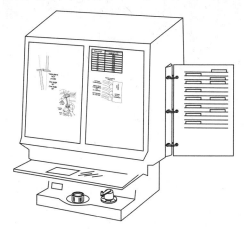

Fig. 8-10 Microfiche film and reader.

speed necessary for maximum horsepower, but on this application the transmission is set at one speed, and a higher speed than 3100 RPM would cause the lawnmower to move too fast for the operator.

The "code"-number is necessary because it is a date code, as shown in figure 8-9. This number is not a sequential serial number, so there could be many engines built on the same day that will have the identical number. Any design changes that occur during the production of the engine may be designated by this number.

ORDER ENGINE PARTS

Parts can be obtained from an authorized dealer or distributor in your area. Dealers for the engine type you are overhauling advertise in the phone book's yellow pages under the section "Engines—Gasoline." If you work on many engines, it is advisable to set up a "trade-account" with the local distributor, which will allow you to purchase parts at a discount. To receive the discount, many distributors require that you order parts by the manufacturer's parts numbers. This will require the use of a parts book or the commonly used microfiche film and reader, as shown in figure 8-10.

The parts manual for the Briggs & Stratton engines is divided into many different sections to account for the various engines. Not only is the model number used to locate the correct section, but a second set of numbers, called the type-numbers, is necessary to identify certain variations of the basic model, as shown in figure 8-11.

Once the correct section in the parts book is located, the picture of the various parts is identified by a reference number, as shown in figure 8-12. The reference number for the part needed is used to find the part number, which is located in the additional pages of the parts section, as shown in figure 8-13. There might be some decisions necessary to choose the correct part number from the reference number. This is where the complete set of engine numbers is necessary.

When one rebuilds an engine, there is a basic list of necessary parts. Additional components may be needed, depending upon the item missing or damaged on the engine, as shown in figure 8-14.

An important step in obtaining the correct part is to check the part number in the price book. The price book will reveal whether the number is "good" (still valid), the price, and the availability. Many part numbers are superseded to a new part number, and the manufacturers update their price list more often than the parts books. The availability is indicated by the stocking code to identify whether a part is in stock locally or if it may have to be ordered.

Engine rebuild gasket kit

The gasket set will have all the common gaskets necessary for the rebuild and more. Many times a gasket set is packaged that will service more than one model number. Thus, the service individual may end up with more gaskets than is needed for a specific engine, but inventory will require fewer gasket sets.

Oil seals

The magneto oil seal is found in the cylinder block on the flywheel side of the engine. Oil seals are used to prevent the crankcase oil from leaking at the point where the crankshaft extends out of the cylinder block.

MAGNETO OIL SEAL A new seal must be used if the crankshaft is removed.

PTO OIL SEAL This power-take-off (PTO) seal is found in the cylinder block, where the crankshaft

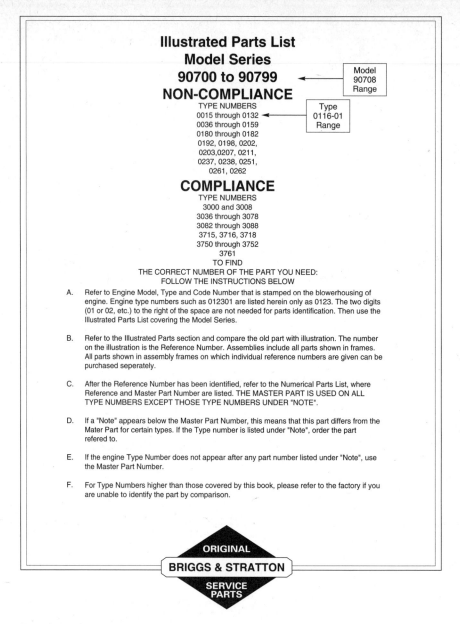

Illustrated Parts List
Model Series
90700 to 90799 ← Model 90708 Range
NON-COMPLIANCE
TYPE NUMBERS
0015 through 0132 ← Type 0116-01 Range
0036 through 0159
0180 through 0182
0192, 0198, 0202,
0203,0207, 0211,
0237, 0238, 0251,
0261, 0262

COMPLIANCE
TYPE NUMBERS
3000 and 3008
3036 through 3078
3082 through 3088
3715, 3716, 3718
3750 through 3752
3761
TO FIND
THE CORRECT NUMBER OF THE PART YOU NEED:
FOLLOW THE INSTRUCTIONS BELOW

A. Refer to Engine Model, Type and Code Number that is stamped on the blowerhousing of engine. Engine type numbers such as 012301 are listed herein only as 0123. The two digits (01 or 02, etc.) to the right of the space are not needed for parts identification. Then use the Illustrated Parts List covering the Model Series.

B. Refer to the Illustrated Parts section and compare the old part with illustration. The number on the illustration is the Reference Number. Assemblies include all parts shown in frames. All parts shown in assembly frames on which individual reference numbers are given can be purchased seperately.

C. After the Reference Number has been identified, refer to the Numerical Parts List, where Reference and Master Part Number are listed. THE MASTER PART IS USED ON ALL TYPE NUMBERS EXCEPT THOSE TYPE NUMBERS UNDER "NOTE".

D. If a "Note" appears below the Master Part Number, this means that this part differs from the Mater Part for certain types. If the Type number is listed under "Note", order the part refered to.

E. If the engine Type Number does not appear after any part number listed under "Note", use the Master Part Number.

F. For Type Numbers higher than those covered by this book, please refer to the factory if you are unable to identify the part by comparison.

ORIGINAL
BRIGGS & STRATTON
SERVICE PARTS

Fig. 8-11 Type numbers identify certain model variations.

extends from the crankcase and connects to the equipment.

AUXILIARY OIL SEAL (OPTIONAL) This oil seal is found on some engines that use an extra shaft extending from the crankcase. The shaft may be used to drive the wheels of a self-propelled lawnmower.

Piston rings

Piston rings are installed on the piston to form a seal between the piston and the cylinder.

STANDARD PISTON RINGS These are the piston rings used on an engine when the cylinder specifications are within the limits of wear. The Briggs & Stratton specification chart indicated that standard rings can be used when the cylinder bore is 0.003-inch or less wear and 0.002-inch out-of-round. Since the engine is within the specification, standard piston rings are used.

CHROME PISTON RINGS If the engine's cylinder wear is greater than 0.003-inch but less than 0.005-inch, chrome piston rings (if available from the manufacturer) may be used to recondition the engine. The chrome piston ring set has the outer edge of the top compression ring chrome-plated. This metal is harder than the aluminum in the cylinder and wears off the irregularities on the cylinder walls to allow better compression sealing. The oil control ring and the middle ring have expanders installed for greater outward pressure and better oil control. The chrome ring is about four times the price of standard rings.

OVERSIZE PISTON RINGS If the cylinder measurements indicate that the cylinder must be bored to the next size, a new larger piston and ring set must be purchased. Engine manufacturers sell piston and rings as standard size, 0.010-inch oversize, 0.020-inch oversize, and sometimes larger sizes. It is only cost effective on the large cast-iron cylinder engines to have the engine bored at a machine shop and an oversize piston and ring set installed.

Ignition Parts

SOLID STATE IGNITION These units are enclosed and not serviceable. If the unit is operational, no parts should be ordered, except for a spark plug. If the unit is defective, the whole module must be ordered.

POINTS AND CONDENSER A new set of points and condenser should be ordered for the overhaul. Briggs & Stratton offers a solid state ignition module that can "retrofit" many of its engines that were built from 1961 to present. The advantages and reliability of the solid state components are attained at a minimal cost. The repair of choice is to install the new Magnetron® module instead of replacing the points and condenser. When ordering the

Magnetron® (B&S # 394970), as shown in figure 8-15, also order the plug (B&S # 231143) that will seal the plunger hole when the points are removed.

> **NOTE:** THERE IS NO NEED TO ORDER THE PLUNGER HOLE PLUG IF THE MAGNETRON IS INSTALLED WHEN THE OLD POINTS AND CONDENSOR ARE NOT REMOVED.

A new harder flywheel key is supplied with the Magnetron® unit.

Spark plug

Check the spark plug manufacturer's catalog for the proper spark plug. The engine we are rebuilding used a Champion CJ-8, which is a "shorty plug" of the longer version of the J-8, as shown in figure 8-16. The "C" at the beginning indicates the "shorty" version. The only difference is the shorter external ceramic length. The CJ-8Y, as shown in figure 8-16, has a "Y" at the end, which denotes an extended center electrode. This spark plug is used in two-cycle engines because it better resists oil fouling.

Carburetor parts

To overhaul a carburetor, it is best to order a carburetor rebuild kit that contains all the common parts necessary. Some manufacturers do not list the parts in kit form. The demonstration Briggs & Stratton engine being rebuilt uses a Walbro carburetor, built for Briggs & Stratton, which allows us to order an overhaul kit.

Starter parts

The starter is used to start the engine components in motion so that the combustion process can begin.

ROPE Inspect the rope and if it is damaged or frayed, a new part should be ordered.

HANDLE Inspect the rope handle and, if defective, order a new one.

SPRING If the starter was sluggish, binding, or the spring area rusted, order a new part.

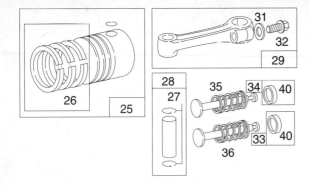

Fig. 8-12 Reference numbers are used to identify parts.

RING SETS:

Ref #	Part #	Description
26	298982	Ring set–standard Piston
	298983	Ring set–.010" O.S. Piston
	298984	Ring Set–.020" O.S. Piston
	298985	Ring Set–.030" O.S. Piston
27	26026	Lock–piston pin
28	298909	Pin assembly–piston– Standard
29	294201 Note:	.005" O.S. Rod assy.–connecting For connecting rod with .020" undersize Crankpin bore–order No. 296079

Fig. 8-13 Part numbers are located in the additional pages of the parts section of the book.

Air cleaner element

The type of element could be constructed of foam or paper or a combination of both. If the element cannot be cleaned, it should be replaced.

Missing/damaged parts

Any additional parts that are missing or broken, e.g., the muffler, should be ordered.

Ref no.	Part no.	Description
358	298989	Gasket set
3	299819	Seal-oil (magneto)
20	391483	Seal-oil (PTO)
26	298982	Ring set-standard
333	398593	Armature-magneto
337	J-8	Spark plug (Champion)
121	398669	Carburetor overhaul kit
769	66734	Rope vertical pull 48"
777	390066	Spring-starter
967	397795	Filter-air

Fig. 8-14 Additional parts may be needed on the engine.

EXAMINE VALVE GUIDES

Use a reject gage, dial indicator, or small hole gage and micrometer to determine if hole size is acceptable. Correct any valve guide problems before continuing with the valve job.

The "common specifications" chart for the Briggs & Stratton engine model 90708 indicates what valve guide reject gage (B&S #19122) to use, as shown in figure 8-17 and figure 8-18. If the end of the gage can be inserted into the hole more than 5/16-inch, then the valve guide hole is too large and must be overhauled before the valve job can be continued.

Recondition valve guides

If the valve guides are not within tolerance, recondition by one of the following methods:

REAM GUIDE AND INSTALL BRONZE BUSHING Install a bronze bushing in the worn aluminum or cast-iron guides. Briggs & Stratton supplies tools to accomplish this procedure.

1. *The B&S model 90708 has a valve guide with an inner diameter of 0.250-inch (1/4-inch), so the counterbore reamer (B&S #19064) is placed into the valve guide and centered by using a pilot bushing (B&S #19191), as shown in figure 8-19 and figure 8-20.*

2. *Set the new bushing (B&S #63709) on the pilot, and mark the reamer with tape at the top of the bushing. This will indicate the depth the reamer will cut to allow for the installation of the new bushing.*

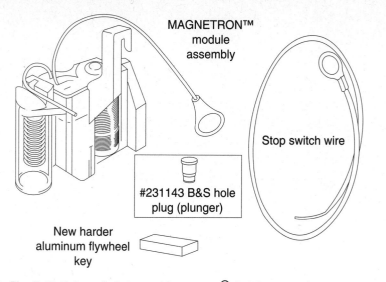

Fig. 8-15 Briggs & Stratton Magnetron® (B&S #394970) electronic ignition. Note Plug (B&S #231143) not included in kit.

MAGNETRON™ module assembly

Stop switch wire

#231143 B&S hole plug (plunger)

New harder aluminum flywheel key

Fig. 8-16 Champion spark plug. Left CJ8Y (extended tip) - right CJ8 (normal).

3. *Enlarge the valve guide hole with the counterbore reamer (B&S #19064). Use some oil during the reaming process and always turn the reamer clockwise as you enter and exit the hole. Ream only to the depth of the new bushing, as shown in figure 8-21. The counterbore reamer will oversize the valve guide so that a bronze bushing can be driven in.*

4. *Use the bushing driver (B&S #19065), as shown in figure 8-22 and figure 8-23, and a lightweight hammer to drive in the new bushing with a light tapping action that will allow installation without mushrooming the bushing. When the bushing reaches the correct depth, a change in the tapping sound will be noticed. Rotate the bushing driver while tapping.*

5. *A finish reamer (B&S #19066) is then used, along with a modified pilot bushing, as shown in figure 8-24, to bring the new valve guide to the proper dimensions. Take care to only "finish ream" the valve guide and not the valve lifter hole.*

NOTE: USE KEROSENE OR ANOTHER "THIN" LUBRICANT WHILE "FINISH REAMING."

6. *Wash the new bushing with solvent and then use compressed air to remove all the chips.*

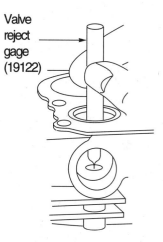

Valve reject gage (19122)

Fig. 8-17 Checking valve guide wear with Briggs & Stratton reject gage (B&S #19122).

7. *Install the valve and make sure it moves without binding.*

REMOVE OLD BUSHING AND INSERT NEW If the old bushing is a bronze insert, it must be removed before the new bushing is installed.

1. *A 7 mm tap (B&S #19273) is used with the modified pilot (B&S #19191) to install threads into the old bushing no more than 3/4-inch in*

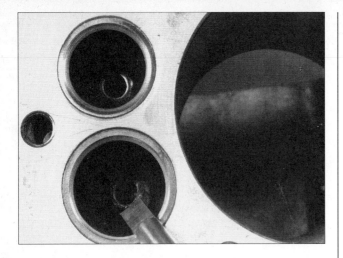

Fig. 8-18 Reject gage.

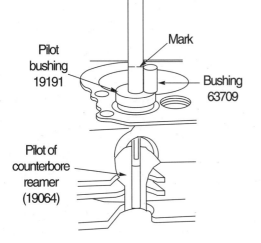

Fig. 8-19 Prepare the cylinder for a new bushing by oversizing the valve guide bore.

depth. A 7 mm tap is used because it will cut a thread into the old bushing without damaging the aluminum of the engine block.

2. Rotate the puller nut (B&S #19272) up to the head of the puller screw (B&S # 19270), as shown in figure 8-25, and insert the puller screw down through the washer (B&S #19270). Thread the puller screw into the tapped bushing until

the screw bottoms in threaded hole. Back off the screw 1/8-turn and place a drop of engine oil on the threads of the puller screw.

3. Hold the puller screw stationary and turn the puller nut down on the washer until the valve guide bushing is removed.

INSTALL NEW VALVE STEM WITH LARGER STEM A valve with a larger stem diameter may be purchased for certain engines. This method is popular with Tecumseh engines. When the valve guides become excessively worn, they are reamed oversize to accommodate a 1/32-inch oversize valve stem.

RECONDITION OR REPLACE INTAKE AND EXHAUST VALVES

Hard starting, loss of power, or loss of compression accompanied by high fuel consumption may be symptoms of faulty valves. Carefully inspect all valve guides, valves, and valve seats for evidence of deep pitting, cracks, or distortion.

Clean valves

Clean the valves with a power wire brush, then inspect for defects, such as warped valve head, excessive corrosion, and worn or bent valve stems, and replace if necessary. The cost of the machinery to reface the valves may make the purchase of new valves the economical choice.

Resurface valve face

A valve, as shown in figure 8-26, can be reconditioned and reused if it is in good shape. An automotive valve refacer machine can be used or the Neway hand valve refacer tool kit can be used, as shown in figure 8-27. It may be more cost effective to routinely replace the valves when the engine is overhauled than to purchase a valve refacer.

The valve angle most commonly used for the valve face is 45°, but on some engines it is possible to have a 30° intake or exhaust valve. The easiest way to identify if both valves are 45° is to put them together, as shown in figure 8-28, and if the stems are in the same plane, the angles are both 45°. If the stems do not follow a straight line, then one valve is 30°, as shown in figure 8-29.

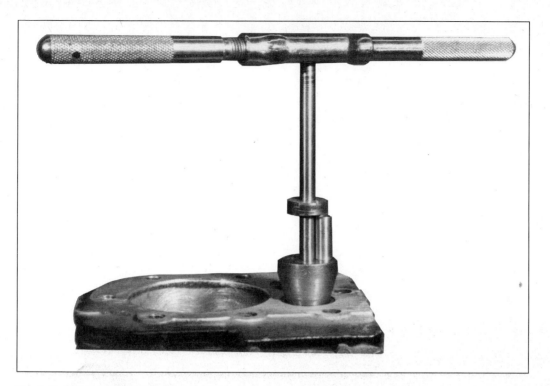

Fig. 8-20 Bushing #63709 and reamer.

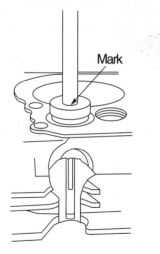

Fig. 8-21 Turn reamer clockwise to the depth of the mark indicating the height of the replacement bushing.

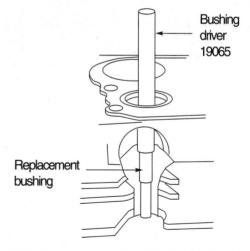
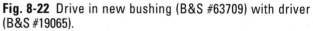

Fig. 8-22 Drive in new bushing (B&S #63709) with driver (B&S #19065).

Check margin width

After reconditioning, if the valve margin is less than 0.031-inch (1/32-inch or 0.8 mm), as shown in figure 8-30, a new valve must be used. The width of the valve margin is important because if it is too thin, the valve can burn and crack.

Fig. 8-23 Drive in new bushing

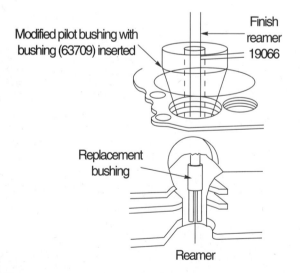

Modified pilot bushing with bushing (63709) inserted

Finish reamer 19066

Replacement bushing

Reamer

Fig. 8-24 Finish reaming the new bushing to final diameter.

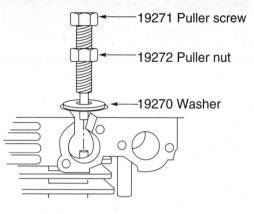

19271 Puller screw

19272 Puller nut

19270 Washer

Fig. 8-25 Tools needed to remove old bronze bushing.

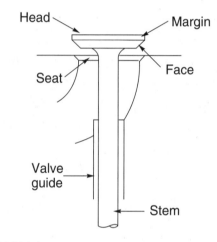

Head

Margin

Seat

Face

Valve guide

Stem

Fig. 8-26 Valve parts.

RESURFACE VALVE SEAT
Insert pilot

Insert the correct pilot into the valve guide from the Neway valve seat cutter kit (B&S # 19237), as shown in figure 8-31. Use the pilot that can be inserted into the hole, but the pilot should not bottom out on the ridge of the valve guide. The pilot is actually a tapered shaft, and if it bottoms out, then you know you have a worn valve guide.

Determine the angle of the valve face and seat

Most engines have 45° angles but some have 30°. The seats are actually cut to 46° (or 31°). This one

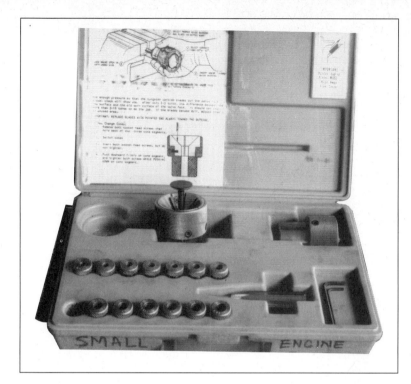

Fig. 8-27 Neway valve refacer.

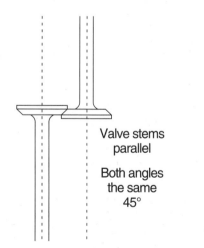

Valve stems
parallel

Both angles
the same
45°

Fig. 8-28 Valve stems parallel—both angles 45°

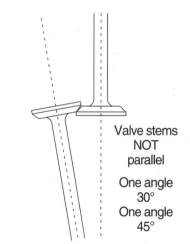

Valve stems
NOT
parallel

One angle
30°
One angle
45°

Fig. 8-29 Valve stems not parallel—one angle 45° and the other 30°.

degree difference between the valve face angle and the valve seat angle is called the "Angle of Interference." It is needed for proper valve seating when the engine is hot.

To determine a valve seat angle, match the valve face against the angle of the cutter. If the valve stem and the cutter pilot are parallel, use that side of the cutter, as shown in figure 8-32.

Cut valve seat

1. Color the valve seat with a black marker for easier identification of the newly cut area.

2. Place the valve seat cutter on the pilot already in the valve guide, with the correct angle facing the valve seat, as shown in figure 8-33 and 8-34. With light pressure, turn the cutter with the T-handle clockwise for approximately three turns.

NOTE: IT IS BETTER TO USE A SPEED WRENCH, IF AVAILABLE, TO IMPROVE THE DOWNWARD PRESSURE AND OBTAIN A MORE CONSISTENT PRESSURE, AS SHOWN IN FIGURE 8-35.

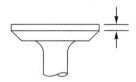

Acceptable if greater than 0.031" (1/32" or 0.8 mm)

Fig. 8-30 Minimum valve margin after refacing.

Remove the seat cutter and observe the valve seat to determine whether metal has been removed evenly around the complete seat, as shown in figure 8-36. If one area does not display cut marks, use the cutter to take off additional metal.

3. The final cut should be made with a piece of waxed or regular paper between the cutter and the valve seat. The paper acts as a shim that allows the cutters to remove only the last high and low spots without any additional valve metal being removed, as shown in figure 8-37.

Check valve seat width

The width of the valve seat is important for obtaining proper valve sealing and cooling, as shown in figure 8-38. The width of the valve seat must be maintained between 0.046-inch to 0.063-inch (3/64-inch to 1/16-inch or 1.1 mm to 1.6 mm), as shown in figure 8-39. If the valve seat width is lower than the specification, additional metal may be removed by using the valve seat cutter. If the valve seat is above the specification, the valve seat can be narrowed using a 31° angle for the 46° valve seat, as shown in figure 8-40.

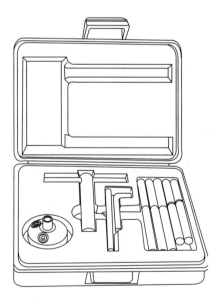

Fig. 8-31 Neway valve seat cutter kit (B&S #19237).

Some seat cutting kits have two additional angled cutters. One is 60° and the other is 15°. These cutters can also be used to narrow the valve seat width.

45° ANGLE VALVE SEAT WIDTH If the seat width needs to be narrowed in a valve with a face and seat angle of 45°, the 60° cutter can be used to lessen the

Fig. 8-32 Determine correct side of cutter by matching the valve face with the tool.

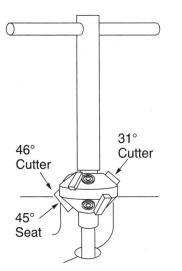

46° Cutter

31° Cutter

45° Seat

Fig. 8-33 Correct angle facing the valve seat.

seat width at the bottom. The 30° cutter can be used to lessen the seat from the top. Narrowing the seat from the top or bottom will depend on where the seat contacts the valve face. Every attempt should be made to keep the touching metal of the seat about midway on the valve face.

30° ANGLE VALVE SEAT WIDTH If the seat width needs to be narrowed in a valve with a face and seat angle of 30°, the 60° cutter can be used to lessen the seat width at the bottom. The 15° cutter can be used to lessen the seat from the top. Narrowing the seat from the top or bottom depends on where the seat contacts the valve face. Every attempt should be made to keep the touching metal of the seat about midway on the valve face.

Lap valves

The valves should be lapped with lapping compound to remove grinding marks and assure a good seal between the valve face and the seat, as shown in figure 8-41. Lapping compound, as shown in figure 8-42, containing a fine grit abrasive, is applied carefully to the valve face. Do not put any compound on the valve stem. Rotate the valve with the lapping tool in a back and forth motion, lifting it off the seat occasionally. After lapping, the area of

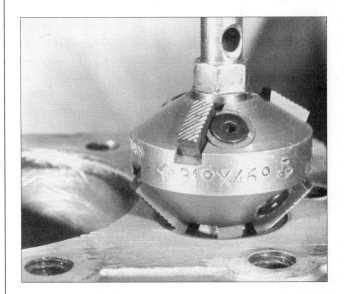

Fig. 8-34 Valve seat cutter.

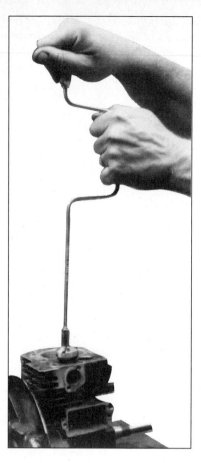

Fig. 8-35 Valve seat cutter and speed handle.

Fig. 8-36 Proper valve seat cutting will remove metal evenly from the complete seat.

Fig. 8-37 The final cut should be made with a piece of paper used as a shim.

Fig. 8-38 Measure the width of the valve seat.

Acceptable if between 0.046" to 0.063"(3/64" - 1/16" or 1.1 mm - 1.6 mm)

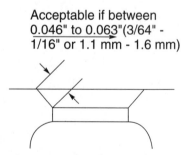

Fig. 8-39 The width must be between the specifications.

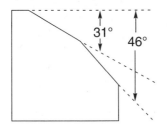

Fig. 8-40 Valve seat can be narrowed by using the other side of the cutter angle.

Fig. 8-41 Valves should be lapped with lapping compound.

Fig. 8-42 Valve lapping compound.

metal contact will be dull, as shown in figure 8-43, and will verify whether the valve seat dimension is correct. It will also break-in the valves so that they will seal better under operation.

Check for leaks

Clean the lapping compound from the valve and seat, and check the valve and valve seat region for any leaks. Hold the valve against the valve seat with your finger and place a thin liquid, like brake cleaning fluid, either in the exhaust port for the exhaust valve or in the intake manifold for the intake valve, as shown in figure 8-44. The seal between the valve seat and the valve face should stop any liquid flow. If the valves do not seal the liquid, continue lap-

ping. If additional lapping does not stop the leakage, then recut the valve face and valve seat, lap the valves, and retest.

CYLINDER PREPARATION

The cylinder must be properly prepared for the installation of the piston and new rings.

Remove upper cylinder ridge

Sometimes the cylinder will have an upper ridge of metal and carbon that must be leveled so that the new rings are not damaged during the installation. The ridge can be sanded with a 300-grit paper, or a ridge remover may be used, as shown in figure 8-45. Caution should be used on aluminum cylinders so that a hard cutter does not destroy the soft aluminum cylinder wall.

Hone cylinder

A cast-iron cylinder must be honed, while on an aluminum cylinder the process is not mandatory, but is advisable. A common honing tool is the flex hone.

CAST-IRON CYLINDER A cast-iron cylinder must be honed to deglaze the shiny low friction area of the cylinder wall. Cylinder honing is necessary for proper break-in between the rings and the cylinder

Fig. 8-43 After lapping, the area of metal contact will be dull.

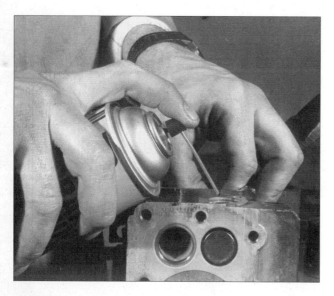

Fig. 8-44 Check for leaks.

wall. Initially, the outer surface of the new rings and the cylinder have uneven metal areas. During the break-in period, the high spots on both surfaces are worn away to obtain a good sealing surface. The friction levels are higher than after the break-in, so it is important to provide adequate lubrication.

ALUMINUM CYLINDER Aluminum cylinders do not have to be honed because the aluminum is soft enough to wear with the rings during the break-in period; however, the honing may be performed on an aluminum cylinder to provide proper etching so that oil will be retained in the cuts and will also provide the necessary lubricant during the break-in period. **Caution:** Unless the proper cleaning steps (described later in the chapter) are followed, honing abrasives remaining in the cylinder will quickly destroy a new set of piston rings. Some of the honing grit can be retained in the etching, and when the engine is operated, it will dislodge, cause abrasions, and increased friction.

FLEX HONE The flex hone can be used in a variable speed drill motor to hone the cylinder wall, as shown in figure 8-46. Lubricate the cylinder surface with honing oil or engine oil throughout the process. The drill motor should run slowly, while the up and down motion of the hone should be rapid. Make about ten up and down strokes in the cylinder and then wipe the cylinder with a paper towel and check the etching pattern, as shown in figure 8-47.

RIGID HONE A rigid hone set, such as the Briggs & Stratton #19205 for aluminum bore engines, or #19211 for cast-iron bore/sleeve engines, can be used on wavy, scuffed, or scratched cylinders. The rigid stones "true" any cylinder irregularities. The

Fig. 8-45 Ridge removing tool.

cylinder should be lubricated with a honing oil, and the hone should be driven by a slow-speed drill motor from 10 to 12 complete strokes.

PISTON PREPARATION

The piston must be properly prepared for the new piston rings and installation into the cylinder.

Clean the piston ring grooves

This is necessary so that the build-up of carbon will not exert a high outward pressure on the new rings and cause premature failure. The old rings should be removed and the ring grooves cleaned with a special tool, or an old piston ring can be broken and used, as shown in figure 8-48. The sharp edge is used as the scraper. **Caution:** The rings are very brittle, and fragments of the ring may propel when fractured. Wrap the ring in a shop towel before breaking.

Clean piston skirt

If the piston perimeter has any rough spots, it should be lightly smoothed with a fine (300+ grit) sand paper in a pattern matching the cylinder wall. Be careful, though, not to use an electric wire wheel or grinder on the piston, for they could remove the thin chrome plating. Even though the surface of the piston appears lackluster, pistons used in aluminum cylinders are chrome plated. Chrome plating does not have to be shiny and bright. Without the chrome plating, the surface of an aluminum piston in an aluminum cylinder would fuse together from the engine heat and cause engine seizure in a short period of time. Dissimilar metals must be used on the piston and cylinder surfaces.

CLEAN ALL DISASSEMBLED PARTS

Removal of all old gaskets, oil seals, and components are necessary in order to clean the parts properly.

Fig. 8-46 Flex-hone in a variable speed drill motor.

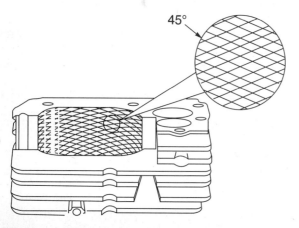

45°

Fig. 8-47 Proper angle for cross hatch after honing.

De-carbonizing formula

Carbon deposits can be softened for easy removal by soaking in the following solution:

Heat mixture to lukewarm (105° F) and place head into the mixture from 30 to 60 minutes. Remove and clean off the carbon.

> 1 CUP   AMMONIA
> 1/2 CUP   VINEGAR
> 1/4 CUP   BAKING SODA
> MIX TO 1 GALLON OF WATER

NOTE: A COMMERCIAL SAND BLASTER SHOULD NEVER BE USED TO CLEAN ANY OF THE ENGINE COMPONENTS. IT IS NEVER POSSIBLE TO REMOVE ALL THE GRIT FROM THE ENGINE DURING THE CLEANING PROCESS, AND ANY RESIDUE WILL DAMAGE THE ENGINE QUICKLY WHEN IT IS STARTED.

Cleaning Steps

The engine components that will be used again must be properly cleaned as close to the assembly time as possible in order to reduce the amount of contamination.

REMOVE OLD GASKET Make sure all the old gasket material is scraped off with a scraper that will not gouge the aluminum alloy surfaces, e.g., a sharpened piece of hardwood or a dull putty knife. It is important to remove all old gaskets. There are sprays that will chemically dissolve the gasket material for easier removal.

DEGREASE PARTS Cleanse all parts in a grease solvent.

WASH WITH SOAP AND WATER Wash all components in a soap and warm water solution, as shown in figure 8-49. It is very important that the engine cylinder be washed with soap and warm water at the final stage because it's almost impossible to remove all the abrasive particles in the cross hatch grooves. A soap solution works best so the dirt and oils will dissolve better in the water. A bottle brush works well for the cleaning. Not only should the cylinder be washed but also wash any elements that could be contaminated with abrasive dirt.

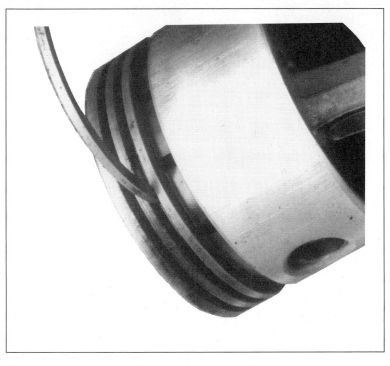

Fig. 8-48 Use a broken old ring to clean piston ring grooves on piston.

WIPE WITH CLEAN DETERGENT OIL After all the pieces have been washed and dried, wipe the metal parts with a clean paper towel that has been dipped in clean detergent engine oil. Continue this process until no black residue appears after the parts are wiped. Any black residue indicates additional dirt that must be removed.

The oil film left from the wiping is sufficient pre-lubrication for the later piston insertion. It is important to also clean the assembly tools in a similar manner to avoid any contamination. The parts should be cleaned immediately before they are assembled into the engine so dirt will not be attracted to the lubricated surface.

ASSEMBLY PRECAUTIONS

The following points are important in the assembly process to maximize engine life.

Assembly area

Establish an area that is free from contaminants to reassemble the engine, as shown in figure 8-50. Put the engine parts together with clean hands and clean tools.

NOTE: THERE IS NO STEP AS IMPORTANT AS A CLEAN WORK SPACE TO REDUCE PREMATURE ENGINE FAILURE. MANY SHOPS HAVE A TEAR-DOWN PLACE THAT IS SEPARATE FROM THE ASSEMBLY AREA AND A SEPARATE SET OF TOOLS TO ACCOMPANY EACH LOCATION. THE DISASSEMBLY REGION IS NORMALLY CONTAMINATED WITH A LARGE AMOUNT OF DAMAGING DIRT, WHILE THE ASSEMBLY SITE IS KEPT CLEAN.

Bolt/nut torquing

One of the prime objectives of the assembly process is to have an operating engine that is trouble-free. Even though the engine may initially operate satisfactorily, it is not cost effective to have a rebuilt engine returned to the serviceperson because of loose bolts or leaking gaskets. Correct assembly techniques can reduce engine failure.

WHO SHOULD USE A TORQUE WRENCH The question always arises concerning how tight the engine's fasteners should be. The experienced technician has developed a kinesthetic sense to know how much twisting can be applied to a bolt so that it will not vibrate loose, and he/she has probably

Fig. 8-49 Wash parts with soap and water.

Fig. 8-50 A clean assembly area is necessary.

learned that an overabundance of twisting can cause "stripped" threads that will produce an insufficient holding potential. With the proper use of a torque wrench and a bolt torque table, the inexperienced technician can greatly increase the probability that the fasteners will work properly. Torque wrenches are used to deliver and monitor the specific limit (torque) to which a fastener can be safely stretched. Four types of torque wrenches can be used: direct reading, dial, clutch, and click.

Certain fasteners, such as the connecting rod bolts, head bolts, and flywheel nut, must always be torqued to the proper specification, as recommended by the engine manufacturer, but the other fasteners may or may not be secured by using a torque wrench, depending on the experience of the technician.

WHY USE A TORQUE WRENCH A deck of baseball cards can be kept together by a rubber band if it is applied properly. The size of the rubber band should be smaller than the deck of cards so that the band must be stretched as it is put around the cards. When the rubber band is released, the rubber molecules attempt to return to the "at rest" size, but this is not possible because the card deck is in the way. The force holding the cards together is determined by the strength of the elastic particles attempting to return.

Metal fasteners work in a similar manner. One of the characteristics of a metal is elasticity. Elasticity is when a piece of metal is twisted or stretched slightly, it will return to its original size when released, comparable to the rubber band. If the metal is stretched beyond its "yield point," it will return when released, but not to the same area, and will have a reduced holding force.

When a bolt is screwed into either a nut or the threads of an engine component, the threads attempt to pull the metal bolt shaft away from the head of the bolt. The greater the twisting force, the greater the stretch of the metal. When the twisting is ceased, the metal of the bolt will attempt to return to the "at rest" position, but it will not be able to do so because of the metal parts sandwiched between it. This is the holding force that fastens parts together.

Bolts are made of different metal mixtures that have different elasticity limits. If the bolt is twisted beyond the limit, it will not have sufficient holding power and may also damage the threads. If a bolt is not twisted enough, the metal will not be stretched enough for proper fastening.

TORQUE CHART A torque chart, included at the end of the chapter, in figure 8-96, is a precalculated table which indicates the amount of twisting required for proper fastening, depending on the bolt's metal characteristics, the bolt's diameter, and the bolt's threads per-inch. The bolt head indicates the metal composition by the markings or lack of markings. This chart can be used to discover the proper torque for all the bolts that the mechanic's repair manual does not indicate.

TORQUE WRENCH The selection of the proper torque wrench is necessary to obtain correct readings. It is not uncommon to have at least two torque wrenches with different graduations handy, because torque wrenches are most accurate at their mid-scale reading. For example, if a bolt needs 90 inch-pounds of torque, a torque wrench that ranged from 0–200 inch-pounds would have a mid- scale reading of 100 inch-pounds, which is very close to the 90 inch-pounds of torque required. This would be more accurate than a wrench that ranged from 0–600 inch-pounds, where the mid-scale is at 300 inch-pounds.

It is very easy to convert from inch-pounds to foot-pounds by using a factor of 12. There are 12 inch-pounds for every 1 foot-pound. If a foot-pound torque wrench is used for a bolt that requires only 90 inch-pounds, the conversion would be attained by dividing the 90 inch-pounds by 12. This would be about 8 foot-pounds, but if your wrench ranges from 0–200 foot-pounds, this measurement would be far from the mid-scale number and, hence, not very accurate.

ASSEMBLY

INSTALL CRANKSHAFT

Lubricate bushings and moving parts. Keep parts oiled and insure that they are lightly coated with lubricant after they are cleaned. Not only does this practice make reassembly easier, but it insures minimal friction of the parts during initial operation of the engine after reassembly. Use the same oil to coat the parts that are used in the engine's crankcase.

INSTALL PISTON RINGS
Check ring end gap

Before positioning the new piston rings on the piston, push one of the rings down the cylinder with the piston head. Check ring end gap for a clearance between 0.005–0.035-inch, as shown in figure 8-51. If the rings have too much or too little gap, the incorrect set of piston rings may have been ordered. One

"rule-of-thumb" states that for every one inch of cylinder bore, 0.003-inch end gap is acceptable.

Put rings on piston

Put the piston rings on the piston in the proper groove and in the proper direction. The use of a ring expander is recommended so that the piston rings will not be overstretched or broken when they are installed, as shown in figure 8-52. Too much outward pressure on the ring during installation could possibly cause premature piston and cylinder wear. Normally, the bottom piston ring is exclusively for oil control and the top ring for compression sealing. The middle piston ring can be for compression sealing and oil control.

When installing these rings, they must go in the proper order on the piston. The general rule of thumb is that all inside chamfers (beveled edges) are aligned to the top of the piston, and any outside notches are aligned toward the bottom of the piston, as shown in figure 8-53. Many times the middle ring has the outside notch to help control the oil. This ring becomes a dual purpose ring, holding the compression in at the top of the cylinder and also scraping off or controlling the oil so that it doesn't get to the top of the cylinder. This outside notch is the oil scraper and must be installed in the downward position.

The gaps can be staggered, but since the rings rotate during operation, all that is being accomplished by staggering is to reduce the odds that all the gaps will align at any one time during operation.

Check for piston wear

After the new rings are installed, the piston can be further evaluated for wear. Check the space between the ring on the groove and the remaining space in the roof of the groove with a feeler gage. If a 0.007-inch feeler gage can be inserted, the piston is worn and should be replaced, as shown in figure 8-54.

The piston should be washed again in soap and water and coated lightly with assembly grease or engine lubricant. Unless the piston will be assembled into the cylinder immediately, place the lubricated piston unit in a plastic bag.

INSTALL PISTON ASSEMBLY

Using proper tools and procedures will minimize any component damage during the installation of the piston.

Compress piston rings

Clean and oil the piston ring compressor tool as described previously. Insert the piston and ring assembly into the ring compressor tool. Align the bands of the tool in the proximity of the rings on the piston and tighten the bands so that the rings are compressed into the piston ring grooves, as shown in figure 8-55.

Position crankshaft

Rotate the crankshaft so that the crankpin, or rod journal, is at the 5 o'clock position in order to avoid having the connecting rod hit the crankshaft when assembled.

Insert assembly into cylinder

Place the piston into the cylinder and notice how the connecting rod aligns with the journal on the crankshaft. Lightly tap the bands of the ring compressor tool with a hammer before the final tightening of the tool's bands. If the device has been used properly, only the pressure of your thumbs is necessary to push the piston into the cylinder. Do not use the wooden handle of a hammer or any other object to move the piston into the cylinder as this may damage the rings.

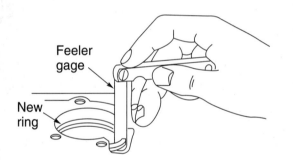

Fig. 8-51 Check ring end gap with a new ring.

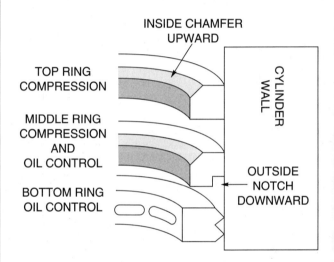

Fig. 8-53 General "rule of thumb" for piston ring installation.

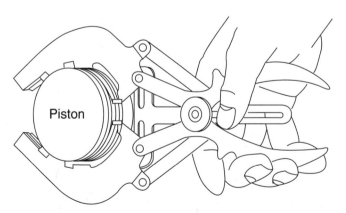

Fig. 8-52 Use a ring expander tool for new ring installation (B&S #19340).

Torque rod nuts

Continue pushing down on the piston so that it mates into the rod journal on the crankshaft and then apply the rod cap, using the aligning marks made during the tear-down. The rod bolt has a crucial torque specification that should be obtained from the manufacturer's specification chart, as shown in figure 8-56. The small Briggs & Stratton (model 92908) requires a torque of 100 inch-pounds. In addition to the torque value, another important consideration is the sequence to be followed when tightening the bolts.

Always tighten the bolts or screws in stages, fastening each approximately the same amount each time. Since the torque on these bolts is 100 inch-pounds, and if one desires to tighten the bolts

in three steps, dividing the 100 by 3 indicates using 33 inch-pounds of torque for each state or degree of fastening. Tighten the first bolt to 33 inch-pounds with a torque wrench, as shown in figure 8-57. Do the same for the second bolt. Return to the first bolt and now torque it to 66 inch-pounds and then apply the additional torque to the next bolt. Finally, bring the torque of each bolt to the final value of 100 inch-pounds.

Check for binding

After the rod is in place and torqued down, rotate the crankshaft to make sure there is no binding and that the rod and piston are installed correctly.

INSTALL VALVE LIFTERS (TAPPETS)

Install the lifters in their proper hole using the following steps:

Invert engine

Invert the engine so the valve lifters will not fall out during the camshaft installation, as shown in figure 8-58.

Lubricate parts

Lubricate the metal surfaces of the lifters and lifter guides.

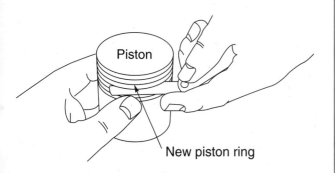

Fig. 8-54 Check for piston wear with new rings installed.

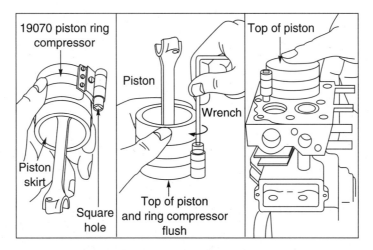

Fig. 8-55 Use ring compressor for proper piston installation.

Basic model series	Idle speed	Armature air gap		Valve clearance		Valve guide reject gage	Torque specifications		
		Two leg	Three leg	Intake	Exhaust		Flywheel nut ft. lbs.	Cylinder head in. lbs.	Conn rod in. lbs.
60000, 6B	1750	.006 .010	.012 .016	.006 .007	.007 .009	19122	55	140	100
80000, 81000 82000, 8B	1750	.006 .010	.012 .016	.005 .007	.007 .009	19122	55	140	100
90000, 91000 92000, 93000 94000, 95000	1750	.006 .010		.005 .007	.007 .009	19122	55	140	100
100200, 100900	1750	.010 .014	.012 .016	.005 .007	.007 .009	19122	60	140	100
100700	1750	.006 .010		.005 .007	.007 .009	19122	55	140	100
110000	1750	.006 .010		.005 .007	.007 .009	19122	55	140	100

Fig. 8-56 Critical torques required by manufacturer.

Insert lifters

Place the valve lifters into their proper holes, noting that the identifying mark made during the disassembly was placed on the exhaust lifter and that the exhaust side can be identified by the exit port which has the threads or mounting hardware for the muffler.

INSTALL CAMSHAFT
Lubricate camshaft

Lubricate all the camshaft surfaces and cylinder camshaft bushings.

Align timing marks

The camshaft, which operates the valves, must be coordinated with the rotation of the crankshaft. Turn the crankshaft so that its timing mark, located on the gear, can be aligned with the timing mark of the camshaft gear, as shown in figure 8-59.

NOTE: IF THERE ARE NO MARKS ON THE CRANKSHAFT GEAR, THE TIMING MARK IS USUALLY IN LINE WITH THE KEYWAY.

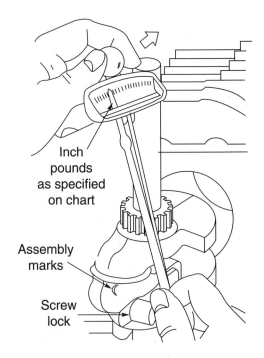

Inch pounds as specified on chart

Assembly marks

Screw lock

Fig. 8-57 Torque the rod nuts.

INSTALL REMAINING CRANKCASE PARTS

After you put in the camshaft and the lifters, assemble the remaining parts located inside the crankcase. The oil slinger is the only part remaining for the small Briggs & Stratton engine and fits on the camshaft with the slinger gears meshed into the camshaft. When the crankcase cover is applied, the slinger will be aligned automatically, as shown in figure 8-60.

INSTALL CRANKCASE GASKET AND COVER

Apply the crankcase cover gasket supplied in the gasket set. Some gasket sets provide many synthetic or fiber sheets of different thickness and color to properly adjust the crankshaft end play. The gasket set for the small Briggs & Stratton engine has four crankcase gaskets with the thicknesses of 0.005-inch, 0.009-inch and two at 0.015-inch, as shown in figure 8-61. At least one gray-colored gasket (0.015-inch) must be used. Additional gaskets are added to adjust the endplay, but never less than one 0.015-inch gray gasket, because the metal surfaces between the crankcase and cover will not be sealed properly; and small oil leaks will be present when the engine is operated.

Fig. 8-58 Invert engine when installing valve lifters.

Rotating the crankshaft slightly will make the installation of the crankcase cover easier. Insert the crankcase bolts and torque them to 90 inch-pounds in increments of 30 inch-pounds, using a crossing sequence until the final torque is obtained.

CHECK CRANKSHAFT ENDPLAY FOR PROPER SPECIFICATION

The crankshaft endplay on an engine is the side-to-side movement of the crankshaft after the crankcase cover has been installed and torqued. The amount of endplay may be measured, as shown in figure 8-62. A dial indicator inserted in a holder can be used to register the measurement, as shown in figure 8-63. The dial indicator, which uses a dial face, is calibrated in thousandths of an inch, and the movable contact arm should touch the crankcase when the gage is mounted on the crankshaft.

The acceptable range will vary, depending on the engine's manufacturer, but for the small Briggs & Stratton engine the endplay must be from 0.002-inch–0.008-inch, as shown in figure 8-64. Too little endplay will produce excessive pressure and wear

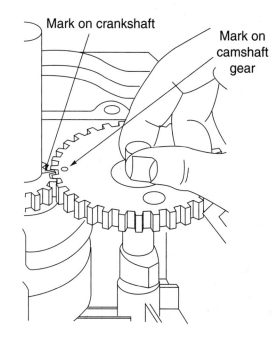

Fig. 8-59 Align the timing marks on the crankshaft and camshaft.

between the crankshaft and cylinder bearing surfaces; excessive endplay will cause piston rod damage.

Many times the measurement is greater than the maximum specification, and the temptation is to remove the 0.015-inch gasket and put in a thinner one. This will reduce the endplay, but it may result in an engine oil leak. The proper procedure to correct the endplay is to purchase a shim for the engine and place it inside the crankcase on the crankshaft, as shown in figure 8-65 and figure 8-66. The thrust

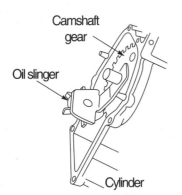

Fig. 8-60 Install oil slinger.

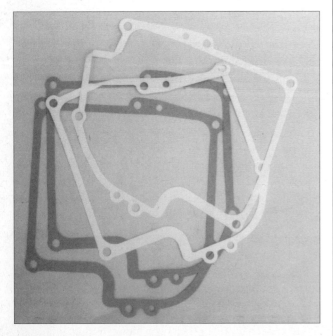

Fig. 8-61 Different gaskets for endplay adjustment.

washer will remove much or all of the crankshaft endplay, so an additional gasket or gaskets must be added to the original gray one (0.015-inch).

An estimation of the thickness of the additional gasket can be made by using the following method: Imagine an Briggs & Stratton small engine that has 0.012-inch endplay which is out-of-spec. If only the thrust washer/spacer is added, the endplay would disappear altogether, so an additional gasket is needed.

Endplay with gray gasket (0.015")	+0.012"	Out-of-spec too much
Thrush washer spacer added	-0.015"	Takes up the space
Subtotal	-0.003"	Too tight
Additional gasket	+0.009"	Adds more space
Subtotal	+0.006"	OK

Recheck the endplay to verify your estimate. If the movement is within specification, rotate the crankshaft completely for two complete revolutions, correcting any binding or unusual noises.

APPLY OIL SEALS TO CRANKSHAFT

Oil seals are used to prevent the lubricant from leaking out of the crankcase and to prevent dirt from entering. The oil seal is usually damaged at the time of removal and should be replaced with each engine rebuild. Whenever a rotating part moves through a stationary part, it is necessary to provide a seal. Basically, all seals consist of a sleeve or wide ring,

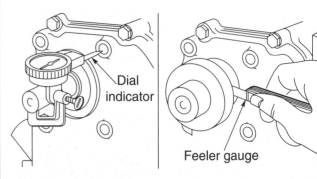

Fig. 8-62 Check crankshaft endplay.

Fig. 8-63 Dial indicator to measure endplay.

which is fitted carefully into the crankcase, and a pliable wiper lip, which is bonded to the sleeve and held against the rotating part by spring pressure, as shown in figure 8-67. Most seals have one pliable wiper lip at one face. It is important that the sleeve is installed in the same direction as the original seal, most commonly with the pliable wiper lip towards the crankcase.

Lubricate oil seals

The seal should be lubricated around the wiper lip before it is installed.

Position with seal protector

A seal protector should be used, as shown in figure 8-68, which can be purchased or made simply by wrapping waxed paper around the crankshaft. Sliding the seal over the crankshaft without the protectors may create unnoticeable rips in the pliable wiper lip that will be followed by an oil leakage problem when crankcase oil is added and the engine

Check chart

All popular engine models

4. Top governed speed: see Briggs & Stratton service bulletin no. 467 engine replacement data

5. Crankshaft end play: .002–.008" except model series 100700 and 120000

Note: on model series 100700 and 120000 crankshaft end play is .002–.030".

Fig. 8-64 Briggs & Stratton endplay specifications.

is operated. After the seal is slid over the crankshaft, the plastic protector may be removed and the seal may be inserted into the flange of the crankcase cover.

Press in seal

A deep socket or special tool may be used to evenly distribute the pressure as the seal is tapped gently in, as shown in figure 8-69. The final seating of the oil seal to the appropriate depth, as shown in figure 8-70, can be accomplished by tapping lightly with a hand punch all the way around the seal. Install all the seals in this manner.

INSTALL VALVES AND VALVE SPRINGS

Install the intake and exhaust valve with the correct spring and retainer using the following steps.

Place the piston at TDC

It is necessary to have the piston at TDC of the compression stroke when adjusting the valve stem clearance. The easiest method to find this position is to place the valves into the valve guides without attaching the springs. Revolve the crankshaft in its normal operating rotation (clockwise from the flywheel side) while watching the valves opening and closing. After the intake stroke (piston moving downward with intake valve open), the piston rises in the cylinder for the compression stroke, and both valves are completely closed when the piston reaches the top of its travel. It is recommended that the

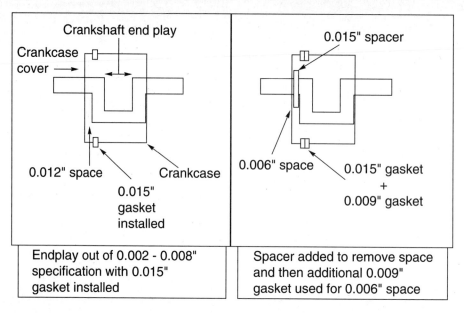

Endplay out of 0.002 - 0.008" specification with 0.015" gasket installed

Spacer added to remove space and then additional 0.009" gasket used for 0.006" space

Fig. 8-65 Proper endplay adjustment.

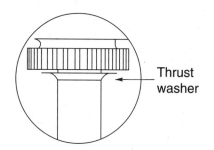

222949 Thrust washer for 1.181" (30.00 mm) dia. crksft.
222951 Thrust washer for 1.367" (35.00 mm) dia. crksft.

Fig. 8-66 Spacer position on crankshaft.

piston be rotated additionally so that it is descending on the power stroke by a quarter-inch (past TDC). This is done to verify that the correct position has been achieved because, if the piston is in the wrong position, the intake would start to open. It is also done to affirm that the compression release system has been bypassed.

Look up valve clearance specification

The clearance specification should be obtained from the manufacturer's manual for the engine, as

shown in figure 8-71. The valve stem to valve lifter clearance must be within the specifications. When the engine is operating, the valve train parts will expand enough (from the engine's heat) to reduce most of the valve clearance so that the engine will operate with minimum valve train noise. If the valve stem to valve lifter clearance is too small, the valves will not seat properly during operation and will cause a loss of compression and premature valve failure.

If the valve stem to valve lifter distance is too large, a tapping sound will be heard when the

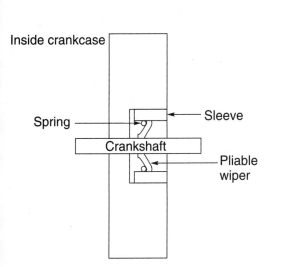

Fig. 8-67 Oil seal parts and position.

Inside crankcase

Spring — Sleeve

Crankshaft

Pliable wiper

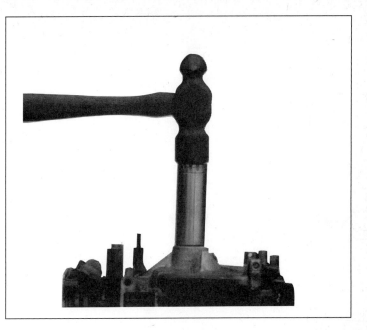

Fig. 8-69 Deep socket used to install oil seals.

engine is running, and the engine's compression release system will not operate. Most engines are designed so that the exhaust valve will open slightly during the compression stroke to allow some of the gases to escape, which will allow less effort in starting the engine. Too much valve gap will make this system inoperable.

Check valve stem to valve lifter clearance

Valve clearance between the valve and valve lifter will probably be too narrow because, when the

valve seat and face were cut, the valve descended lower into the crankcase and the prior clearance was changed. Since the Briggs & Stratton specification for the engine's intake valve is between the range of 0.005-inch–0.007-inch, the 0.005-inch feeler gage is used. The lowest measurement is used so that if too much metal is ground off, the gap will still be within the specification range.

Measure the clearance before the valve springs are installed by placing the feeler gage between the valve lifter and the valve stem. Press down on the valve top so that the feeler gage is sandwiched between the valve parts. While pressing down, pull

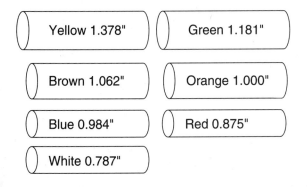

Yellow 1.378"

Green 1.181"

Brown 1.062"

Orange 1.000"

Blue 0.984"

Red 0.875"

White 0.787"

Fig. 8-68 Oil seal protector set.(B&S #19334).

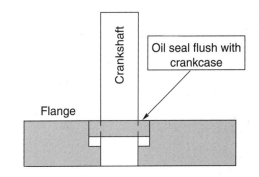

Crankshaft

Oil seal flush with crankcase

Flange

Fig. 8-70 Oil seal should be flush with the crankcase.

Basic model series	Idle speed	Armature air gap		Valve clearance		Valve guide reject gage	Torque specifications		
		Two leg	Three leg	Intake	Exhaust		Flywheel nut ft. lbs.	Cylinder head in. lbs.	Conn rod in. lbs.
60000, 6B	1750	.006 .010	.012 .016	.006 .007	.007 .009	19122	55	140	100
80000, 81000 82000, 8B	1750	.006 .010	.012 .016	.005 .007	.007 .009	19122	55	140	100
90000, 91000 92000, 93000 94000, 95000	1750	.006 .010		.005 .007	.007 .009	19122	55	140	100
100200, 100900	1750	.010 .014	.012 .016	.005 .007	.007 .009	19122	60	140	100
100700	1750	.006 .010		.005 .007	.007 .009	19122	55	140	100
110000	1750	.006 .010		.005 .007	.007 .009	19122	55	140	100

Fig. 8-71 Valve clearance specification.

Fig. 8-72 Press down on valve when checking clearance.

Fig. 8-73 Use a grinder and a proper holder to take metal off squarely when adjusting valves.

out the feeler gage. There should be a slight friction between the valve train and the gage, but there should be no "clicking" sound as the gage is finally removed, as shown in figure 8-72. If the valve clearance is too small, metal must be taken off from the bottom of the valve stem and then the clearance rechecked. It is important to use a grinder and holder device, as shown in figure 8-73, that will take off the metal squarely from the bottom of the valve. If too much metal is taken off the valve, use the feeler gage with the highest allowable gap. If the gap is still too wide, then the valve seat or valve face can be cut

so as to lower the valve in the cylinder and lessen the valve gap.

Lubricate the valve stems

An "antiseize" compound should be applied to the valve stems before assembly to prevent any future restriction in movement of the valve, as shown in figure 8-74.

Fig. 8-74 An "antiseize" compound should be applied to the valve stem.

Put in valve springs and retainer clip

If the valve springs differ, the spring that provides more tension is used on the exhaust valve. The valves are held in the engine with a spring and retaining devices, as shown in figure 8-75. A tool may be used to compress the valve spring so that the pin, collar, or retainer may be installed, as shown in figure 8-76.

Put valve cover on

After the valve springs are installed, the valve cover can be assembled. The valve cover on the Briggs & Stratton engine also serves as a breather valve for the engine. The valve allows the creation of a crankcase vacuum during operation, which provides better oil control and performance. The fiber disk in the breather valve must operate properly. It can be checked by using a paper clip that is made of metal wire with the diameter of about 0.045-inch. This is the maximum clearance that should be between the fiber and the breather body, as shown in figure 8-77. When installing the valve cover, use a new gasket. The use of the old gasket with a liquid sealer can plug the oil return holes.

INSTALL CYLINDER HEAD AND GASKET

Install the cylinder head along with a new cylinder head gasket using the following steps.

Prepare cylinder head

It is important to check the gasket surface area of the head to determine whether it is warped or distorted. An easy method is to attach a piece of 300+ grit sand paper on a flat surface, using masking tape or rubber cement. If a metal surface plate is not available, an economical substitute is a sheet of 3/8-inch thick or greater Plexiglass, as shown in figure 8-78 and 8-79. Place the gasket surface of the cylinder head against the sand paper and, with a light pressure, move the head in a circular or "figure eight" motion. Continue this for about 10 seconds and then examine the etched marks on the surface of the cylinder head. If the gasket area is not completely sanded, there are areas that are not the same level. Continue sanding until all the gasket area is etched with the sand paper. Aluminum is soft enough so that the high spots will be honed. Clean the head in soap and water before assembling.

Lubricate head bolts with "antiseize"

The head bolts should be lubricated not only on the threads, but more importantly on the under-surface of the bolt head. The bolts, when torqued to the proper value, will not loosen because of the oiling, which will keep the fasteners from rusting and seizing in the engine.

Torque head bolts

Check for the proper torque specification, as shown in figure 8-80. Each manufacturer publishes the sequence that the head bolts should be tightened, as shown in figure 8-81. The proper method for torquing a sequence of bolts should be followed. For example, since the final torque value for the Briggs & Stratton engine will be 140 inch-pounds, the final value is divided by 3 to approximately 50 inch-pounds. The tightening sequence is followed to tighten all the head bolts to 50 inch-pounds. The next increment would be 50 additional inch-pounds to a subtotal of 100 inch-pounds. The sequence would again be followed at this new level. The steps will be repeated for the third time, finally reaching the required torque.

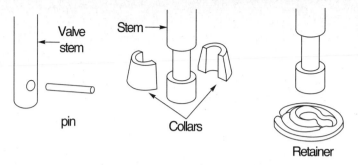

Fig. 8-75 Valves are held in the engine with different retaining devices.

Install the spark plugs

The last step involves installing the correct, new spark plug to prevent any contaminants from entering the assembled cylinder. The spark plug can be installed finger tight and torqued to 15 ft.-lbs. after the ignition parts are installed and tested. Set the spark plug gap to the specification listed in the engine repair book. The plug gap used for the small Briggs and Stratton engine is 0.030-inch.

INSTALL IGNITION PARTS

The points and condenser or the solid state ignition components should be installed.

Points and condenser ignition

Refer to the ignition chapter for proper assembly steps. Remember to set the ignition timing after the points are installed. Some systems require only the correct adjustment of the points to obtain the correct ignition timing.

When installing the electrical components, it is important that any future problems be anticipated, e.g., loose wires or wires that may be severed by rubbing on metal engine parts.

Solid state ignition

When installing the solid state modules, be careful not to overtighten the mounting screws; most can be applied properly by torquing them to 35 inch-pounds.

Fig. 8-76 Valve spring installation tool (B&S #19063).

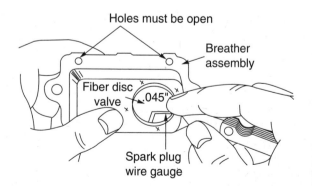

Fig. 8-77 The fiber disk in the breather valve can be checked with a paper clip.

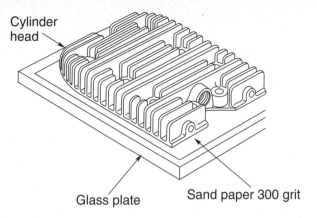

Cylinder head

Glass plate

Sand paper 300 grit

Fig. 8-78 An economical substitute is a sheet of 3/8" thick or greater Plexiglass.

INSTALL FLYWHEEL

Rotate the crankshaft so that the keyway is easily accessible. Wipe the contact area of the flywheel and the crankshaft with a small amount of lubricant to prevent any future rust formation. Place the flywheel on the crankshaft and then insert the key into the flywheel. Place the washer provided (with the curved side away from the engine) on the flywheel and install the flywheel nut.

Look up the flywheel nut torque in the specification chart, as shown in figure 8-82. Using a torque

Fig. 8-79 Observe sanding pattern.

wrench and a strap wrench, tighten the flywheel nut to the specification value, as shown in figure 8-83.

INSTALL AND ADJUST COIL/ARMATURE AIR GAP

Only an coil/armature that is found outside of the flywheel may be adjusted. Refer to the specification

Basic model series	Idle speed	Armature air gap		Valve clearance		Valve guide reject gage	Torque specifications		
		Two leg	Three leg	Intake	Exhaust		Flywheel nut ft. lbs.	Cylinder head in. lbs.	Conn rod in. lbs.
60000, 6B	1750	.006 .010	.012 .016	.006 .007	.007 .009	19122	55	140	100
80000, 81000 82000, 8B	1750	.006 .010	.012 .016	.005 .007	.007 .009	19122	55	140	100
90000, 91000 92000, 93000 94000, 95000	1750	.006 .010		.005 .007	.007 .009	19122	55	140	100
100200, 100900	1750	.010 .014	.012 .016	.005 .007	.007 .009	19122	60	140	100
100700	1750	.006 .010		.005 .007	.007 .009	19122	55	140	100
110000	1750	.006 .010		.005 .007	.007 .009	19122	55	140	100

Fig. 8-80 Cylinder head torque specification.

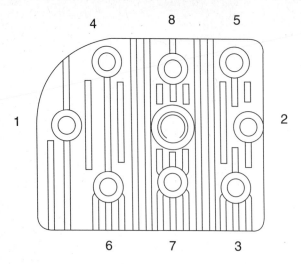

Fig. 8-81 The proper sequence for tightening cylinder head bolts.

tion. Lightweight flywheels have weights added to counterbalance the mass of the magnets and are often mistaken for the magnets.

Preset coil

Loosen the coil's mounting screws and pull the coil the maximum distance from the flywheel. Lightly tighten the fasteners while holding the coil in this position.

Insert measuring tool

Rotate the flywheel magnets so that they are directly below the legs of the coil and insert a gage (old microfiche or feeler gage) between the coil's legs and the magnets.

Establish final adjustment

Loosen the coil's mounting screws. The magnets will pull the coil to the flywheel, but the distance will be limited by the gage that is inserted. Tighten the screws again (35 inch-pounds) while the gage is inserted and then rotate the flywheel so the gage can be removed.

Check for proper spark

The ignition system can be checked for spark at this point. It is easier to correct any problems before

chart for the air gap range, as shown in figure 8-84. A stronger spark will be obtained by setting the gap to the lowest value in the specification range.

Identify the magnets

Rotate the flywheel so that its magnets are not aligned with the legs of the coil. Place a screwdriver near the magnets and observe the magnetic attrac-

Basic model series	Idle speed	Armature air gap		Valve clearance		Valve guide reject gage	Torque specifications		
		Two leg	Three leg	Intake	Exhaust		Flywheel nut ft. lbs.	Cylinder head in. lbs.	Conn rod in. lbs.
60000, 6B	1750	.006 .010	.012 .016	.006 .007	.007 .009	19122	55	140	100
80000, 81000 82000, 8B	1750	.006 .010	.012 .016	.005 .007	.007 .009	19122	55	140	100
90000, 91000 92000, 93000 94000, 95000	1750	.006 .010		.005 .007	.007 .009	19122	55	140	100
100200, 100900	1750	.010 .014	.012 .016	.005 .007	.007 .009	19122	60	140	100
100700	1750	.006 .010		.005 .007	.007 .009	19122	55	140	100
110000	1750	.006 .010		.005 .007	.007 .009	19122	55	140	100

Fig. 8-82 Flywheel nut torque specification.

Fig. 8-83 Use a strap wrench and proper tool with a torque wrench.

the remaining engine pieces are assembled. Normally the flywheel can be carefully spun fast enough by hand to generate a spark through the

spark tester. If no spark is apparent, refer to the ignition or troubleshooting chapter for the correct testing procedure.

INSTALL SHEET METAL PARTS AND ROUTE WIRES

Any sheet metal parts that direct the cooling air over the cylinder fins should be installed. Since the engines are painted after they are completely assembled, the "shadows" of bare metal will expose any missing parts.

The ignition wire must be routed properly to the stop-switch. The wire from the coil is used to ground the primary circuit to the engine frame so that the engine will stop, as shown in figure 8-85. Make sure that it will not cross any sharp edges that may cause the insulation to wear.

INSTALL CARBURETOR
Clean carburetor parts

Refer to the carburetor chapter for in-depth carburetor rebuilding. The carburetor should be cleaned and the passages cleared with compressed air.

Basic model series	Idle speed	Armature air gap		Valve clearance		Valve guide reject gage	Torque specifications		
		Two leg	Three leg	Intake	Exhaust		Flywheel nut ft. lbs.	Cylinder head in. lbs.	Conn rod in. lbs.
60000, 6B	1750	.006 .010	.012 .016	.006 .007	.007 .009	19122	55	140	100
80000, 81000 82000, 8B	1750	.006 .010	.012 .016	.005 .007	.007 .009	19122	55	140	100
90000, 91000 92000, 93000 94000, 95000	1750	.006 .010		.005 .007	.007 .009	19122	55	140	100
100200, 100900	1750	.010 .014	.012 .016	.005 .007	.007 .009	19122	60	140	100
100700	1750	.006 .010		.005 .007	.007 .009	19122	55	140	100
110000	1750	.006 .010		.005 .007	.007 .009	19122	55	140	100

Fig. 8-84 Air Gap Range specification.

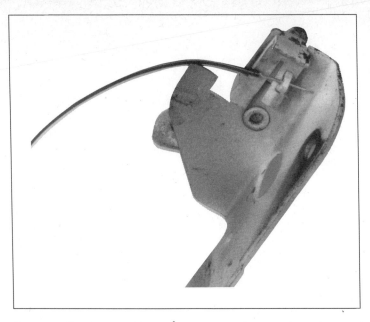

Fig. 8-85 The ignition wire must be routed properly to the stop switch.

Assemble carburetor

Assemble the carburetor using the new parts ordered. Review all the precautions and procedures that pertain to the carburetor.

Preadjust carburetor

Preset the adjusting screws according to the engine's service manual.

Attach governor controls

Refer to the governor chapter for the theory of the solid link attachment and the spring connection. The sketch made during the disassembly will help in the correct assembly, or refer to the governor section in the engine's repair manual, as shown in figure 8-86. In the small Briggs & Stratton engine, the governor parts are installed when the carburetor is mounted to the engine. The governor link is already attached to the air vane governor, so the other end of the link is inserted into the throttle plate. Notice that there is a solid link between the throttle and the governor vane and a spring link between the air vane and the speed control.

Install all the bolts that hold the carburetor to the engine, but do not tighten them to the final torque until all have been partially bolted.

NOTE: CARE MUST BE USED WHEN INSTALLING GOVERNOR LINKS SO THAT THEY ARE NOT BENT FROM THEIR ORIGINAL SHAPE.

REBUILD REWIND STARTER

Refer to the engine's repair manual for the steps in rebuilding the rewind starter. The starter should be cleaned and lubricated before it is assembled. A lubricant is silicone spray, which will not attract dust and dirt like other petroleum-base lubricants.

INSTALL BLOWER HOUSING

After any cooling, air filter screens have been installed. The blower housing can then be attached by starting all the fasteners in their proper positions before the final tightening is done. Since the bolts are low carbon steel with a 1/4-inch–20 pitch, the final torque should be 90 inch-pounds. This is not one of the mandatory torques, but it is advisable to torque as many fasteners as possible. After the housing is in place, inspect the governor controls for any binding or rubbing against the housing cover. A throttle, stuck in the wide open position, can have a disastrous effect on a rebuilt engine. Also check for any wires that may have been pinched between two pieces of metal.

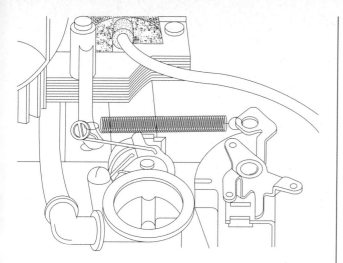

Fig. 8-86 Attach governor controls.

INSTALL MUFFLER

Apply an antiseize compound to the muffler threads or bolts to insure easy future removal.

PAINT ENGINE (OPTIONAL)

If it is desirable to paint the engine after assembly, perform the tasks described next:

Sand engine

Remove all the loose paint and rough up any glossy surface paint to allow proper adhesion of the new paint.

NOTE: DO NOT USE A BEAD BLASTER TO REMOVE ANY OF THE PAINT.

Degrease engine

Degrease the entire engine with rubbing alcohol or commercial surface preparation solution. Do not handle the engine with your hands after it has been degreased.

Suspend engine

Hook the engine and suspend it from a rope in the area where it will be painted. This will allow the engine to be rotated during the final spraying process.

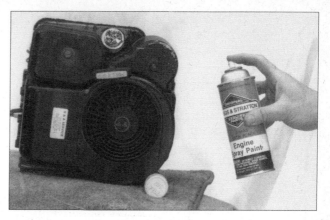

Fig. 8-87 Paint engine after assembly.

Spray primer coat

Apply one light coat to the engine and allow to dry for 20–30 minutes.

Spray tack coat

Follow the directions on the paint container and spray a thin tack coat over the complete engine.

Spray final coat

After a few minutes when the tack coat is sticky, apply the final coat to the complete engine sufficiently to "wet" the complete surface so that all parts are covered and there are no areas with excess paint, as shown in figure 8-87. If any spot has excess paint, wipe it off with a clean cloth immediately and reapply a "wet" coat to the area.

MOUNT ENGINE

The engine should be mounted on the machinery that it will activate. The best final adjustment can be achieved when the engine is in its normal position and operating with a load. If it is not possible to mount the engine to the application, a test stand can be made, as shown in figure 8-88, that will safely allow the use of a blade, necessary on some applications to start the engine, or a flywheel plate added to the PTO-side of the crankshaft, as shown in figures 8-89 and 8-90. Frequently, the blade is the flywheel of the engine and, consequently, without a flywheel, the engine will not start and may cause a kickback if an attempt is made to start it, causing damage to the engine.

Sharpen blade

If the blade is present, sharpen it, as shown in figure 8-91, and balance it, as shown in figures 8-92 and 8-93, before installation. When balancing the blade, remove additional metal from the heavy side.

Add oil

Place a sufficient amount of oil in the crankcase. Use the oil viscosity recommended by the engine manufacturer (usually SAE 30). It is advisable to change this oil after only 30 minutes of operation so that any dirt remaining from the assembly process will be suspended in the oil and removed before it causes internal damage.

Add fresh fuel

Fresh fuel must be used when starting the engine. It may be advisable to mix in some two-stroke-cycle oil with the first tank of fuel. This oil/gas mixture will provide additional lubrication during the initial break in period when friction levels are greatest.

START AND ADJUST THE ENGINE

The engine should be started and warmed up before any final adjustments are made to the carburetor.

Warm up the engine

After the engine is started, run it at moderate speed for about five minutes. During this initial time period, vary the speed of the engine from idle to mid-speed. The accelerations and decelerations are necessary for the initial break-in of the metal parts.

Adjust the carburetor

After the engine has warmed up (five minutes) adjust the carburetor according to the engine repair manual.

Fig. 8-88 Engine test stand.

Set high and idle speed

A vibra-tachometer, as shown in figure 8-94, should be used to accurately adjust the high speed governor setting for safe operation. Many times the length of the mower blade determines the maximum safe governed speed. The chart, shown in figure 8-95, is used to prevent the blade tip speed exceeding 19,000 feet per second, which could lead to the blade breaking. The longer the blade length, the slower the engine must be operated. The idle speed must be set properly so that the engine will accelerate smoothly. The settings for the fictitious 3.5 HP Briggs & Stratton engine is limited to 3100 RPM for the top speed, not because of the blade length, but because if the speed is greater, the self-propelled unit would move the mower too fast. The idle is set at 1750 RPM.

Break-in period

The break-in period can last for about 10 hours of operation. During this stage, the metal surfaces will wear against each other, causing the high and low spots to disappear, and a good sealing of the rings to the cylinder will be achieved. The engine may

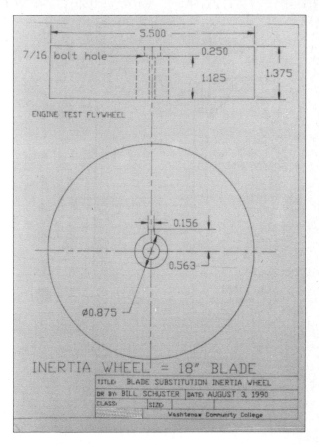

Fig. 8-90 Inertia wheel (substitute flywheel).

Fig. 8-89 Flywheel simulator attached.

Fig. 8-91 Sharpen the blade.

Chapter 8, Engine Assembly **273**

Fig. 8-92 Balance the blade.

Fig. 8-93 Balance the blade.

smoke and use excess oil until this phase has finished, so it is advisable to check the oil level often. The engine will break in faster if the engine is operated at normal speed with a moderate load for the time period. An example of a poor break in procedure would be to operate an engine at one speed with no load for a long period of time, such as operating a lawnmower on a cement surface. It would be

Blade length	Recommended maximum rotational r.p.m.
18"	3800
19"	3600
20"	3400
21"	3250
22"	3100
23"	2950
24"	2800
25"	2700

Fig. 8-95 Safe blade speed chart.

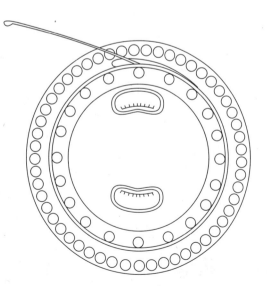

Fig. 8-94 Vibra-Tach, used for speed indication.

better to operate the engine at maximum governed speed with a light load, such as normal cutting of short grass.

It is recommended that the engine oil be drained after about one to two hours of operation to remove any contaminates remaining from the assembly process.

The cylinder head torque should be checked after the engine has operated at least 1/2 hour and after it has cooled.

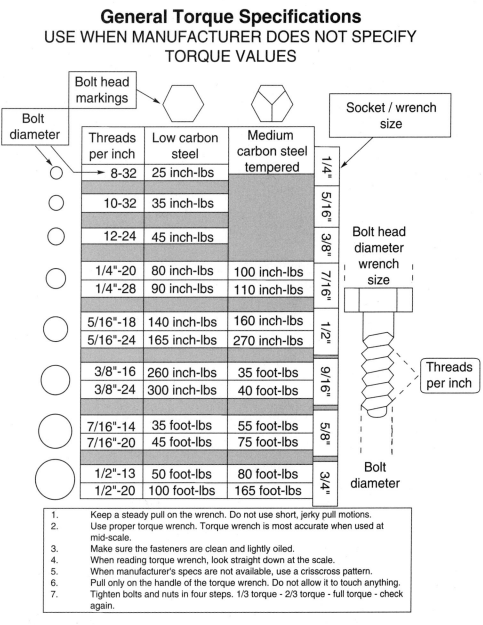

General Torque Specifications
USE WHEN MANUFACTURER DOES NOT SPECIFY TORQUE VALUES

Threads per inch	Low carbon steel	Medium carbon steel tempered	Socket / wrench size
8-32	25 inch-lbs		1/4"
10-32	35 inch-lbs		5/16"
12-24	45 inch-lbs		3/8"
1/4"-20	80 inch-lbs	100 inch-lbs	7/16"
1/4"-28	90 inch-lbs	110 inch-lbs	7/16"
5/16"-18	140 inch-lbs	160 inch-lbs	1/2"
5/16"-24	165 inch-lbs	270 inch-lbs	1/2"
3/8"-16	260 inch-lbs	35 foot-lbs	9/16"
3/8"-24	300 inch-lbs	40 foot-lbs	9/16"
7/16"-14	35 foot-lbs	55 foot-lbs	5/8"
7/16"-20	45 foot-lbs	75 foot-lbs	5/8"
1/2"-13	50 foot-lbs	80 foot-lbs	3/4"
1/2"-20	100 foot-lbs	165 foot-lbs	3/4"

1. Keep a steady pull on the wrench. Do not use short, jerky pull motions.
2. Use proper torque wrench. Torque wrench is most accurate when used at mid-scale.
3. Make sure the fasteners are clean and lightly oiled.
4. When reading torque wrench, look straight down at the scale.
5. When manufacturer's specs are not available, use a crisscross pattern.
6. Pull only on the handle of the torque wrench. Do not allow it to touch anything.
7. Tighten bolts and nuts in four steps. 1/3 torque - 2/3 torque - full torque - check again.

Fig. 8-96 General Torque specifications.

SUMMARY

The small 3.5 horsepower air-cooled engine is rarely rebuilt for a customer because the cost of the labor and parts are more than the cost of a short-block or new engine, but for the hobbyist and student, it provides a good value. Parts can be obtained from an authorized dealer or distributor in your area. Dealers for the engine type you are overhauling advertise in the phone book's yellow pages under the section "Engines–Gasoline."

Using the step-by-step sequence described in this chapter, a successful engine rebuild will result. The assembly area must be clean of all dirt and the engine components must be washed to avoid the possibility of dirt contaminating the rebuilt engine.

Questions

1. **Engine model numbers are helpful in obtaining the correct parts. Technician A says that for the Briggs & Stratton engine, all that is needed to order a gasket set is the model number. Technician B says that for a Briggs & Stratton engine, the model, type, and code number are needed to order a crankshaft. Who is correct?**

   A. Only Technician A
   B. Only Technician B
   C. Both Technician A and B
   D. Neither Technician A or B

2. **When rebuilding an engine, some parts need to be replaced only when they are defective. Technician A says that some of these parts would include the oil seals, piston rings, and gaskets. Technician B says some of these parts would be the air cleaner, piston, and connecting rod. Who is correct?**

   A. Only Technician A
   B. Only Technician B
   C. Both Technician A and B
   D. Neither Technician A or B

3. **Valve service is necessary during the rebuild process. Technician A says that the order to follow is (1) cut valve seats, (2) cut valve face, and (3) recondition valve guides. Technician B says the order to follow should be (1) lap valves, (2) cut valve face, and (3) cut valve seat. Who is correct?**

   A. Only Technician A
   B. Only Technician B
   C. Both Technician A and B
   D. Neither Technician A or B

4. **Technician A says that most valve seats will be cut to a 46° angle. Technician B says that most valves seats are cut to 46°, but some will have a 31° angle. Who is correct?**

   A. Only Technician A
   B. Only Technician B
   C. Both Technician A and B
   D. Neither Technician A or B

5. **Technician A states that "as long as the piston rings fit into the grooves, it does not matter which side is the top or bottom." Technician B states that "the general rule for new piston rings is that any outside notches should be facing downward and be installed on the middle ring groove." Who is correct?**

   A. Only Technician A
   B. Only Technician B
   C. Both Technician A and B
   D. Neither Technician A or B

6. It is important that the engine be assembled in a clean area. Technician A says that it is best to wash the parts closely to the time they will be assembled. Technician B says that all the parts should be covered with grease after they are washed so they will not rust.
   Who is correct?

   A. Only Technician A
   B. Only Technician B
   C. Both Technician A and B
   D. Neither Technician A or B

7. The fasteners used on the engine are designed so that parts will not separate under normal operating conditions. Technician A says that after each fastener is torqued down, it should be further turned 1/8- to 1/4-turn with a box end wrench. Technician B says that it is necessary to use a torque wrench on every engine bolt or nut.
   Who is correct?

   A. Only Technician A
   B. Only Technician B
   C. Both Technician A and B
   D. Neither Technician A or B

8. The crankshaft endplay for the Briggs & Stratton engine should be between 0.002-inch–0.008-inch. If the crankcase endplay is less than the lowest specification, Technician A says to add a gasket. Technician B says to remove a gasket or put a thinner one in.
   Who is correct?

   A. Only Technician A
   B. Only Technician B
   C. Both Technician A and B
   D. Neither Technician A or B

9. The cylinder head should be prepared before it is installed. Technician A says to scrape the gasket area with a screwdriver and apply a gasket sealer to the gasket area. Technician B says to sand the gasket surface, wash it, and install on the engine with a new cylinder head.
   Who is correct?

   A. Only Technician A
   B. Only Technician B
   C. Both Technician A and B
   D. Neither Technician A or B

10. The following statements are made about an engine after it has been assembled. Technician A says to change the oil only after about a half-hour of operation so that any suspended contaminations from the assembly process are removed before excessive wear occurs. Technician B says to check the cylinder head torque after a short time of operation, but only after the head has cooled.
    Who is correct?

    A. Only Technician A
    B. Only Technician B
    C. Both Technician A and B
    D. Neither Technician A or B

CHAPTER 9

Engine Maintenance

INTRODUCTION

Engines are built to operate from 300 to 2500 hours before an overhaul. This can range from 6 to 20 years when the equipment is used approximately 50 hours a year. The difference depends upon the design and materials used in the manufacturing process. The two factors that affect longevity are cylinder construction, (e.g., aluminum cylinders wear faster than cast-iron cylinders), and how efficiently engine heat is dissipated.

The type of engine maintenance helps determine the maximum life of the engine. A good maintenance program includes keeping the engine clean and well-oiled to reduce friction and allow adequate heat transfer. In order to get the maximum operating hours from an engine, a proper maintenance schedule must be followed.

HOURS OF OPERATION

Many times, the automobile's mechanical life is rated by the number of miles possible before a major overhaul. The air-cooled engine's life is rated by the hours of operation. Maintenance schedules are set-up so that certain procedures are performed at intervals of operation hours. The intervals in hours of operation can be converted to days, weeks, or months by estimating the number of hours of use per day and the number of days of operation per week. It is easier to remember to change engine oil every month than it is to change it after every 25 hours of operation.

Typical consumer engine

The average mowing time for a homeowner is about two hours per week. Since the average mow-

ing season is six months, this can be converted into about 50 hours of operation for the year. An aluminum alloy cylinder engine that is rated for an operating life of 300 hours will operate for approximately six years with proper maintenance before a major overhaul is necessary.

Typical commercial engine

A commercial purchaser may use the engine 40 hours a week for a total of 1000 hours a year. For the commercial consumer, the aluminum alloy cylinder would not be an economical choice. The better engine would be one with industrial quality, such as a cast-iron cylinder, that could operate for 2500 hours and then have the economic capability to be rebuilt for an additional 2500 hours of operation. Cast-iron cylinders are either fully cast-iron or a cast-iron liner that is installed in an aluminum block casting.

The industrial/commercial (I/C) engine has a larger oil capacity, operates at a lower temperature, and has more efficient impellers on the flywheel to move a greater volume of air across the engine's cooling fins, as shown in figure 9-1. This translates to longer life. The valve train materials and construction are able to resist heat distortion, insuring proper valve seating. Many exhaust valves include a rotator that turns the valve slightly on its up and down movement to reduce carbon formation and engine failures, as shown in figure 9-2.

The I/C engine's air cleaner is commonly a two-piece construction using a combination of a paper automotive-type filter along with a lubricated foam filter which is used as a pre-cleaner. Since the engine intake of dirt is a great contributor to engine failure, the dual element cleaner promotes longer life.

Model Series 114900
4 HP I/C

Displacement 11.39 cu.in. (187 cc)

Bore 2.78 in. (70.6 mm)

Stroke 1.88 in. (47.6 mm)

Features:

- Cast iron sleeve bore for extended life

- Pulsa-Jet carburetor with all-temperature choke for easier starting under changing temperatures

- Maintenance free Magnetron® electronic ignition

- Dual element air cleaner with paper cartridge and foam pre-cleaner provides double air filtration

- Compression release for smooth, quick starting

- Mechanical governor for quick power response

Fig. 9-1 The Industrial/Commercial (I/C) engine has special qualities to prolong the engine life.

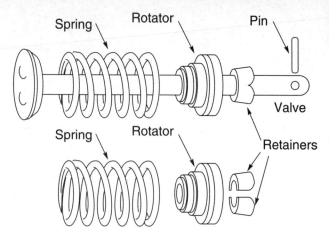

Fig. 9-2 Many I/C engines include a rotator on the exhaust valve that turns it slightly with every up and down movement.

AUTOMOBILE ENGINE VS. AIR-COOLED ENGINE

The average automobile engine is rated at about 100 horsepower and at approximately 5000 RPM, while the a typical air-cooled engine is rated at about 4 horsepower and at approximately 3600 RPM. The automobile needs only about 40 horsepower (40 percent of power) when cruising on a level road at about 55 MPH at 2200 RPM. At this level of operation, the stresses on the engine are minimal, and the cooling system is able to remove the "heat of operation" effectively. Engine life is long and maintenance is low because of the low power requirements.

The air-cooled engine, on the other hand, functions near the top RPM, and many times it is operating at about 80 percent maximum horsepower. The margin of unused power is much less and the heat of operation is greater. Greater precautions

must be taken to maintain the small air-cooled engine because of its power requirements and the dirty environment in which it is used.

PREVENTATIVE MAINTENANCE

Problems and breakdowns occur most frequently when the engine and equipment are not maintained properly. A rule in the business is that the probability of a mechanical failure is greatest when closest to the middle of a project or when you are at the maximum distance from the necessary tools. The old adage, "an ounce of prevention is worth a pound of cure," can be translated into, an hour spent on preventive maintenance can save many hours of downtime and prevent many problems.

Setting up a maintenance schedule can assure timely lubrication and repairs and also help establish an equipment record-keeping system, as shown in figure 9-3. A schedule should be formulated in accordance with the manufacturer's suggested intervals, which can be found in the owner's manual. A file that includes the owner's manual and the maintenance records should be made for each piece of equipment. This kind of system will help in ordering necessary maintenance supplies before they are needed.

DATE	OPER-ATOR	HOURS USED	CUMU-LATIVE HOURS	ADD	OIL CHANGED	AIR CLEANER SERVICED	REMARKS

EQUIPMENT USED ON _____

MODEL _____   TYPE _____ SERIAL NO. _____

TYPE OF EQUIPMENT _____ MAKE _____

Fig. 9-3 Setting up a maintenance schedule can assure timely lubrication and repairs and helps establish an equipment record.

DAILY MAINTENANCE

Before the daily operation of an engine, certain steps should be followed to detect and correct any problems:

Daily Maintenance

1. Check oil level

2. Replenish fuel supply

3. Clean air intake screen

4. Check air cleaner mounting

5. Check safety items

Check oil level

Before checking the oil level, clean the area where the oil level is checked so that debris and abrasive dirt do not have a chance to enter the engine. Any dirt that enters the crankcase offsets the benefits gained by maintaining a correct oil level.

Check the oil level and fill to the "full" mark, as shown in figure 9-4. An engine that operates with the oil at the maximum level position will be less susceptible to overheating than an engine with a low crankcase supply. One of the functions of the oil is to cool the engine. As the oil is splashed or pumped to the engine components, the engine heat is absorbed and returned to the crankcase. The heat is then dissipated to the metal crankcase, which radiates it to the atmosphere. The closer the oil level is to the "full" mark, the greater the amount of heat that can be absorbed by the oil.

Do not overfill the crankcase with oil. "If a little is good, a lot must be better" is **not** the slogan that applies in this situation. The engine oil level is calibrated so that it is high enough for the slinger,

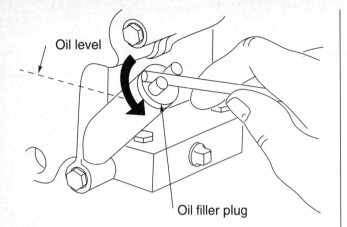

Oil level

Oil filler plug

Fig. 9-4 Check the oil level and fill to the "full" mark.

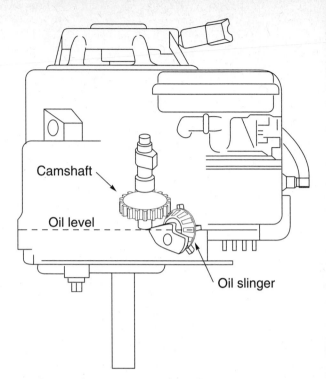

Camshaft

Oil level

Oil slinger

Fig. 9-5 The engine oil level is calibrated so the oil slinger can work properly in this Briggs & Stratton.

dipper, or oil pump to work properly and deliver oil to the internal components, as shown in figure 9-5. If the level of oil is too high, the piston rod assembly and crankshaft might dip into the reservoir and cause the oil to foam.

As the parts revolve, air may be beaten into the lubricant in the same way that cream is whipped. Unless the air bubbles escape from the lubricant as fast as they form, foaming occurs. Foaming causes an increase in volume that may allow the foam to be splashed to the load-carrying components instead of the oil. Foam is not an efficient load-carrying medium and will reduce the lubrication at crucial parts.

Replenish fuel supply

Use clean, fresh, lead-free gasoline that has no more than 10 percent alcohol (Ethanol). Remember: any alcohol in the fuel will remove corrosion from the inside of the fuel tank, and the particles will remain suspended in the fuel. The fuel filter should be changed more often when alcohol is blended with gasoline.

Examine the interior of the fuel tank for any signs of debris or water that may cause engine operating problems. Do not fill the fuel tank to the point of overflowing. Provide approximately 1/4-inch of tank space for fuel expansion.

NOTE: DO NOT FILL THE FUEL TANK WHEN THE ENGINE IS RUNNING OR WHEN IT IS HOT.

Clean air intake screen

The engine cooling is at 100 percent when the engine is operating at 3600 RPM and the air intake screen is free of adherent grass, leaves, etc. The screen is used to stop any large items from being injected into the cooling system which may plug the cooling fins around the cylinder. However, an air screen that is coated with enough debris so that air cannot get through will lead to marked overheating of the engine. Clean the air intake screen by wiping it or brushing it off, as shown in figure 9-6. Avoid directing a burst of compressed air at the area which may propel some debris into the engine.

Check air cleaner mounting

There must be no leaks around the air cleaner where it attaches to the carburetor. Check the fasteners that hold the air cleaner in place and make sure they are secure.

Air intake screen

Fig. 9-6 Clean the air intake screen with a brush or with compressed air.

Check safety items

All safety devices installed by the manufacturer must be working properly. Do not operate the engine with any switches or components bypassed.

MAINTENANCE EVERY 25 HOURS

This interval is approximately twice a mowing season for the consumer who uses the mower for an average of two hours a week. In a normal mowing season, the two best times to perform these operations are midway and at the close of the mowing season. When conditions are dirtier than normal, these procedures should be performed every twelve operating hours.

The end of the mowing season is recommended for maintenance, rather than the beginning, because when the oil is changed, all the foreign particles that are suspended in oil will not have time to settle to the bottom of the crankcase and adhere to the internal walls while the engine is stored. The oil will be warm after the last mowing, and all the dirt that has been suspended will flow easily from the crankcase. The first mowing of the year usually takes place while the grass is quickly growing, and the stresses to the engine are greatest because of the long blades of grass. There is usually not enough time to perform all the necessary preventive maintenance steps properly.

Maintenance every 25 hours

1. *Change oil*
2. *Inspect and clean air cleaner element*
3. *Perform daily maintenance steps*

Change oil

Operate the engine until it is warm so that the dirt particles are suspended and the oil will flow quicker. Remove the oil drain plug from the bottom of the engine and allow the old oil to flow into a container. Clean the drain plug and, after all the oil has been drained, insert and tighten it. Fill the crankcase immediately with the proper new oil and adjust the level so that it is full.

Oil can be drained from some engines through the fill tube attached to the crankcase. This method allows less contamination during the oil change because as the oil is drained out, a good deal of any external dirt around the fill tube is also flushed away. This cleans the area so that when the new oil is added, very little dirt will enter.

The plastic oil container that allows a direct pour into the filler tube is one of the best improvements in maintenance for the consumer. Before the new container was popular, the oil change process sometimes added more dirt to the engine than it removed. First, a dirty oil spigot may have been used to penetrate the top of a dusty can so that the oil could be poured into the engine. Second, the area around the oil filler opening may not have been cleaned and, as a soiled funnel was inserted into the filler tube or opening, the surrounding debris was transported into the channel. When the oil was finally poured into the funnel, the contaminants from all of the mentioned sources were carried into the crankcase. The situation could have been made worse if the oil filler was opened to enhance the draining of the oil prior to new oil being added.

Inspect and clean air cleaner element

Protecting an engine from the intake of dirt is very important to the total life of the engine. The air

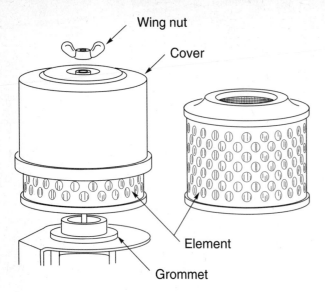

Fig. 9-7 Clean the dry element cartridge by tapping it gently on a flat surface.

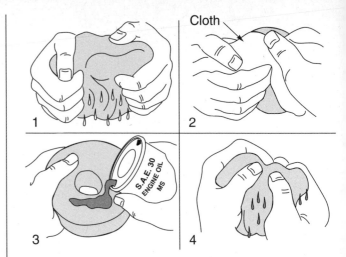

Fig. 9-8 The "oil-foam" or sponge type of air cleaner is only effective if it is lubricated after it has been cleaned.

cleaner can be made from a variety of materials and are classified as:

1. *Dry element air cleaner*
2. *"Oil foam"-type air cleaner*
3. *Dual element-type air cleaner*

DRY ELEMENT AIR CLEANER

Remove the dry element cartridge, as shown in figure 9-7, and clean it by tapping it gently on a flat surface. If the cartridge is very dirty, it should be replaced or washed in a soap and water solution. A cartridge that is washed must be dried before installing (without the use of pressurized air). Compressed air should never be employed as it may damage the paper element and cause dirt to enter the engine. The air cleaner may be inspected for any rips in the surface area by holding a light bulb or flashlight behind the paper surface and observing any perforations. Check for filter distortion of the molded shape.

Do not use any petroleum solvents, such as kerosene, to clean the cartridge as they may deteriorate quickly, and do not oil the cartridge after it is cleaned.

"OIL FOAM"-TYPE AIR CLEANER

The "oil foam" or sponge type of cleaner is effective only if it is lubricated after it has been cleaned, as shown in figure 9-8.

1. *Remove the element and wash it with liquid detergent and water rather than in a petroleum solvent such as kerosene.*

   NOTE: WHEN THE ELEMENT IS WASHED WITH A PETROLEUM SOLVENT, SOME OF THE SOLVENT REMAINS AFTER IT IS SQUEEZED OUT, WHICH WILL CAUSE THE OIL THAT IS ADDED TO IT TO SETTLE TO THE BOTTOM RATHER THAN BEING DISPERSED THROUGHOUT THE ELEMENT.

2. *After the rinse water has been squeezed out into a dry, unsoiled cloth, clean fresh engine oil is applied to the element. Enough oil should be added so that the sponge is saturated. Knead the oil through the element until all the pieces are lubricated.*

3. *Take a clean dry towel and squeeze the foam element in it until all the excess oil has been removed and then install the element into its holder that has also been cleaned.*

DUAL ELEMENT-TYPE AIR CLEANER

The dual element-type is the most efficient for filtering the

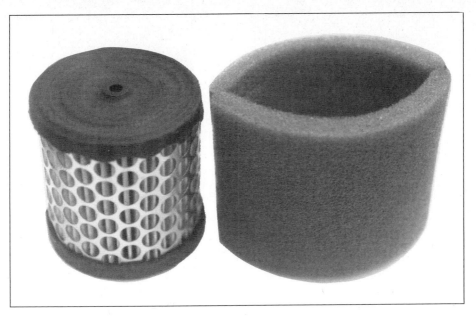

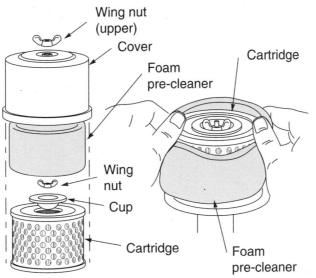

Fig. 9-9 The dual element air cleaner is most efficient. The precleaner filter should be cleaned often to preserve the dry element cartridge.

incoming air, as shown in figure 9-9. Wash the pre-filter in liquid detergent and water, squeeze it dry, add oil to the foam, and squeeze it in a towel to remove the excess oil. If the sponge is cleaned regularly, the paper element will accumulate very little dirt. It is possible to obtain 100 hours of operating life from the paper filter if the foam pre-cleaner is maintained properly.

MAINTENANCE EVERY 50 HOURS

This interval is approximately once a mowing season for the consumer who uses the mower for two hours a week, on average. A good time to perform this is at the end of the mowing season when more time is usually available. If the engine is stored where small animals may make their nest inside the

Fig. 9-10 An engine that is stored where small animals may make their nest inside the blower housing should be inspected at the beginning of each season.

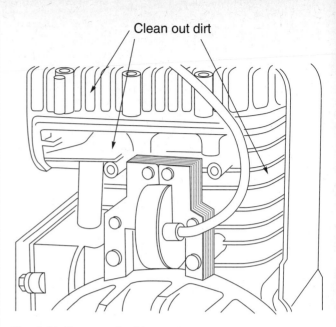

Clean out dirt

Fig. 9-11 Remove the blower housing and clean out the cooling fins.

blower housing, then it should be inspected at the beginning of the season, as shown in figure 9-10.

Maintenance Every 50 hours

1. Clean engine

2. Change fuel filter and flush tank

3. 25-hour cleaning steps

Clean engine

Dirt and debris can enter the blower housing with the cooling air and clog the cooling fins. Continued operation with a clogged cooling system will cause severe overheating and engine damage. Remove the blower housing and clean out the cooling fins, as shown in figure 9-11.

It is also necessary to remove all the oil film and dirt from the exterior of the crankcase. Most of the engine oil heat is removed through this area, and a "blanket" of dirt and oil can insulate some of the heat inside the engine.

Change fuel filter and flush tank

The fuel filter should be replaced, if utilized, and all the fuel should be removed from the tank so that any contaminants are removed.

MAINTENANCE EVERY 100 HOURS—MINOR TUNE-UP

Many steps of the minor tune-up are included at different time intervals in the maintenance schedule, but the steps are listed again because often the minor tune-up is the only maintenance the engine receives. Frequently, it is not performed until a problem arises and the mower is brought in for service.

One hundred hours can convert to every other mowing season and, preferably, it should be performed by a technician at the service repair shop.

Minor tune-up—Every 100 hours

1. Change oil

2. Ignition test

3. Inspect starter mechanism

4. Compression check

5. Replace spark plug

6. Remove blower housing and clean fins

7. Inspect flywheel

8. Wash engine

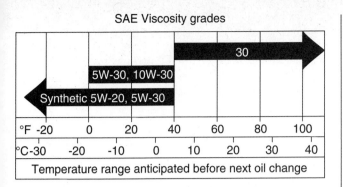

Fig. 9-12 Select the correct viscosity engine oil.

Fig. 9-13 API label for a SAE 30 oil.

9. Clean or replace air filter

10. Check fuel system

11. Inspect muffler

12. Adjust carburetor

13. Sharpen the blade

14. Service the equipment

15. Service electrical components

Change oil

The old oil is removed from the engine and new oil with the proper API rating and viscosity grade is added, as shown in figure 9-12.

1. *Operate the engine. Oil has detergent and dispersants added. The detergent acts like soap: it keeps the engine free of varnish and sludge deposits, it prevents carbon build-up around the piston rings, and it neutralizes acids. The dispersants keep various contaminants in suspension, preventing them from settling on vital engine parts. The internal engine contaminants are dumped when the oil is changed. The engine should be started and operated until the oil is warm before the oil change so that the suspended contaminants will drain easier.*

2. *Drain the oil. The oil may be removed from the bottom of the crankcase or it may be drained from the filler tube. If the drain plug is used, the filler cap should be removed to vent the crankcase in order to increase the speed of the draining oil. It is important to remove any*

debris from the filler area before it is opened up so that dirt will not enter with the fresh oil.

3. *Add new oil. The air-cooled engine operates at a higher temperature than its water-cooled counterpart. A good oil must lubricate, cool, seal, and clean the engine. A high-quality lubricant is needed with many additives to protect the air-cooled engine. The oil's viscosity rating indicates its resistance to flow. A cold oil flows slower than a hot oil, so the oil must maintain a determined viscosity at high temperatures to ensure that it will lubricate properly. The oil must also be thin enough to ensure that when it is cold, the engine parts can move easily to start the engine. Small air-cooled engines should use a straight **SAE 30** oil with an **API rating of SE, SF, or SG** be used if the anticipated temperature range of 40° F or greater, as shown in figure 9-13 and 9-14. If the engine is used at a lower temperature than 40° F, a **SAE 10W-30** multi-viscosity rating is desired. Most manufacturers advise that a **SAE 10W-40 oil should never be used** because the greater proportion of viscosity index improvers and the added heat of the air-cooled engine can lead to oil breakdown and eventual engine failure, as shown in figure 9-15. Tests have shown that an air-cooled engine operating at full-rated horsepower will run for about 7.5 hours before engine failure occurs when a SAE 10W-40 oil is used. The small air-cooled engine stores from one to two quarts of oil in its crankcase. The additives break down with use,*

Fig. 9-14 A typical brand and packaging of one quart of oil.

Fig. 9-15 Never use a 10W-40 viscosity or weight oil.

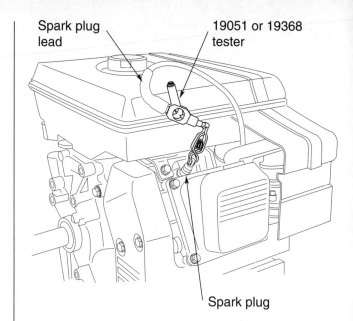

Fig. 9-16 The spark should jump the gap on a spark tester such as the one sold by Briggs & Stratton.

and dirt is suspended in the oil. The oil should be checked and changed often. Most manufacturers give an interval for this change. Briggs & Stratton recommends 25 hours of operation. When performing an oil change, be certain that you do not cause more harm that good. The operator knows that it is important to change the engine oil often, but may not be aware of the dangers of introducing dirt and other foreign matter into the crankcase during the oil change.

Ignition test

The spark should jump the gap on a spark tester like the one produced by the Briggs & Stratton, as shown in figure 9-16. If there is no or intermittent spark, a major tune-up must be performed. This includes replacement of points and condenser on engines so equipped or conversion to solid state ignition, if available.

Inspect starter mechanism

Inspect the starter system and repair any sticky mechanisms or replace any worn parts, such as the starter rope.

Compression check

A compression gage can be used to check the compression. The engine should be operated until warm and the carburetor choke and throttle valves held in the open position during the test in order to obtain the maximum reading.

A compression gage should be installed and the starter activated for about three fast pulls of the starter rope, or for about ten compression strokes with an electric starter. The reading should range between 50–90 PSI for most small air-cooled engines. If the reading is less than 50 PSI, then further tests should be performed to isolate the problem. If the reading is higher than 90 PSI, combustion deposits have built up and should be removed. The cylinder head should be removed so that the piston and head can be scraped with a soft wooden tool. Be careful not to gouge any of the metal surfaces if a metal scraper is used.

Replace spark plug

Remove the spark plug and check to see if it is the correct type. Note the color and condition of the electrode. The plug should be light tan in color to denote correct combustion. In a multi-cylinder engine, it is a good practice to remember from which cylinder the spark plug has been removed so that when all the plugs are compared together, any deviations of color or wear can indicate the problem cylinder.

It is not cost effective to clean an old plug. Obtain a new spark plug designed for the engine. The plug that is recommended is for average running conditions. Different heat ranges or electrodes may be used for abnormal conditions. An engine that operates at a high speed for an extended period may have a spark plug with a lower heat range than specified. An engine that operates near idle speed may have a higher heat range. The heat range of a spark plug is the measure of how fast the heat dissipates from it. The hotter the plug, the slower the heat loss. A certain amount of heat is necessary at the electrode so the combustion deposits will be burned off. Gap the new plug to the proper dimension.

After the new spark plug is gapped, it should be installed and torqued from 15 to 20 ft.-lbs., as shown in figure 9-17. Avoid cleaning the spark plug with any abrasive material, such as a sand blaster or sand paper, because some of the abrasive grit will remain tightly packed within the electrode of the plug after the cleaning operation no matter how carefully the sand is rinsed. The grit that is introduced into the engine as a result of a soiled spark plug will account

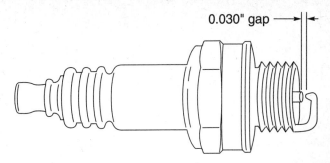

Fig. 9-17 A new spark plug should be gapped and be torqued from 15 to 20 foot-pounds.

for unexpected and premature wear. Tests have shown that all the grit from a sand blasting procedure cannot be removed with compressed air or solvents.

Remove blower housing and clean fins

The cylinder cooling fins must be uncovered and inspected for any obstructions. Dirt and debris can enter the blower housing with the cooling air and clog the cooling fins. Engine operation with a clogged cooling system will cause overheating and engine damage. The pneumatic governor may also be affected and cause the engine to operate at an excessive speed. Clean out the cooling fins with compressed air or when the engine is pressure washed.

Inspect flywheel

Check for any cracks or damaged fins. If any fins are missing, vibrations may occur that can loosen some of the fasteners. A flywheel that has a broken fin should be discarded.

Wash engine

Customer satisfaction and apparent running condition seem to increase with a clean engine. The clean engine will transfer heat better and prolong engine life. The engine can be pressure washed or simply sprayed with a grease solvent that will mix with the dirt on the engine and wash away with water. A running engine attracts dust, if there is any oil mist or leak around the engine crankcase, that will build up into a thick layer, causing the engine to retain harmful heat.

Steam and high-pressure washers work well for removing grease, debris, and grass stains, but use care. Water from the high-pressure washer can contaminate some components, which may cause early failure. Do not use the pressure washer near electrical components such as interlocks, switches, gages, starters, alternators, relays or electric clutches.

When cleaning the painted exterior, use a mild detergent mixed with water. An industrial soap or degreaser may clean the area, but may also contain chemicals that can cause premature paint oxidation and fading.

Take care to cover any decals that may be damaged by the cleaning process. The decals provide important operation or user information and, if they become damaged or painted over, they must be replaced in the same spot.

Clean or replace air filter

10,000 gallons of air passes through the air cleaner for every gallon of gasoline used, and the air filter must remove all the debris from it.

Clean the air filter about every 25 running hours, but under dusty conditions, clean more frequently. Most engines use the "oil foam," or dry type, of air cleaner, but the most effective method of air filtration is the combination of both, where the oil foam is used as the pre-cleaner.

The oil foam filter can be cleaned in liquid detergent and water. The filter is then wrapped in an absorbent material and squeezed dry. Clean, fresh engine oil is added throughout the element, and the element is again squeezed in absorbent material to remove excess oil.

Cleaning with liquid detergent and water is the most efficient method. The dirty oil is completely removed and, when the fresh oil is added, the oil residue adheres to the element better and prevents gravity from pooling the oil at the bottom of the filter.

The dry element is cleaned by tapping it lightly on a hard surface. If the element is very dirty, replace the cartridge or wash in a low or non-sudsing detergent and warm water solution. Do not use petroleum solvents to clean the element as they may cause the cartridge to deteriorate. Do not oil the car-

tridge or use pressurized air to clean or dry. A light bulb may be inserted into the center of the element, and any holes will be evident when you observe it from the outside.

Check fuel system

Gasoline is subject to oxidation, which results in the deposit of fuel gum in the carburetor. This can precede many problems, the most common of which is a sticky intake valve. The ideal preventative procedure would be to buy gasoline by the tankful, as you do with an automobile, or only the quantity of fuel that can be used in a one-month period.

Flush the fuel tank and carburetor with fresh fuel. The old fuel may be siphoned or dumped from the tank and disposed in a proper container. The fuel line and the carburetor should be flushed by adding fresh fuel to the fuel tank and running the engine until the fresh fuel has purged the carburetor of the old. Some carburetors have drain plugs on the bottom of the float chamber that can be used to clear the system.

Inspect muffler

The muffler should be replaced if defective. A muffler is necessary to protect the operator from excessive noise levels. Furthermore, it protects the exhaust valve from overheating by providing the proper back pressure.

Backpressure is necessary to slow down the exit of exhaust gases so that the heat of combustion does not overheat the exposed parts of the exhaust valve. The muffler produces a resistance to the exhaust gas. A spark arrestor screen is located in some two-stroke-cycle engines that must be inspected for any blocked areas.

Coat the muffler threads with an antiseize compound that will permit easy future removal. Check for a damaged muffler deflector and replace it if bent, damaged, or no longer effective.

Adjust carburetor

1. *Preset carburetor adjustment needles, according to the specifications in the repair manual, and install the air cleaner before adjusting the carburetor.*

2. *Start the engine and allow it to warm up to a*

normal running temperature for five minutes at a moderate speed.

3. *Increase the engine running speed to the maximum recommended RPM and adjust carburetor high speed circuit with the procedures found in the repair manual or the carburetor chapter.*

4. *After the main system is adjusted, move the speed control lever to the idle position and follow the same procedure for adjusting the idle system. The idle speed should be kept at about 1800 RPM while adjusting.*

5. *Repeat the adjustment procedure between the main mixture adjustment screw and idle at least three times. This confirms that the carburetor will work properly under load after it leaves the service area.*

6. *Test the engine by running it under normal load. Make sure that the engine accelerates from idle speed to high speed without hesitation. It should respond to load pick-up immediately; one that "dies" is too lean. An engine which ran rough before picking up the load is adjusted too rich.*

Sharpen the blade

Caution: Before any work or inspection of the rotary blade, remove the ignition cable from the spark plug and anchor it securely against any metal

Fig. 9-18 Remove the ignition wire before any work is done under the mower deck. Anchor the spark plug wire securely against any metal on the engine.

of the engine, as shown in figure 9-18. The spark plug wire has a "memory" that will move it back to the plug area unless secured.

In applications where a cutter blade is used, the blade should be inspected, removed, cleaned, sharpened, balanced, and installed. When the edge is sharp, the grass is cleanly cut, but when a dull blade is used, grass leaves or blades are torn away. This will cause the formation of a brown tint to the lawn in a couple of days. The ragged ends of the torn grass heal slower than a clean cut, and valuable nutrients are lost.

The mower blade should be removed, cleaned, and sharpened on a grinder, or it can be secured in a table vise and filed to clean the cutting edge. The cutting edge should be close to 35°.

After the blade is sharpened, it must be balanced before installation. Remove the debris adhering to the blade, and place the blade on a balancing tool or simply use a straight edge. Remove metal from the heavy side (the side that moves downward) until it balances. Metal can be ground from the end of the blade to quickly reduce the weight.

The under side of the blade shroud should be scraped and cleaned of any debris that will hinder the discharge of cut material. Check for any cracks in the housing and possible safety problems.

Coat the blade fasteners with a light oil or an anti-seize compound that will facilitate future removal. Tighten the blade bolt or nut securely, turn the machine upright, and replace the spark plug cable.

Service the equipment

Check the following, if equipped:

1. *Inspect all belts for fraying and cuts and replace, if worn.*

2. *Test the adjustment and tension of all belts.*

3. *Lubricate all wheels and drive mechanisms. Check the pressure of the inflatable tires on a riding mower and inspect for wheel damage on mowers with solid wheels.*

4. *Look for grease fittings and lubricate with a grease gun.*

5. *Inspect all gear boxes for proper lubricant levels.*

6. *Check the engine mounting bolts for tightness*

and test the other small bolts and screws on the engine for security.

Service electrical components

1. Evaluate the electric starter.

2. Check for proper battery installation and electrolyte level.

3. Look over the battery terminals for excessive corrosion, and clean with a wire brush. When disconnecting battery cables, remove the negative cable first and reconnect it last.

4. Clean the battery top with a solution of water and baking soda.

MAINTENANCE EVERY 200 HOURS—MAJOR TUNE-UP

The procedures in the major tune-up are identical to those in the minor tune-up, with the addition of certain areas. The valve train and the ignition system are renovated. The major tune-up is the recommended procedure for most engines that are brought in for service. Many times, a customer returns shortly after the engine was serviced because of an operation problem. The problem may not be associated with the previous service, but to the customer, the serviceperson is responsible for the problem. Regardless of who is responsible for the failure, the customer has an unproductive feeling about the situation. The individual brings his machinery to the engine center for service and also for assurance that it will operate correctly. The major tune-up is the technician's way of offering this implied warranty.

Maintenance every 200 hours—Major tune-up

1. Valve service

2. Cylinder and head carbon cleaning

3. Ignition component inspection

4. Fuel system inspection and flush

5. All minor tune-up steps (100 hours)

Valve service

A valve job consists of the following:

Valve service

1. Remove valves from engine

2. Wire-brush valves and valve stems

3. Clean carbon from the valve guides

4. Evaluate the valve guides and stem clearance

5. Cut valve seats

6. Reface, replace, or reseat valves

7. Check seat articulation

8. Check valves for leakage

9. Check clearance between valve stem and lifter

10. Install springs and valve keepers

A complete description of the valve service can be found in the assembly chapter.

1. **Remove valves from engine**. There are a variety of tools that may be used to compress the spring and remove the retainers. The springs should be marked so that they will be replaced on the same valve stem.

2. **Wire-brush valves and valve stems**. The valves can be inspected better when they are clean. Before you clean the carbon, inspect the valves for indication of leaks.

3. **Clean carbon from the valve guides**. Use a small caliber rifle cleaning brush or reamer.

4. **Evaluate the valve guides and stem clearance**. By using a micrometer and a hole gage, you can evaluate the clearance for the valve stems. A dial indicator may also be placed on the head of the valve and the side-to-side movement measured. Movement less than 0.003-inch is acceptable. Briggs & Stratton offers a reject gage. If the gage can be inserted into the hole a measured distance, the valve guide must be serviced.

5. **Cut valve seats**. Cutting new seats is very important. Use the proper tools to obtain the correct angle. Remove the minimal amount of metal necessary to "true" the valve seat.

6. **Reface, replace, or reseat valves**. The operation performed will depend on the condition of the valve. Some shops have the means to reface the valves. If refacing is not economically feasible, then new valves may be purchased.

7. **Check seat articulation**. The valves may be lapped to the seat with a grinding compound, or

bluing may be used to indicate the width of the touching surfaces. Many technicians believe that lapping the valves will make them seal better when the engine is operating. Regardless of the truth of this statement, either method is important for certifying proper contact.

8. **Test valves for leakage**. *A low viscosity fluid, such as kerosene, can be poured into the exhaust and intake port of the engine. When the valves are held in place, the fluid should not leak past the metal-to-metal contact surface. It is acceptable if the area looks "wet" from the fluid, but does not form droplets.*

9. **Evaluate clearance between valve stem and lifter**. *Check the manufacturers' specifications for the clearance. Some valves are not adjustable, except for grinding metal off the stem to open the gap. If the gap is too great, the valve seats may be cut further so a new valve can be used.*

10. **Install springs and valve keepers**. *The springs are installed and the gap is again checked.*

Cylinder and head carbon cleaning

Remove the cylinder head of the engine (note the position of the bolts so that they can be returned to their original position) and scrape off the deposits in the combustion area. Make sure that the piston is at TDC with both valves closed to prevent any of the loose carbon from entering the valve train.

An engine will lose power from a build-up of combustion deposits. The deposits can lodge between valve and seat, resulting in burned valves. Inspect the cylinder head for any distortion, as shown in figure 9-19, and correct it by sanding the head on a flat surface. You may re-use the head gasket if it is in good condition and it shows no signs of leakage but it is better to use a new one. Torque the head bolts back to specification.

Ignition component inspection

If points and condensers are used in the engine, they should be removed and replaced with new components or the coil retrofitted with an electronic ignition module, if possible. Refer to the ignition chapter or service manual for precautions and procedures.

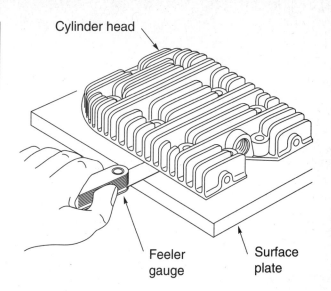

Fig. 9-19 Inspect the cylinder head for any distortions.

Solid state ignition systems require no service except to check the strength of the spark. If a spark tester is installed and there is voltage to jump the tester's expanded gap, the spark strength is sufficient and no further service is necessary.

Fuel system inspection and flush

The fuel system should be inspected and flushed. The fuel tank should be emptied and cleaned of all contaminants. A new fuel filter should be installed, and any fuel lines replaced if damaged or deteriorated.

The carburetor should be disassembled and inspected. All dirt should be cleaned from the passages, and new components and gaskets should be used if damaged parts are found.

MAJOR OVERHAUL

If the piston rings need to be replaced, the engine should have a major overhaul. The major overhaul procedure will include the disassembly and evaluation of the engine parts so that an economical choice of repairs can be made. If it is economically feasible to repair the engine, then the assembly chapter should be referred to for the appropriate steps.

OFF-SEASON STORAGE

In the case of a lawnmower, one winter of bad storage can do more damage than several seasons of operating. Serious damage from rust and corrosion, annoying damage from plugged carburetor passages, and stale fuel can lead to problems the next time the mower is used.

Off-season storage

1. Wash equipment
2. Change oil
3. Inspect and clean air cleaner element
4. Add fuel system stabilizer
5. Clean air intake screen
6. Check air cleaner mounting
7. Sharpen blade and clean under deck
8. Fog engine
9. Replace spark plug
10. Store machinery

WASH EQUIPMENT

The equipment should be washed with soap and water or a degreaser that is soluble with water. Flush off the equipment with a stream of water. A dirty engine may attract moisture that can cause corrosion.

CHANGE OIL

The engine should be operated until it is warm, then the oil can be drained easier with much of the internal dirt suspended in the old oil. Fill the engine to the proper level with new oil which has the correct API rating and viscosity.

INSPECT AND CLEAN AIR CLEANER ELEMENT

The air cleaner should be inspected and cleaned properly or replaced.

ADD FUEL SYSTEM STABILIZER

The debate whether to store the equipment with a full or drained fuel system can be answered by determining where the equipment is to be stored. If it is to be stored inside a heated garage or in a basement, the fuel system must be drained completely. If it is to be stored in a moist area, the fuel tank can be full. Filling the system is the preferred method because, when the system is stored with the fuel tank full, there is less opportunity for condensation to form in the fuel tank or for the internal components in the carburetor and fuel pump to "dry" out and become brittle.

Fuel additives can be bought and added to the fuel to decrease the oxidation process. These additives are recommended for all fuels if they are stored seasonally or for a long period of time. The additive is only effective if it is added to the fuel, and the engine is then operated for a period of time to bring the protected fuel into the fuel lines and carburetor.

Some of the additive manufacturers claim that the fuel can be protected from oxidation for up to two years. Throughout this extended period, a vented fuel cap will cause many of the lighter molecules to be lost and may also cause starting problems even though the fuel is protected from oxidation breakdown. Briggs & Stratton markets a gasoline additive (part #5041) that can be used for gasoline stabilizing.

CLEAN AIR INTAKE SCREEN

The intake screen is used to stop any large items from being injected into the cooling system, which may plug the cooling fins around the cylinder. Clean the screen by wiping or brushing off the debris. Avoid directing a burst of compressed air at the area, which may drive some of the debris into the engine.

CHECK AIR CLEANER MOUNTING

There must be no leaks around the air cleaner where it attaches to the carburetor. Check the fasteners that hold the air cleaner in place and make sure they are secure.

SHARPEN BLADE AND CLEAN UNDER DECK

Remove the blade properly after the spark plug wire has been disconnected from the spark plug and properly grounded by connecting it or attaching it to the body of the engine or machine. A sharpened blade will cause less stress on the engine the next time it is used which, for a lawnmower, is most likely in a fast grass-growing season. Removal of the debris under the mower deck will lessen the chances for oxidation to occur during the storage period and prevent premature destruction of the mower deck.

FOG ENGINE

With the old spark plug in the engine, the engine should be "fogged" to coat the internal parts with oil so as to prevent any problems with corrosion or rust.

"Fogging" the engine can be accomplished by removing the air cleaner and starting the engine. While the engine is operating at a moderate speed, engine oil or fogging oil is squirted into the carburetor air horn with the air cleaner removed. The oil is introduced fast enough to stall the engine. This process will coat the intake system, cylinder and rings, and the exhaust system. When the engine is started the following season, it will smoke for a period of time until all the excess oil is removed.

An alternative method to "fogging" is to take a teaspoon of oil and pour it into the spark plug hole so that the piston area is coated with lubrication. Although this coats the cylinder and piston, "fogging" the engine provides a more complete protection.

REPLACE SPARK PLUG

The old spark plug should be removed after the "fogging" and the new one installed after it is gapped to specification.

STORE MACHINERY

The machinery should be coated with a thin coat of engine oil to prevent rusting. Dip a cloth into some oil and rub it over the components. If the equipment will be stored in a damp area or area that is exposed to the elements, a plastic bag should cover it as much as possible. The plastic should be secured down, but should not be air-tight. The purpose of the covering is not only to keep the equipment away from the direct exposure of rain or dust, but also to allow it to "breath" and vaporize any condensation that has accumulated.

Whether the equipment is stored inside or outside, the spark plug wire must be detached and anchored to the metal of the engine to prevent it from accidentally starting if the starter is activated.

SUMMARY

The type of engine maintenance helps determine the maximum life of the engine. A good maintenance program includes keeping the engine clean and well-oiled to reduce friction and allow adequate heat transfer. Maintenance schedules are set up so that certain procedures are performed at intervals of operational hours.

An aluminum alloy cylinder engine that is rated for an operating life of 300 hours will operate for approximately 6 years with proper maintenance before a major overhaul is necessary. A cast-iron cylinder can operate for 2500 hours and then have the economic capability to be rebuilt for an additional 2500 hours of operation. Cast-iron cylinders are either fully cast-iron or a cast-iron liner that is installed in an aluminum block casting.

Questions

1. **The construction material used in the engine cylinder determines the potential operating life. Technician A states that the cast-iron cylinder has a longer operating life than the aluminum cylinder. Technician B states that the potential life of cast-iron and aluminum cylinders is about the same. Who is correct?**

   A. Only Technician A
   B. Only Technician B
   C. Both Technician A and B
   D. Neither Technician A or B

2. Proper air filtering on the carburetor intake is important to a long-lasting engine. Technician A says that the paper filter is the best filtering system. Technician B says that the combination of the paper filter with the foam pre-cleaner filter is the best.
Who is correct?

A. Only Technician A
B. Only Technician B
C. Both Technician A and B
D. Neither Technician A or B

3. The oil level should be checked often to observe if the level is proper. Technician A says that it is better to overfill the oil level in order to provide a reservoir. Technician B says that overfilling the oil level is just as harmful as a low oil condition.
Who is correct?

A. Only Technician A
B. Only Technician B
C. Both Technician A and B
D. Neither Technician A or B

4. Changing the oil often is vital to long engine life. Technician A states that if certain steps are not taken, some customers may add more dirt than they remove during the oil change. Technician B says that any oil that meets the API rating of SC or higher can be used.
Who is correct?

A. Only Technician A
B. Only Technician B
C. Both Technician A and B
D. Neither Technician A or B

5. When performing the minor tune-up, Technician A says to always replace the spark plug with a new one. Technician B says that the spark plug can be used again if it is sand blasted properly.
Who is correct?

A. Only Technician A
B. Only Technician B
C. Both Technician A and B
D. Neither Technician A or B

6. A flywheel has one fin broken off the air impeller. Technician A says that an engine vibration may occur. Technician B says that the flywheel can still be used if there are no engine vibrations.
Who is correct?

A. Only Technician A
B. Only Technician B
C. Both Technician A and B
D. Neither Technician A or B

7. Washing the engine is important. Technician A says that this makes the customer feel good about the service. Technician B says that washing is necessary because the engine will dissipate the heat of combustion better.
Who is correct?

A. Only Technician A
B. Only Technician B
C. Both Technician A and B
D. Neither Technician A or B

8. A lawnmower blade is out of balance. Technician A says that this will cause the blade to smash the grass rather than cut it. Technician B says that the unbalanced state may vibrate the lawnmower.
Who is correct?

    A. Only Technician A
    B. Only Technician B
    C. Both Technician A and B
    D. Neither Technician A or B

9. The major tune-up is necessary to return the engine to full power. Technician A states that the cylinder must be scraped to remove the deposits in the major tune-up. Technician B says that the piston rings should be replaced in the major tune-up.
Who is correct?

    A. Only Technician A
    B. Only Technician B
    C. Both Technician A and B
    D. Neither Technician A or B

10. Fogging the engine is a good method to protect the metal components during the storage period. Technician A says this means to spray the complete outside of the engine with a lubricating mist. Technician B says that this means to spray the complete underside of the mower deck to prevent rust.
Who is correct?

    A. Only Technician A
    B. Only Technician B
    C. Both Technician A and B
    D. Neither Technician A or B

CHAPTER 10

OVERVIEW

THE PROFESSIONAL TECHNICIAN

Stories are numerous about customers who have been dissatisfied with the expense or quality of small air-cooled engine repairs. Many technicians who are just "part changers" have given the profession a black eye. Parts may be changed needlessly with the hope that the problem will be solved. If the difficulty still persists, additional parts are interchanged. The expense of the parts and labor may equal or exceed the value of the engine itself.

A professional technician is an individual who can identify the problem efficiently and solve it with minimal expense. The charge to the customer is excessive when the parts replaced do not solve the problem and costly to the technician when the customer returns with the machinery so that additional time is required to correct the problem. The ability to troubleshoot a problem quickly and accurately is one of the most valuable talents or "tools" that the repairman can have.

CAUSE VS. EFFECT

The cause of the problem must be corrected, not just the effect. Many times, the cause is overlooked and the malfunction corrected, only to have the engine trouble recur. Therefore, it is important to recognize the cause and effect relationship.

If a connecting rod breaks through the crankcase wall (**effect**), what is the **cause** of this? The over-speeding or lack of lubrication that **caused** the problem must be corrected as well as the repairs made to the engine components.

For example, a customer complained that her tractor did not have enough power to cut the grass.

The technician cleaned and adjusted the carburetor, and the tractor tested-out properly. The woman returned the next day with the same complaint, so the technician asked her to demonstrate the problem. As she mounted the tractor, she pulled the choke lever out. The engine started, but ran poorly (**effect**) because she never realized that the choke lever needed to be pushed in after the engine started (**cause**). She was advised to push in the choke lever after starting the engine, which corrected the cause as well as the effect.

Another customer complained about poor performance when cutting the grass on his riding lawnmower. The technician knew to check the sharpness of the blade and the condition of the mower bearings for a **cause**, realizing that its **effect** on the engine would be poor performance, rather than assuming that a worn engine was the difficulty.

The final example is about a lawnmower starter that was brought into the repair facility with a broken starter rope. Following the starter rope repair, the customer assembled it to the engine and attempted to start the engine. He pulled the whole machine a foot off the ground because, even though the rope was repaired (effect), the seized engine (cause) was not.

PROBLEM SOLVING

The technician increases the probability of efficiently correcting any problem when he logically collects correct, relevant information and interprets it properly. Troubleshooting is a systematic approach of problem solving that requires a special plan of attack that is used from one engine to another. The utilization of this approach diminishes the chances that a valuable piece of information will be overlooked. The good technician will work in

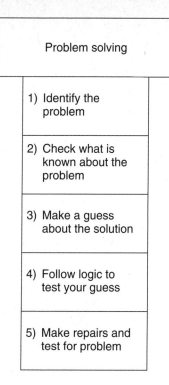

Problem solving
1) Identify the problem
2) Check what is known about the problem
3) Make a guess about the solution
4) Follow logic to test your guess
5) Make repairs and test for problem

Fig. 10-1 There are five steps in the problem solving procedure.

Fig. 10-2 "Won't run" tag fastened to the engine may be the only communication between the customer and service person.

an orderly fashion and never assume or leap to conclusions.

There are five steps in the problem solving procedure, as shown in figure 10-1.

1. *Identify the problem to be solved.*

2. *Determine what is already known about the problem.*

3. *Form an educated and experienced guess about what the answer might be.*

4. *Follow your troubleshooting logic to verify your assumption.*

5. *Make the repairs and then test for the problem. If the problem is still unresolved, try a new approach.*

1. Identify the problem

Often the "won't run" tag fastened to the engine is the only communication between the customer and

service person, as shown in figure 10-2. It is important to gather as much information as possible from the user of the machine, whether it be the customer, a neighbor, or an acquaintance. Troubleshooting becomes a time consuming process when limited data concerning the engine is available.

It is beneficial to have the operator demonstrate the problem to the technician. For example, a woman brought in a snowblower that was hard to start, and the technician asked her to demonstrate her starting procedure. She was embarrassed when it started on the second pull of the rope. The same snowblower was returned the following week by the owner, who was the woman's father. He again complained about starting problems. When asked to duplicate the problem, the technician noticed that the man's strength was so reduced that he was not able to pull the starter rope fast enough. The technician advised that an electric starter be added.

2. Check what is known about the problem

This step is often skipped. Most manufacturers distribute bulletins that notify the technician regarding problems encountered with their products. Check the literature for possible solutions.

The experienced technician may make better judgements because of past experiences with similar problems. Communication among technicians in the shop and attendance at local update schools is important to keep repair times down to a minimum.

3. Make a guess about the solution

Get to the problem by making a qualified guess based on the facts presented and your prior knowledge and experiences. Numerous levels in the troubleshooting flow chart can be bypassed when the appropriate facts about the problem are known.

The importance of customer communication in the diagnostic process is shown when a customer enters a repair shop with a lawnmower that won't start during the spring season. The customer states, "When I put the mower away last fall, the engine was working properly. I used up the fuel in the tank and stored the mower in my shed. When I took the mower from the shed last week and filled the gas tank with fresh fuel, it would not start." The technician observes that the owner had maintained his lawnmower properly, but remembered that what is said is not always what is meant. Further inquiry uncovers that the fresh fuel was poured from a fuel container that had been stored on the garage shelf since the previous fall. The technician quickly determines that since the fuel in the can had exceeded acceptable shelf life limits and the customer had not put any fuel stabilizer in it, the problem likely was caused by the stale fuel. This example demonstrates the importance of acquiring a good history in the troubleshooting process.

4. Follow logic to test your guess

The importance of following a logical sequence when attempting to isolate a problem cannot be overemphasized. If little is known about the problem, a systematic procedure must be followed to acquire the correct diagnosis.

A customer complained that his engine stopped and would not start again after he hit an object while mowing. In attempting to locate the problem, the technician removed the spark plug, laid it touching the metal of the engine, and pulled the starter rope. No spark was produced. The technician suggested that the flywheel key may have been sheared, with the result that the magnets were no longer timed to the legs of the armature. The sheared key caused a reduction in available spark plug voltage. In addition, the sudden stoppage of the engine also could have bent the crankshaft. The conversation allowed the technician to properly estimate the cost and type of repairs.

5. Make repairs and test for problem

After the repairs are finished, test the machinery to see if the malfunction has been corrected. If the problem is still unresolved, try a new approach. Be careful that your personal frustrations at this point do not blind your judgements. Sometimes the best solution is to put the job "on hold" and proceed to another engine.

NOTE: TIME AWAY FROM THE PROBLEM MAY BE ALL THAT IS NECESSARY, SO THAT WHEN YOU RETURN, A NEW APPROACH CAN BE DEVELOPED.

SYSTEMATIC CHECK SEQUENCE—PART 1 (WITH CUSTOMER)

The **Systematic Check** is an efficient, simple routine that is a logical approach to troubleshoot starting problems. The three main areas of ignition, compression, and carburetion, as shown in figure 10-3, will be checked for possible problems. This routine will identify most problems and should be performed with the customer present. The communication between the service person and the customer can save hours of frustration and allow a reasonable estimate of the time and cost of repairs. When the **Systematic Check** identifies a problem, further tests, found in this chapter, will isolate and correct it. Always start your troubleshooting with the easiest or least complicated possibilities, as shown in figure 10-4.

The time spent with the customer can be used to identify parts that may be broken, and authorization for the repairs can be made immediately. The customer is present to answer any of the technician's questions about the problem. It is not unusual for the technician to ask the customer to start the engine in order to observe any problems.

This opportunity with the customer can be used to educate him/her on simple maintenance procedures concerning the machine and instill confidence in the professionalism of the shop and the technician.

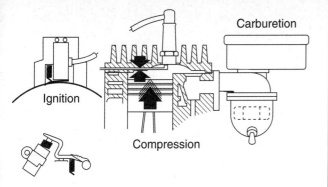

Fig. 10-3 Is the problem ignition, compression, or carburetion?

PROBLEM HISTORY

Find out the details of the problem. Skillful use of questions can disclose hints about the problem. This step can be used as a "launching board" to other parts of the troubleshooting process. Document the important information on the repair order.

Visualize people in any profession whom you consider as competent, and the common denominator is their skillful use of questions and clues to discover the problem.

Some questions that can be pertinent are:

- What are the symptoms?
- When and how did the problem occur?
- Has the problem happened before?
- How old is the fuel in the tank?
- Who is the operator of the equipment?
- Who was the operator when it failed?
- What does the customer mean by "fresh fuel"?
- Ask the customer what he/she thinks is the problem?
- Who was the last technician to work on the machine?

CHECK THE OIL LEVEL

Fill, as necessary, before proceeding. The oil is needed to reduce friction and to seal the piston for proper compression. An insufficient oil level and/or contaminated oil could suggest excessive internal

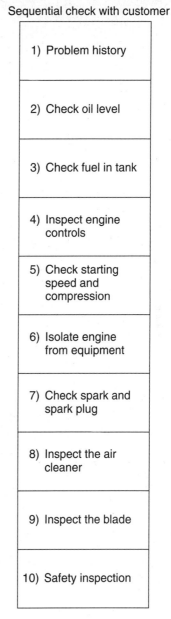

Sequential check with customer

1) Problem history

2) Check oil level

3) Check fuel in tank

4) Inspect engine controls

5) Check starting speed and compression

6) Isolate engine from equipment

7) Check spark and spark plug

8) Inspect the air cleaner

9) Inspect the blade

10) Safety inspection

Fig. 10-4 Systematic check sequence part 1 (with customer).

engine wear. The condition of the oil may provide a clue to the mystery. An oil sump that is filled with clean oil could indicate proper maintenance or, more likely, the oil was added after the engine failed. If the oil is low, fill before additional testing in order to avoid seizure of the engine and accusation regarding the blame.

CHECK FUEL IN TANK

Determine that there is fuel in the tank and that the fuel shut-off is in the "on position." Verify that the fuel is fresh, using the smell and volatility tests. A varnish odor indicates that the gasoline has deteriorated, necessitating a complete fuel system flush, which will include the carburetor. The volatility test checks the combustibility of the fuel and is performed by dipping a finger into the tank, removing it, and waving it around. When a liquid gasoline changes to a gaseous phase, it absorbs heat energy, and this rapid evaporation of fuel will "cool" your finger and indicate a good amount of volatile molecules. Gasoline is a blended hydrocarbon containing a ratio of smaller, lighter molecules to the heavier, longer ones. The proportion is changed by the refiner from one season to the next to ensure easier starting and operation in extreme climatic conditions. In the winter, the fuel is blended with more light, volatile molecules to facilitate cold weather starting. In summer, the fuel is less volatile to prevent vapor lock. Eventually, the lighter molecules, combine to form larger molecules, and gasoline is subject to oxidation, which results in deposits of fuel gum in the carburetor. The shelf life for gasoline is about three months unless the gasoline has been mixed with oil, then it diminishes to one month.

INSPECT ENGINE CONTROLS

On an engine with a remote control throttle cable or "kill" switch, inspect cables, links, wires, and springs for binding, sticking, bends, or modifications which can affect control action. Inspect the speed control setting to ensure proper contact of the grounding switch and choke lever. Make certain that the cutout, or stop switch, is not grounding out the primary circuit, preventing the engine from starting.

CHECK STARTING SPEED AND COMPRESSION

Ensure that the starting speed is not hindered by any excessive load force or an underpowered starter. Evaluate the strength of the operator and the length of the rope on a recoil start model and the battery charge on an electric start. A lawnmower should be taken out of the grass and started on the sidewalk to minimize starting resistance.

A quick check of the compression can be made by noticing a pulsing or intermittent resistance to the movement of the piston as the engine is started. Little or no resistance may indicate a compression problem.

ISOLATE ENGINE FROM EQUIPMENT

It is important to differentiate whether the problem is an engine or an equipment problem. Remove the equipment load from the engine and observe if the problem disappears. An example would be a idler pulley bushing that has seized and impedes the engine from operating correctly or a damaged blade bearing that produces an immense load while engaged.

CHECK SPARK AND SPARK PLUG

Check the condition of the ignition system and the spark plug.

Ignition tester with spark plug

Remove the high tension lead or wire attached to the spark plug terminal and insert a spark tester between it and the spark plug while the spark plug is in the engine. The air gap in the tester can be modified for the different types of ignition systems. The point ignition requires a smaller gap than the solid state ignition. A 0.250-inch air gap is sufficient for the point ignition, and a 0.500-inch air gap is sufficient for the solid state ignitions systems. An adequate ignition system produces enough voltage for the spark to jump both the tester and the spark plug gap. Rotate the flywheel or impeller vigorously with "pull rope" or starter, if present. If a spark is observed to be jumping the tester gap, the ignition system and spark plug are adequate. If no spark is noticed at the tester, then bypass the plug and use the tester alone. A spark tester can be bought or made.

Ignition tester without spark plug

Remove the spark plug from the engine and look for the presence of gasoline on the electrodes. Then insert the tester between the high tension lead and the engine metal, e.g., cooling fins, and revolve the flywheel rapidly. If spark is produced, one can theorize that the ignition set-up is sufficient to produce a spark, but that the spark plug is defective. If no spark is noted and the plug electrodes are moist, a more detailed evaluation of the ignition system is indicated.

If no spark appears and the plug electrodes are dry, both carburetion and ignition elements need to be examined minutely.

Solid state ignition

In solid state ignitions like the Magnetron®, the spark may be good, but the engine timing may be incorrect because of a sheared flywheel key.

INSPECT AIR CLEANER

The condition of the air cleaner is a good indication regarding the general maintenance given to the engine by the operator. Look for debris that might obstruct the intake of air.

Determine if dirt has entered the engine through the air intake system. Examine the air cleaner stud threads for indications of dirt and also check beneath the cover of the air cleaner. Dirt found on these parts indicates possible abrasive damage within the engine. Inspect the air cleaner element for damage or holes. Verify that the foam element contains the proper amount of oil.

Attempt to start the engine after detaching the air cleaner in order to observe the choke operation.

INSPECT THE BLADE

NOTE: ALWAYS REMOVE THE SPARK PLUG WIRE AND WEDGE IT ONTO A METAL ENGINE PART WHEN INSPECTING UNDER THE LAWNMOWER DECK. THE IGNITION WIRE HAS A "MEMORY" IN IT AND WILL MOVE BACK TO THE SPARK PLUG UNLESS IT IS IMMOBILIZED.

Make sure that the blade is not loose or installed "upside-down." The blade is the flywheel on many applications and if not attached properly, the engine will not start. A flywheel is heavy mass that spins with the crankshaft and develops a considerable amount of inertia, which keeps the crankshaft rotating through the non-firing stroke of the four-stroke-cycle-engine. The expense of the engine is reduced when the lawnmower blade can be used as the flywheel. If the blade is loose, insufficient rotating motion will be developed during the non-power strokes of the engine.

Notice the condition of the blade to observe if it has hit any hard objects that may have sheared the flywheel key. Remember that the absence of any marks on the blade does not rule out that the engine was stopped abruptly, because hitting a wooden implement will leave very little evidence.

Observe the center bolt that fastens the blade as the blade is rotated for any signs of a bent crankshaft, e.g., wobbling movement of the bolt.

SAFETY INSPECTION

Since 1982, all consumer lawnmowers used in the United States have had a requirement that the blade must stop rotating within three seconds when the operator releases the push handle. The repair shop must make certain that these functions are operating before the machine exits the shop. The customer must be advised that any tape which has been applied to the handle to eliminate any safety features or any safety switches which have been modified or are "out-of-order" will be reinstated before the unit is returned to them.

SYSTEMATIC CHECK SEQUENCE—PART 2 (WITHOUT CUSTOMER)

The preliminary examination of the engine with the customer will lead to a guess about the cause of the malfunction. The next step involves a methodical or well-ordered testing to verify the assumption and arrive at the solution. After each step in the progression, check the engine to determine if the problem has been corrected, as shown in figure 10-5. Sequential steps are performed after the work order is written up and the unit is "prioritized" in the service area.

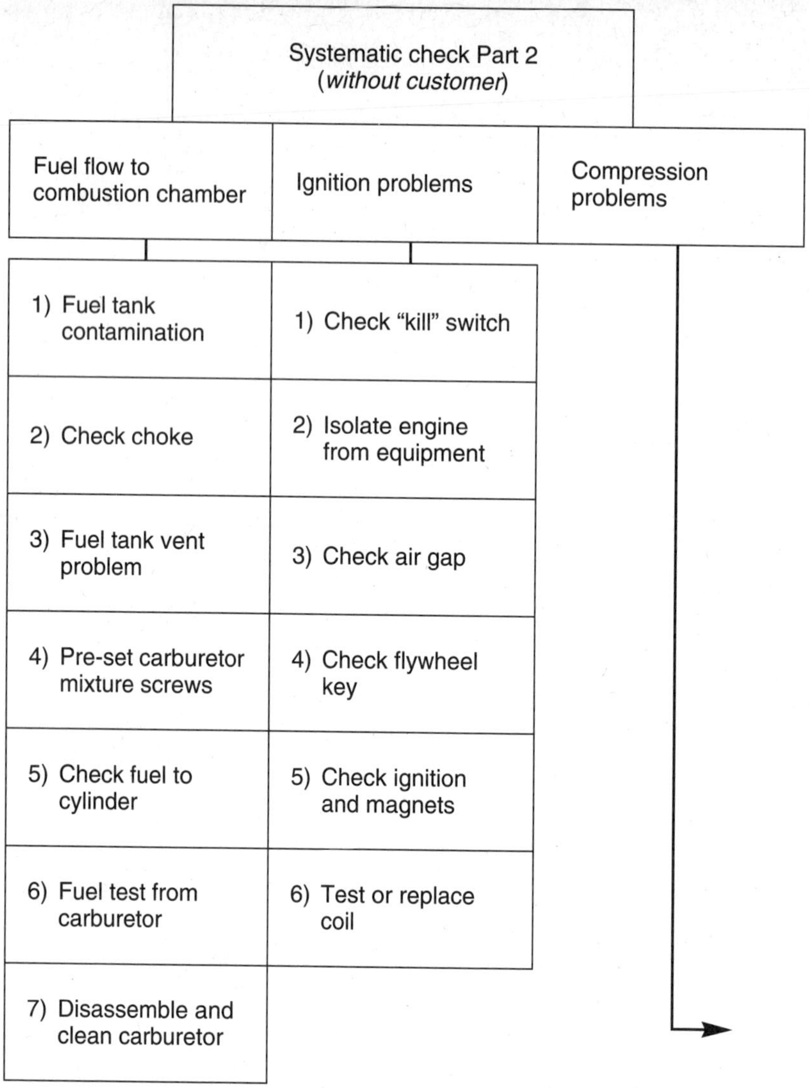

Systematic check Part 2 (without customer)		
Fuel flow to combustion chamber	Ignition problems	Compression problems
1) Fuel tank contamination	1) Check "kill" switch	
2) Check choke	2) Isolate engine from equipment	
3) Fuel tank vent problem	3) Check air gap	
4) Pre-set carburetor mixture screws	4) Check flywheel key	
5) Check fuel to cylinder	5) Check ignition and magnets	
6) Fuel test from carburetor	6) Test or replace coil	
7) Disassemble and clean carburetor		

Fig. 10-5 Systematic check sequence part 2 (without customer).

FUEL FLOW TO COMBUSTION CHAMBER CHECK

Fuel must be present in the combustion chamber for proper combustion.

1. Fuel tank contamination

Check for water in the tank. Since water's density is greater than gasoline and they will not mix, water accumulates at the bottom of the tank. The system should be emptied and cleaned with air to rid the tank, lines, and carburetor of any water or debris. If only the water must be removed from the tank, a suction device can be used.

Debris in the tank could indicate possible blockage in the system. After the system is cleaned, it should be filled with fresh fuel.

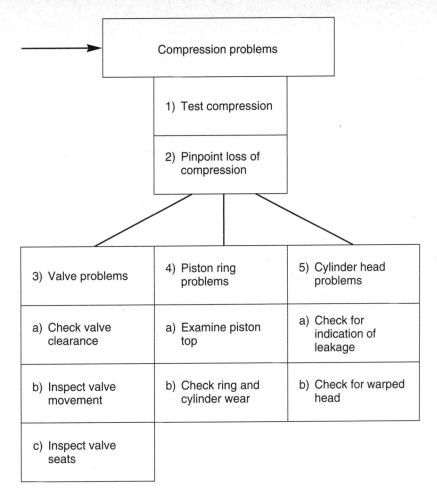

Fig. 10-5 (continued)

2. Check choke

If a choke is present on the unit, make sure that it is functioning.

3. Fuel tank vent problem

Look at the tank vent. Make sure that it is open and that air can enter the tank as gasoline is consumed by the engine. Some engines have tank vents that route through the carburetor. These passages should be inspected for obstructions.

4. Pre-set carburetor mixture screws

The mixture screws, if present, should be rotated clockwise until a slight resistance is encountered

and then backed out the specified number of turns, as indicated in the service manual.

5. Check fuel to cylinder

Crank the engine several times, remove the spark plug, and check to see if fuel is present on the plug electrode. If the plug is wet, the problem may be poor fuel or an ignition problem; but if the spark plug is dry, put 1/2 teaspoon of fuel into the combustion chamber, replace the spark plug, and attempt to start the engine. If the engine starts and then quits, the problem is with the carburetor; no fuel has flowed from the carburetor to the combustion chamber.

6. Fuel test from carburetor

Put a small amount of fuel or spray some W-D 40 into the carburetor's air horn, install the air cleaner, and again try to start the engine. If the engine attempts to start, the problem is inside the carburetor.

NOTE: WD-40 LUBRICATES THE COMBUSTION CHAMBER, BUT IS COMBUSTIBLE. THE USE OF STARTING FLUID (ETHER) IS NOT RECOMMENDED.

Move the fuel shut off valve, if equipped, to the open position. Check the fuel filter for any obstructions. The fuel line hose can be disconnected from the carburetor's inlet and checked for the proper flow of fuel. If flow of fuel to the carburetor is in question, a transparent fuel line can monitor fuel flow. If the engine does not attempt to start when the fuel is put into the carburetor, check the carburetor mounting gasket or O-ring for an air leak.

7. Disassemble and clean carburetor

The carburetor should be disassembled, inspected, and cleaned completely. New parts and gaskets should be installed and necessary adjustments made according to the service manual. If it is a float-type carburetor, the float level should be adjusted properly.

IGNITION PROBLEMS

If there is a no spark condition, the following areas should be checked:

1. Check "kill switch"

The "kill" switch is the most frequent cause for not having spark. The magneto engine is turned off by grounding out the primary circuit. If the "kill" wire has worn through its sheath and is touching the metal of the engine, the primary current will be short-circuited and ignition will not occur. The engine may also be outfitted with a variety of interlock sensors that prevent the engine from starting in an unsafe condition. These devices should be temporarily disconnected to evaluate their effect.

2. Isolate the engine from the equipment

Equipment problems could be a defective safety interlock, misadjusted controls, shorted connections, corroded wires, or parasitic loads, such as an engaged blade deck or auxiliary PTO unit.

3. Check air gap

The air gap is the distance between the legs of the armature and the flywheel magnets. This gap must be set to the manufacturer's specifications. Theoretically, the closer the magnets are to the armature, the stronger the spark. When the engine rotates, the clearance in the bushings of the crankshaft allows the flywheel to move up and down while it is spinning. There must be a minimum opening, or air gap, so that the magnets will not rub on the metal of the armature. On external coils, set the air gap to the lesser measurement of the specified range of numbers given by the service manual. Coils that are beneath the flywheel are not adjustable, but the clearance can be checked by placing a piece of electrical tape on the metal legs of the coil when the flywheel is removed. When the flywheel is replaced and spun, the magnets will move past the legs, just barely touching the tape if the clearance is acceptable.

4. Check flywheel key

The flywheel key should be inspected to determine if it has been damaged or sheared. The key aligns the magnets in the flywheel with the legs of the armature for maximum voltage in the spark system when the points open. The use of a soft key protects the expensive parts, like the crankshaft and flywheel, from damage when the engine is stopped suddenly. A sheared or fractured key will cause the magnets to move away from the optimal alignment, and the voltage produced will not be sufficient to jump the spark plug's gap. The advent of the solid state ignition system requires that a harder key be used. In the electronic system, the spark plug will fire even though the key is sheared, but the spark timing to the piston's position will be affected. A slightly damaged key will produce a good spark, but the "out-of-kilter" timing will cause the engine not to start or to "kick-back" or will "pull the rope from your hands." Certain applications are more apt to shear the key. A rotary lawnmower with a long blade will shear the key if it hits an object and the engine is suddenly stopped.

5. Check ignition and magnets

Even though it is uncommon for magnets to be faulty, it is advisable to evaluate their intensity. Compare the magnetic pull on an article of iron with that of another set of good magnets. Many contemporary magnets are glued to the flywheel and may loosen or break off. In this case, the flywheel assembly should be replaced, rather than gluing the magnets to the flywheel.

A. ENGINES WITH SOLID STATE
Solid state units are constructed and sealed at the factory so that no repairs or adjustments can be made. Determine that the Magnetron® or armature is not installed backwards. The component should be replaced with a new module if there is no spark.

B. ENGINES WITH POINT AND CONDENSER
The point activated engine should have the points and condenser replaced. The price of new points and condenser is minute compared to the labor charge for cleaning and gapping old points. Briggs & Stratton offers a solid state (Magnetron®) unit that can be superimposed on older point ignition systems with minimal cost and which provide certain distinct advantages. The Magnetron® has a five-year warranty that is indicative of the manufacturer's confidence in the product.

6. Defective coil

The coil can be checked with a coil tester, but small businesses might not be able to justify the expense. If the coil cannot be tested, it should be replaced with a new coil. The probability of the coil being the cause of the failure at this point is great.

COMPRESSION PROBLEMS

The engine must have a minimum compression pressure to start. The low compression can be due to leaking valves, bad rings, worn cylinder, or a leaking head gasket. The following are ways to test compression:

1. Test compression

A. ACTIVATE STARTER
The simplest method is to just pull the starter rope or activate the electric starter. While the engine is rotating, a resistance to the rotation will be felt during the compression stroke. This is the most unreliable method because the resistance will not be great; most engines have a compression release.

B. USE COMPRESSION GAGE
A better method to evaluate the pressure is to utilize a compression pressure gage with the spark plug removed. Very few engine manufacturers will give specifications for the minimum and maximum pressures. However, experience has indicated that a minimum of 60 PSI to a maximum of 90 PSI is desirable.

C. REVERSE FLYWHEEL SPIN
Another method used to test compression involves spinning the flywheel reverse to its normal rotation in order to bypass the compression release found on most engines. Quickly spinning the engine will cause the flywheel to stop and reverse direction when the compression stroke is reached. Caution is advised so that your fingers are not cut when spinning the flywheel.

D. CYLINDER LEAKAGE TEST
A cylinder leakage test may be performed if the test equipment is available. The test will indicate the condition of the cylinder and valve train. The cylinder is filled with air by a special connection that screws into the spark plug opening. The engine is rotated until the piston is at the top of its travel on the compression stroke. At this position, both valves are closed and very little air should escape the combustion chamber. A cylinder leakage of 25 percent or less is acceptable. The test will also disclose the origin of a leak. By listening for the escaping air, the source of the leak can be detected. Leaking air from the air cleaner area of the carburetor indicates a defective intake valve. Leaking air from the muffler indicates a bad exhaust valve. Leaking air from the breather tube area indicates worn piston rings.

2. Pinpoint loss of compression

The loss of compression can be localized to facilitate the repair procedure.

LOW COMPRESSION If the compression is low, find the source of the difficulty. The compression problem can be narrowed to the valves, cylinder and rings, or cylinder head. The "wet" test is used for this. Take about a teaspoonful of engine oil and put it into the cylinder through the spark plug hole. Place the pressure gage in the spark plug opening and test the compression. If the reading on the gage is now higher than it was before, the cause of the low compression probably arises from the piston rings and cylinder. If the compression did not change very much from the original pressure, the fault is with the valves. The oil added to the cylinder gives a temporary seal between the piston ring and the cylinder wall; however, it does not affect a leaking valve. A solution of soapy water can be brushed on the spark plug and the head gasket areas. Any loss of compressed air from these areas is revealed by the formation of bubbles. See if the head bolts are secure and tighten any that are loose.

3. Valve problems

The problem of low compression can be caused by a valve train failure. The following areas must be checked:

A. VALVE CLEARANCE If a valve problem is indicated, remove the valve cover and check the clearance between the valve lifters and the valve stems. Refer to the service manual for proper specifications. Valve gap problems are common when installing a shortblock[1]. A shortblock, as shown in figure 10-6, arrives from the factory with the valve components installed, but these engines are not tested on the assembly line. Thus, it is possible that the valve adjustment could be out of specification. Always check and adjust the valves of a new shortblock.

B. INSPECT VALVE MOVEMENT If the problem still persists, then remove the cylinder head and

Fig. 10-6 An engine shortblock.

rotate the crankshaft while observing the valve movement. A broken camshaft will be identified by this test.

C. INSPECT VALVE SEATS Check the valve and its seat and observe the surfaces that touch together. A leaking area will exhibit a different texture or color. If so, a complete valve job should be performed.

4. Piston rings problems

The piston rings must limit the combustion pressures to the top of the piston as well as preventing any oil from entering the combustion chamber from the crankcase.

A. EXAMINE THE PISTON TOP A piston's top surface area that is half clean indicates that the cylinder and rings are not sealing properly. A piston with good sealing between the rings and cylinder wall will show various deposits with only a small clean area.

An engine that has run for a moderate period of time will have adherent deposits. Wipe off any loose material. Notice that there is one area unsoiled by any deposits. All engines push oil from the crankcase, beyond the rings, and into the combustion chamber. As the rings wear, more oil escapes, proportional to the wear of the engine. The oil that is commonly used in modern air-cooled engines is a detergent type and will clean the piston top; thus, the more oil that reaches the combustion chamber by being forced beyond defective rings, the greater the clean surface area of the piston top.

---

[1]shortblock is a new factory assembled cylinder, crankshaft, piston, piston rings, valves, springs, and oil seals.

B. CHECK RING AND CYLINDER WEAR When there is a substandard seal of the rings and cylinder, determine whether it is due to worn rings or to both rings and cylinder. Measure the cylinder and compare it to the specifications. Find out if new rings can be installed without oversizing the cylinder. If the cylinder must be oversized, then a decision must be made as to whether or not the engine should be rebuilt, shortblocked, or a new engine installed. This judgment should be based on a comparison of the final proposed cost of each option.

5. Cylinder head problems

A leaking cylinder head can be caused by a warped head, improper bolt torque, defective gasket, or excessive combustion deposits. It is advisable to periodically clean the threads for the cylinder head bolts using a bottom tap. This is a good practice when you find a warped head or a blown gasket.

A. CHECK FOR INDICATIONS OF LEAKAGE Check the cylinder head for signs of abnormal exhaust leakage. The area of the leakage will be a different color than the remaining gasket area.

B. CHECK FOR WARPED HEAD Low compression can be due to a leaking head gasket. If so, the gasket should be replaced. Inspect the cylinder head to ensure that it has a flat surface. The cylinder head can be "milled" by using a piece of sand paper attached to a flat surface. Move the head in a circular motion on the sand paper and continue until the sanding marks cover the complete gasket surface on the head. The best flat surface to use is a plate of window glass.

PERFORMANCE PROBLEMS

A knowledge of operation is essential to troubleshoot an engine problem properly. The first suggestion in each group represents the probable solution that is easiest to check.

ENGINE VIBRATES

The following are the steps from most common to least common solutions, as shown in figure 10-7.

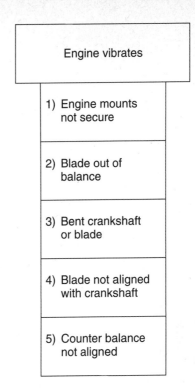

Fig. 10-7 The following are the steps from most common to least common solutions when an engine vibrates.

1. Engine mounts not secure

Every force in one direction has an equal force in the opposite direction. When the power stroke moves the piston downward in the cylinder, the movement is converted to a rotating force by the crankshaft. The twisting of the output shaft in one direction forces the mass of the engine in the opposite direction. The engine bolts and nuts keep the cylinder and crankcase from turning. If the attachments are loose, the engine parts will rotate every power stroke. This will cause a noise in the engine. Inspect all the fasteners that attach the engine to the machinery for proper torque.

2. Blade out of balance

Remove the blade and check its balance. The blade should not produce a vibration when spinning. It must be balanced so that the masses equalize each other by having equal mass on each side of the center. This is accomplished by grinding metal from

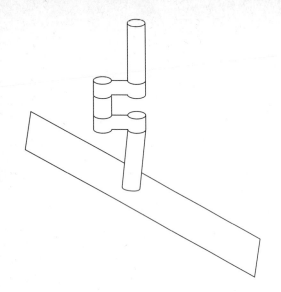

Fig. 10-8 Check for a bent crankshaft.

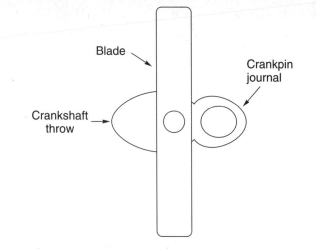

Fig. 10-9 Blade adapter is keyed to the crankshaft so that the alignment is correct.

the heavy side of the blade until it does not teeter. The blade must be able to remain level when the center of the hole on the blade is placed on a knife edge or a special tool. Longer blades will magnify the vibrations caused by any imbalance.

3. *Bent crankshaft or blade*

While the blade is removed, check for a bent crankshaft, as shown in figure 10-8, by removing the spark plug from the cylinder head and activate the starter while watching the PTO side of the crankshaft. Any wobble indicates a bent crankshaft. The crankshaft can bend when a lawnmower blade hits a hard object. It is not advisable to have a bent crankshaft straightened. Crankshafts are manufactured with a "grain" at the molecular level. This grain runs parallel to the center of the shaft. If the metal bends and is then bent back, the grain lines are disrupted and could cause the metal to fracture in the area of the disruption. This possible breakage and the consequences of damage that may occur makes it unwise to repair a bent crankshaft.

4. *Blade not aligned with crankshaft*

Machinery vibrations can be eliminated by adjusting the blade to counter-balance the crank-

shaft throw weights. Normally, the blade adapter is keyed to the crankshaft so that the alignment is correct, as shown in figure 10-9. The crankshaft has weights on it that counteract the power stroke of the engine. If the blade adapter is damaged, proper positioning can be accomplished by marking the crankshaft so that the position of throw weights are known, and then the blade can be placed perpendicular to the mark.

5. *Counter-balance not aligned*

Some large engines have extra masses (weights) geared to the crankshaft to reduce engine vibrations. When the weights are assembled in the engine, they must be timed properly to the crankshaft for proper operation.

ENGINE KNOCKS

The following are the steps to take, from most common to least common solutions, to correct engine knocks, as shown in figure 10-10.

1. *Blade loose*

The bolt or nut that holds the blade may not have been properly tightened or torqued. As the crankshaft accelerates during the power stroke and decelerates on the other strokes, a knocking sound is

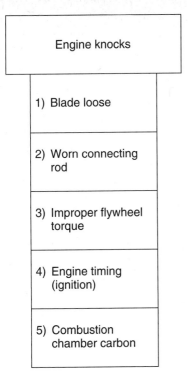

Engine knocks
1) Blade loose
2) Worn connecting rod
3) Improper flywheel torque
4) Engine timing (ignition)
5) Combustion chamber carbon

Fig. 10-10 The following are the steps from most common to least common solutions when an engine knocks.

generated by a loose connection between the blade and crankshaft. The solution is to tighten the blade or adapter.

2. *Worn connecting rod*

The normal clearance between the piston rod and crankshaft journal is 0.002-inch to 0.003-inch. The engine oil occupies this gap while the engine is running to provide a shock absorber action that minimizes the impact of metal against metal. When the rod to the crankshaft clearance becomes excessive, the oil can no longer keep the metal surfaces apart, and the sound transmits throughout the engine. A test for this condition is made by grasping any part of the crankshaft that protrudes from the crankcase and slightly move it clockwise, then counter-clockwise. Any play indicates the excessive rod clearance. To further verify this condition, remove the spark plug or the cylinder head and move the crankshaft, as described earlier. Movement of the

crankshaft without any movement of the piston confirms the diagnosis.

The solution to the problem may be a new engine, a new shortblock, or repair of the engine. If the decision is made to repair the engine, it must be disassembled and the crankpin journal of the crankshaft measured for wear. If the crankpin journal is within specification, then the rod must be replaced.

3. *Improper flywheel torque*

If the physical connection between the flywheel and the crankshaft is not strong enough to hold the acceleration of the crankshaft, a knock may be heard in the engine. There is the probability of damage occurring to the components, as well. A flywheel key that has been sheared, smeared, or distorted to the left side of the keyway (looking down on the flywheel) is the result of an unsatisfactory torqued flywheel nut. Replace the key and properly torque the flywheel nut.

4. *Engine timing (ignition)*

Detonation can cause engine knock. Detonation is the sound caused by the improper ignition of the combustion mixture. The engine is calibrated so that the spark starts the fuel burning at precisely the correct time. The combustion gases burn and cause a pressure wedge that begins as a small force and builds as it moves across the piston top surface. The engine is timed so that the full force occurs just when the piston is starting its downward movement. The spark must be coordinated with the piston and the crankshaft in order to have the flame front start while the piston is moving up on the compression stroke. This allows for the time necessary to build up the maximum force of the combustion. If the time is advanced too much, the flame front will start early and the force may oppose the piston's upward movement. As a result, great opposing pressures are exerted upon the rod and crankshaft journal, and the oil buffer will no longer keep the two metal surfaces separated.

The solution is to check the engine timing and correct it, if necessary, or replace and adjust the ignition components, such as the points and armature timing, to the proper specifications.

5. Carbon in combustion chamber (pre-ignition)

The carbon deposits can glow red from the heat generated in the cylinder and cause pre-ignition. These glowing carbon pieces can ignite the fuel mixture before the spark plug does or can ignite the fuel in other parts of the cylinder after the spark plug does. Either one of these situations causes the wedge-type of burning to be distorted. Considerable

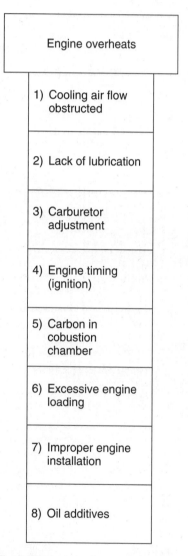

Fig. 10-11 The following are the steps from most common to least common solutions when an engine overheats.

pressure is exerted on the rod and crankshaft, and noise is produced.

Every 100 hours of operation, or sooner if a problem is suspected, the cylinder head should be removed and the combustion deposits removed with a wooden or plastic scraper. A metal scraper may damage the soft aluminum engine parts.

ENGINE OVERHEATS

The following are the steps for engine overheating from most common to least common solutions, as shown in figure 10-11.

1. Cooling air flow obstructed

The air-cooled engine is built with fins attached to the cylinder head to obtain more surface area. Air that is cooler than the engine is moved across this surface to increase the transfer of heat. It is important that the fins be cleared of anything that will diminish the surface area. Make sure that the engine fan is in good shape and that the area where air enters the engine is free from any restrictions. Often, there are screens that filter the air before it enters the fin area. These must be kept clean. The engine's sheet metal (shroud housing) must be totally intact so that the air is directed around corners to gather the maximum amount of heat. Remove the sheet metal before washing the engine at least once a year. Remove the grease and wash clean. Remember to clean the bottom of the oil sump as well as the part of the cylinder without any fins. Engines that have a mist of oil over them gather dust, and this dust and oil film will insulate the engine's heat.

2. Lack of lubrication

Oil must lubricate moving parts in the engine. When friction cannot be minimized or prevented, the heat that friction produces will cause the metal parts to melt. This condition will quickly destroy the engine. It is very unusual to have an engine used without first putting oil into the crankcase unless it was not refilled when drained. The more common problem is low engine oil. One of oil's functions is to cool the engine. Small engines have limited reservoirs of lubricant, and when the oil is splashed to the hot engine parts, the oil absorbs the heat and falls back to the sump. The heat is transferred to the metal surface of the sump and radiates from the engine to the surrounding air. When the oil level is low, the oil has not lost enough of its heat before it is hurled or pumped back to collect more heat. Wear in the engine increases, and life of the engine is reduced. If the metal parts overheat, the engine may stop running.

3. Carburetor adjustment

The leaner the air-fuel ratio is, the hotter the combustion temperatures are. An engine that is running too lean will overheat the exhaust part of the engine. An engine that is adjusted too lean will cause the muffler to glow, and sparks will be discharged, which is especially noticeable in the evening. The sparks are the carbon deposits in the engine that begin to glow and break loose. The exhaust valve cannot handle the excessive heat and may fail. Adjust the carburetor for proper operation.

4. Engine timing (ignition)

An engine in which the timing is too advanced or retarded will cause excessive heat. Less of the combustion energy will be converted to useful work with an increase of heat to be removed from the engine. The heat will cause combustion deposits to glow and ignite the fuel mixture unevenly. Damage can occur to the piston and valves. To counter this condition, properly time the engine.

5. Carbon in combustion chamber

Any addition of a material to the combustion chamber will raise the compression ratio, causing the engine to operate at a higher temperature. For every 100 hours of operation, the cylinder head must be removed and the combustion deposits cleaned out.

6. Excessive engine loading

An engine that is operated under maximum power may produce more heat than it can remove. The cooling is dependent upon the flow of air from the impeller on the flywheel to the metal engine components. An application that requires more power than the engine can provide will cause the engine to slow and will result in a decrease in cooling air delivered to the engine.

The solution is to educate the consumer about the best operating conditions for maximum engine life. A riding lawnmower that is operated in thick, long grass at full engine speed with the transmission in the highest gear will probably cause an overheating condition and reduce the engine's life.

7. Improper engine installation

Many times an engine is installed by the manufacturer or technician on an application that obstructs proper cooling.

8. Oil additives

It is not advisable to utilize any oil enhancers in the air-cooled engine. Most people think that if a little is good, then a lot is better. Under the extreme heat of the air-cooled engine, these additives break down to a sludge that insulates the heat inside the engine.

ENGINE RUNS, BUT STOPS AFTER A PERIOD

The following are the steps to take in order to rectify this problem, from most common to least common solutions, as shown in figure 10-12.

1. Fuel tank vent problem

The fuel tank must be vented to the atmosphere. A defective vent will cause a vacuum in the fuel tank, and the fuel supply to the carburetor will cease.

2. Carburetor float vent problem

The float bowl is vented. If the vent is plugged, the vacuum formed in the float area will lean out the mixture or stop the engine. Many manufacturers have proceeded to design internal vents that are located inside the carburetor. These do not plug easily because the air is filtered before entering the vent. Air passage through a conduit, whose opening is on the outer surface of the engine, will be readily obstructed by dirt and debris coming from the mowing, sawing, etc.

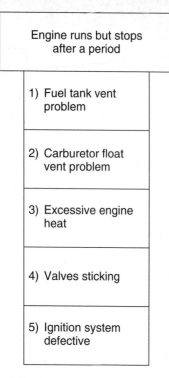

Fig. 10-12 The following are the steps from most common to least common solutions when an engine runs but stops after a period of time.

3. Excessive engine heat

Heat can change the operation characteristics of many components in an engine through thermal expansion. High temperatures may result in engine stoppage. Check for causes of excess heat, e.g., blocked air intake screens, dirt build-up between cooling fins, absent sheet metal engine housing (shroud), etc.

4. Valves sticking

Some valve stems will stick when the engine heats up. The valve may stick in the open position and compression will be lost. As the engine cools, the valve metal will cool and contract. The valve will snap back into place. Question the operator if a click was ever heard in the engine when it was cooling.

The valve should be removed and the valve guide cleaned with the proper reamer. A valve stem lubricant can be used when replacing the valve.

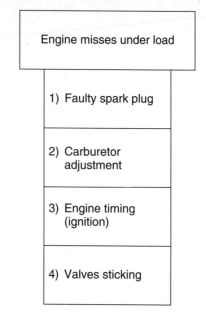

Fig. 10-13 The following are the steps from most common to least common solutions when an engine misses under load.

5. Ignition system defective

Electrical components are very susceptible to damage by heat. Components that operate when the engine is cool will prove defective under operating conditions. This is a difficult problem to isolate. When electrical components are checked with different testers, it is a good practice to heat the components, when possible, in an oven or with a heat gun, and retest. If the test equipment is not available, then replacement of the probable impaired part is necessary after all other possible causes have been ruled out. This can be a very expensive troubleshooting process.

ENGINE MISSES UNDER LOAD

The following are the steps to troubleshoot this problem, from most common to least common solutions, as shown in figure 10-13.

1. Faulty spark plug

Remove the spark plug and check the wear and condition of the electrodes. The distance or air gap between the center and side or ground electrodes should be measured. If the gap is too small, the heat from the spark across the plug's gap may not be sufficient to ignite the compressed air-fuel mixture in the compression stroke. If the gap is too large, the voltage may not be sufficient to jump the spark plug's gap. Remove the spark plug, and gap the plug with a wire feeler gage.

2. Carburetor adjustment

A "lean" carburetor adjustment will cause excessive heat to build up in the combustion chamber. Deposits may glow and ignite the fuel improperly, causing a "miss" in the engine. A lean carburetor mixture requires a higher voltage at the spark plug to ignite the air-fuel mixture. The additional voltage needed may not be supplied by the ignition system at lower engine speeds.

A "rich" carburetor adjustment will create excessive combustion chamber deposits that will interrupt the operation of the spark plug, causing a "miss" in the engine. Adjust the carburetor according to the manufacturer's specifications.

3. Engine timing (ignition)

The timing is set simultaneously or automatically on numerous engines when the point gap is adjusted. If the adjustment is not accurate, the timing will not be correct. A faulty condenser can cause the timing to vary. The structure of the condenser permits the primary current to cease when the points move apart. When the points open, the inertia of the electrons causes arcing or sparking across the small air gap. The primary circuit is not interrupted, and the spark plug is not fired. The condenser is designed into the circuit to prevent this arcing. A defective condenser will allow the engine to have a different timing sequence when the voltage in the primary is low at slow magnet speeds and when the voltage is high at rapid magnet speeds. A method of verifying that a condenser is faulty is to hook up a strobe timing light to the engine. The spark plug firing is sensed by the inductive pickup that is connected to the spark plug wire. While the engine is running, shine the light on any moving metal part. The strobe light will freeze the action if the spark is consistent. If the spark timing is inconsistent, the features on the moving parts will appear to "jump around."

4. Faulty ignition system

The voltage in the primary circuit of the ignition system varies with the speed of the magnets. The faster the magnets move by the coil, the higher the voltage. The points open and close, thus regulating the firing of the spark plug. Eventually they become pitted, corroded, or dirty. As a result, the spark plug voltage begins to deteriorate. Note when the miss occurs. A high-speed miss could be caused by a defective coil or solid state unit. At high speeds, the voltage is high in the primary and any weak area may short the system. A low speed miss could be caused by a high resistance in the primary circuit, e.g., dirty points. The voltage is weak at slow magnet speeds and may not charge the primary circuit.

5. Valves sticking

A valve that occasionally sticks will occur mainly while the engine is in a cool state. The "engine miss" can be recognized at low and high speeds.

A valve that is not seating properly will cause an engine miss, primarily at low engine speed, because the minimum compression necessary will be marginal, while at high engine speeds the slight valve leakage will not affect compression pressures.

A sticking valve may also be detected when a backfire is heard through the carburetor. The valves should be removed and a valve overhaul completed.

ENGINE LACKS POWER

The following are the steps to take in order to determine whether an engine lacks power, from most common to least common solutions, as shown in figure 10-14.

1. Dirty air cleaner

A dirty or plugged air cleaner will restrict the amount of air available to the engine and will cause a lack of power. An engine running at full throttle consumes an enormous amount of air. The volumetric efficiency of an engine has a direct correlation to the available power.

Properly cleanse and install the air cleaner.

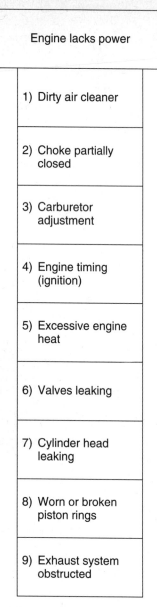

Engine lacks power
1) Dirty air cleaner
2) Choke partially closed
3) Carburetor adjustment
4) Engine timing (ignition)
5) Excessive engine heat
6) Valves leaking
7) Cylinder head leaking
8) Worn or broken piston rings
9) Exhaust system obstructed

Fig. 10-14 The following are the steps from most common to least common solutions when an engine lacks power.

2. Choke partially closed

Inadvertently, the operator may leave the engine choked. Improper choke linkage adjustment may restrict the choke's opening and cause it to work improperly.

3. Carburetor adjustment

Proper adjustment of a carburetor may not be possible because of foreign material contaminating the fuel as well as the carburetor. Make sure that there are no gasket leaks that may allow air to enter the cylinder without going through the carburetor. Check for a worn throttle or choke shaft. A method for detecting leaks involves spraying starting fluid in the leaking areas while the engine is idling. Any change in the sound of the running engine will indicate improper sealing. Adjust the carburetor properly.

4. Engine timing (ignition)

If the timing of the engine is retarded too much, the combustion pressure will not have its full consequence. The power output will be reduced. Correct any improper engine timing.

5. Excessive engine heat

The excessive heat caused by poor lubrication will cause the metal parts to expand, and the fit between the moving parts will be too tight. Binding will cause a resistance and apparent loss of engine effectiveness. Inspect the metal surfaces of the engine for any obstructions to the cooling air flow.

6. Valves leaking

Any valve that is not sealing the combustion chamber will reduce the power output of the engine. The compression pressure of the hot engine should be at least 60 PSI.

Establish that the valve lash adjustment is within specifications. Valve lash is the gap between the valve stem and the valve lifter when the piston is at TDC of the compression stroke. Many times, the valve stem will be too close to the head of the valve lifter, thus not permitting the valve to close tightly (especially true on an engine that has been short-blocked). Even though the valves are already installed when received from the factory, the valve lash adjustment may not be acceptable.

7. Cylinder head leaking

Inspect the cylinder head gasket for possible leaks. Any escape of combustion gases in this area will cause a reduction in the compression with diminishing power.

8. *Worn or broken piston rings*

Worn rings will no longer seal the combustion chamber to harness the maximum power. An indication of worn rings will be an excessive amount of oil exiting the breather valve area. The oil will flow through a tube connected to the breather, to the atmosphere (open system), or routed back to the air cleaner area of the carburetor (closed system).

9. *Exhaust system obstructed*

Assure that the exhaust flow is not inhibited by a kink, dent, or by carbon in the muffler. Two-stroke-cycle engines are likely to have a carbon build up in the exhaust area. Normal maintenance includes inspection of the exhaust port, spark arrester screen, and muffler openings.

ENGINE SPEED FLUCTUATES (SURGES)

The following are the steps to rectify engine speed fluctuations, from most common to least common solutions, as shown in figure 10-15.

1. *Carburetor adjustment*

The idle and high speed mixture adjustments must be correct for the different carburetor circuits to respond properly when the throttle plate is opened or closed.

2. *Governor linkage adjustment*

The governor must also be set so that there is no play in the system. Mechanical governor arrangements have an adjustment procedure found in the engine's service manual. This procedure removes all the gaps between the linkage components so that the throttle will respond quickly to changing engine conditions.

3. *Governor linkage binding*

The governor linkage and spring must move freely when the speed control lever is set at idle while the engine is not operative. Governor parts

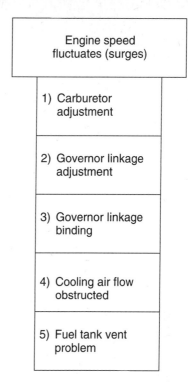

Fig. 10-15 The following are the steps from most common to least common solutions when an engine speed fluctuates (surges).

cannot rub against other portions of the engine because any friction creates an improper force that will disturb the balance of the system. Many times, when an engine is rebuilt, the newly applied paint job fills the holes, causing a binding at the pivot points. This must be avoided.

4. *Cooling air flow obstructed*

The pneumatic (air vane) governor may be affected by inappropriate air flow across the engine fins. Any debris that is caught in the fins or shroud, may cause a restriction in the air flow, resulting in a false sensing of engine speed. The spring tension will increase the engine's speed and lead to rapid deterioration.

5. *Fuel tank vent problem*

Unless atmospheric air is allowed to enter the fuel tank, the flow of fuel to the carburetor will be

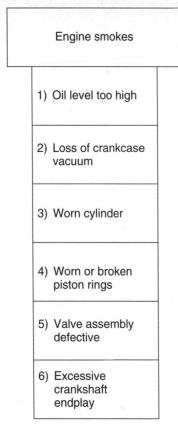

Engine smokes
1) Oil level too high
2) Loss of crankcase vacuum
3) Worn cylinder
4) Worn or broken piston rings
5) Valve assembly defective
6) Excessive crankshaft endplay

Fig. 10-16 The following are the steps from most common to least common solutions when an engine smokes.

impaired. A plugged or restricted fuel tank vent can cause a lean condition in the carburetor.

ENGINE SMOKES

The following are the steps to take in order to rectify engine smoking, from most common to least common solutions, as shown in figure 10-16.

1. *Oil level too high*

If the oil level is too high, the additional agitation by the internal moving parts will create excessive oil droplets that will pass out the breather valve and return to the combustion chamber through the carburetor. The splashed oil might be excessive in the cylinder area, resulting in an inefficient oil control from the piston rings.

2. *Loss of crankcase vacuum*

The loss of crankcase vacuum will allow excessive oil mist to exit the breather area. In a closed system, the breather tube routes back to the carburetor, where the crankcase vapors are burned. If too many oil droplets are sent back to the carburetor, the engine will "smoke" from the exhaust area. Areas to inspect for causes of a low crankcase vacuum are the oil fill gasket, damaged breather, or worn rings and cylinder.

3. *Worn cylinder*

The cylinder must be within specifications for proper sealing of the piston rings. Remove the cylinder head and inspect the piston top for the quantity of carbon. Since detergent oil is used in the engine, a large cleaned area reveals either cylinder or piston ring wear. Measure the cylinder and determine any excessive wear, out-of-round distortion, or taper.

Proper cylinder preparation and break-in is necessary to obtain a good seal between the cylinder and the piston rings. All cast-iron cylinders must be deglazed by a honing process when new piston rings are installed. The cylinder must have the proper cross hatch etched.

4. *Worn or broken piston rings*

If the cylinder is within specifications, the piston rings can be checked for wear by removing them from the piston and placing them into the cylinder bore, about one inch from the top. The ring end gap can be compared to a reject specification or a new piston ring's gap.

It is also possible that the piston rings have not been installed in their proper position or order.

5. *Valve assembly defective*

Remove the intake valve and check under the head of the valve and the valve ports for an oil or carbon deposit. A deposit will indicate that oil is being pulled into the combustion chamber due to a defective breather system, or past the valve stems.

6. *Excessive crankshaft endplay*

Excessive endplay will cause the piston and rings to have abnormal pressures applied to it and may not seal properly.

ENGINE HARD TO RESTART WHEN WARM

The following are the steps to take when an engine is hard to restart when warm, from most common to least common solutions, as shown in figure 10-17.

1. Flooded engine

When the warm engine is difficult to restart, the primary cause is a rich or flooded condition. An engine with a hot restart condition will have the following symptoms: engine starts and runs well, but when shut off, is difficult or impossible to restart until the engine cools off. If an engine has a hard hot restart symptom, first check the engine's spark plug to determine if a flooded condition exists. This is accomplished by removing the plug and observing the tip to see if it is covered with fuel. Hard hot restart can be caused by an improperly adjusted engine and/or equipment controls, or by a partially restricted air filter. Perform all initial adjustments and recheck for a hard restart condition before attempting more repairs.

2. Damaged carburetor parts

A damaged adjusting needle, O-ring, or loose needle/seat will cause a rich condition which contributes to hard hot restart problem. Inspect for missing, damaged, or loose parts.

A damaged diaphragm can cause a rich condition. If the diaphragm has been perforated, fuel may bypass the carburetor's metering system and be pulled directly into the engine through the air compartment and hole in the air horn.

The condition of the diaphragm link, diaphragm, and diaphragm spring can affect proper operation of the choke. If any of these components are damaged, a hot restart problem can occur. The use of non-original parts, which have not been manufactured to the manufacturer's specifications, can cause a hot restart condition.

3. Engine idle speed too low

Another critical factor is the engine's idle speed. When the engine's idle RPM is set too low, hesitation occurs during acceleration. To eliminate this

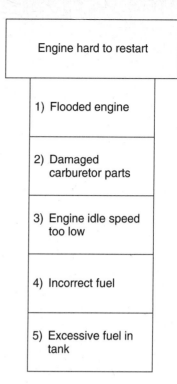

Fig. 10-17 The following are the steps from most common to least common solutions when an engine is hard to restart when warm.

situation, some technicians make the air-fuel mixture richer, which will cure the hesitation but cause a hot restart condition. The mixture adjustment should be on the lean side so that there is a slight hesitation when accelerating without stopping the engine.

4. Incorrect fuel

Winter grade fuel used in warm weather may vaporize too readily and cause a flooding (hot restart) condition.

5. Excessive fuel in tank

Too much fuel added to a warm engine equipped with a tank mount-style carburetor assembly will cause a rich condition resulting in difficult hot restart. The flooded condition will be even worse if the engine is bumped, tipped on an angle, or if the fuel tank cap is not venting properly.

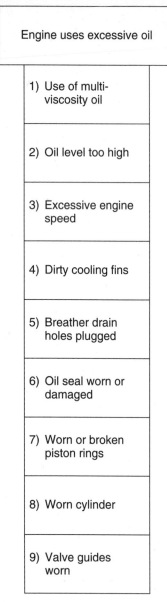

```
┌─────────────────────────────────┐
│   Engine uses excessive oil      │
└─────────────────────────────────┘
│
┌─────────────────────────────────┐
│ 1) Use of multi-                 │
│    viscosity oil                 │
├─────────────────────────────────┤
│ 2) Oil level too high            │
├─────────────────────────────────┤
│ 3) Excessive engine              │
│    speed                         │
├─────────────────────────────────┤
│ 4) Dirty cooling fins            │
├─────────────────────────────────┤
│ 5) Breather drain                │
│    holes plugged                 │
├─────────────────────────────────┤
│ 6) Oil seal worn or              │
│    damaged                       │
├─────────────────────────────────┤
│ 7) Worn or broken                │
│    piston rings                  │
├─────────────────────────────────┤
│ 8) Worn cylinder                 │
├─────────────────────────────────┤
│ 9) Valve guides                  │
│    worn                          │
└─────────────────────────────────┘
```

Fig. 10-18 The following are the steps from most common to least common solutions when an engine uses excessive oil.

USES EXCESSIVE OIL

The following are the steps to take to determine if an engine uses excessive oil, from most common to least common solutions, as shown in figure 10-18.

1. Use of multi-viscosity oil

Most manufacturers recommend the use of a straight weight oil, e.g., SAE 30. The use of multi-viscosity oil increases oil loss. This phenomenon is especially observed in an engine that has some internal wear.

2. Oil level too high

The engine oil level is checked by a "dip stick" on many engines. When the dip stick is lost or damaged, the replacement, although it appears to fit, may not be the proper length. Now the oil level may be too high, even though it is in the safe area on the dip stick. Excessive oil is splashed over the engine parts, and when the oil-saturated air exits the crankcase breather, it reduces the level in the crankcase. Adjust the oil to the proper level.

3. Excessive engine speed

The faster the engine operates, the less efficient the seal between the piston rings and the cylinder walls. Adjust the engine's top speed to the recommendation of the manufacturer.

4. Dirty cooling fins

If the flow of air is obstructed around the cylinder, excessive heat will be retained and cause uneven expansion and contraction of cylinder metal. The circular cylinder bore will be distorted and will allow an inefficient sealing of the piston rings. Remove all debris from the cylinder fins.

5. Breather drain holes plugged

When the air and oil mist is passed through the filter in the breather, the oil is collected and allowed to drain back to the crankcase. If these holes are plugged, the oil will build up in the breather and eventually be pushed out the breather tube. The oil may be expelled to the atmosphere or back to the carburetor, in a closed system, and burned. Oil is lost from the system. Disconnect the breather hose and run the engine. Observe if oil is blown from the breather tube. Also notice if the signs of oil burning in the exhaust gases cease.

6. Oil seal worn or damaged

Oil may be lost from an oil seal that is not properly sealing the crankshaft where is exits the crankcase. Replace any defective oil seals.

7. Worn or broken piston rings

When the rings wear, the oil may escape to the upper part of the combustion chamber. The oil is then burned and the operator must add oil often to the crankcase. The bottom piston ring is the primary oil control ring. If the engine oil is contaminated with abrasive particles, this ring will wear quickly. Inspect the ring wear and replace, if necessary.

8. Worn cylinder

Cylinder walls wear and weaken the seal between the rings and cylinder, resulting in loss of large quantities of oil. If the cylinder is within tolerances, then new piston rings may be installed to correct the problem. If the cylinder bore is out of specification, it must be oversized or replaced with a new one.

9. Valve guides worn

Oil will escape from the crankcase through the clearance between the valve guides and stems. This clearance is a greater problem with the modern engine. The clearance has been widened with the use of unleaded fuels. There is not as much lead in the fuel to lubricate the valve train. The additional clearance allows for some of the crankcase oil vapor to pass through. Replace the valve guides or ream the hole larger and install a new valve with a thicker valve stem.

EXCESSIVE OIL OUT OF BREATHER

The following are the steps to take to determine whether excessive oil is passed out of the breather, from most common to least common solutions, as shown in figure 10-19.

1. Oil level too high

The crankcase air contains an oil mist from the splashing of the engine oil. The breather mechanism is capable of filtering out this mist and returning it

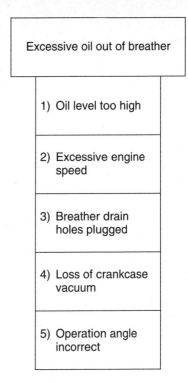

Fig. 10-19 Excessive oil out of breather.

to the crankcase. If the oil level is too high, the oil mist will be excessive, resulting in the breather being unable to filter all the oil from the exiting air. The excessive oil is passed out of the breather tube to the atmosphere or routed to the carburetor's air cleaner.

2. Excessive engine speed

An engine operated for a period of time at a speed greater than recommended will force too much oil mist out of the breather. The piston rings become less effective at this higher speed and allow more blow-by-gases into the crankcase.

With a vibra-tachometer, adjust engine maximum speed to the manufacturer's specifications.

3. Breather drain holes plugged

The oil removed from the vacating crankcase air flows back to the crankcase by a passage in the breather element. If this circuit becomes obstructed, the oil will fill the breather and eventually pass out the breather hose. On some applications the breather cover is installed upside-down, preventing the normal flow back to the crankcase. A new gasket must be used when the breather is removed. A liquid gasket sealer should not be used to prevent the drain holes in the breather from plugging.

4. Loss of crankcase vacuum

It is important for an engine to maintain a crankcase vacuum while it is operating. The crankcase breather valve allows crankcase air to escape when the piston is moving downward, but prohibits entry of external air when the piston is moving upward. This vacuum reduces the air mass in the crankcase so that, during the power stroke, blow-by gases do not add to it to an abnormal degree. If the mass of air in the crankcase is excessive, then too much oil mist will be expelled through the breather causing the excess oil to continue through the breather tube.

The crankcase pressure can be tested by an open-ended manometer that can be purchased or made. A good crankcase vacuum at idle would range from 9–15 inches of water. The vacuum should reduce slightly with an increase in speed. A vacuum gage used for automotive purposes cannot be used since 7 inches of water equals 1 inch of mercury, and it would not be sensitive enough to identify any problems.

A. DAMAGED BREATHER The breather is an important one-way valve that allows air to be pushed from, but not sucked into, the crankcase. Inspect the breather valve for any damage.

B. DAMAGED SEALS The most common leak that causes crankcase vacuum loss is from the oil filler area. This area must not allow any air into the crankcase. The problem is magnified when the extended plastic fill tubes are used. The O-rings become damaged or lost, and many times an old rag is used to plug the filler hole.

Oil seals are primarily designed to prevent oil from exiting the engine around the crankshaft. A secondary but important function is to stop any air from entering during the period of crankcase vacuum. Oil seal failure may occur without any external leaking indications. Temporary seal can be obtained by applying some oil to the outside of the seal when the engine is operating. Then observe any changes.

C. PISTON RINGS NOT SEATING The piston rings may not be sealing the combustion pressures and, thus, allow excessive blow-by. Either the rings may be worn or both the cylinder bore and rings are beyond specifications.

5. Operation angle incorrect

Four-stroke-cycle engines are not as adaptable as are the two-stroke-cycle engines when used on steep angles. The crankcase oil in a four-stroke-cycle engine will flood to one side during steep application and may cause oil to be expelled from the breather.

OIL SEAL LEAK

The following are the steps to take to rectify an oil seal leak, from most common to least common solutions, as shown in figure 10-20.

1. Oil seal worn or damaged

The oil seal may wear and be no longer capable of retaining the oil. It is common for fishing line to accumulate on the power output side of the crankshaft. The string will move on the crankshaft toward the engine, lodging in the oil seal and causing failure.

2. Bent crankshaft or blade

Always look for a bent crankshaft, which can damage an oil seal. Check it out by removing the spark plug from the engine and pulling the starter rope. Watch the end of the shaft for any wobble.

3. Worn crankcase bushing

When replacing a defective oil seal, check the clearance between the crankshaft and the PTO or the magneto bushing or bearing. Move the crankshaft from side to side. If the play is too large, the new oil seal will quickly wear out.

4. Oil seal improperly installed

The oil seal may be damaged during the installation. It is important to protect the soft edges from any abrasion. If a seal protector is not available, then put some wax paper around the shaft before the oil seal is installed. The seal must be put on without damage. Use a deep socket to drive the bushing in. Make careful taps around the outer edge while installing.

5. Defective breather

The crankcase breather must let air out when the piston moves downward and prevent air from entering while the piston is moving upward. If the breather valve is not working correctly, the blow-by gases from combustion will create a positive pressure rather than a vacuum. With this set-up, and the piston moving downward, the crankcase air cannot escape fast enough. The crankcase pressure is great enough to push the oil and air past the oil seals.

6. Loss of crankcase vacuum

The most common cause for loss of crankcase vacuum is the oil filler not being properly sealed. When the piston is moving upward, there is a vacuum formed in the crankcase. If there is not a good seal at the oil filler, unfiltered air will enter and contaminate the oil. In addition, a vacuum is not formed, and the engine will force air and oil out the oil seals. If the engine is equipped with a mechanical governor, the governor shaft that protrudes from the engine will also show signs of leaking oil.

ADDITIONAL TEST
1. Crankcase vacuum

A partial crankcase vacuum is necessary for proper engine operation. Excessive oil consumption or a "smoking" engine can be improperly diagnosed as defective piston rings when actually the cause may be a poor crankcase vacuum. The crankcase vacuum should be checked on an engine that is consuming excessive oil, especially an engine that is relatively new or recently rebuilt.

Low crankcase vacuum is caused by a defective breather valve or other air leaks in the crankcase, such as the oil fill plug or the crankshaft oil seals.

As the piston moves down, the air molecules in the crankcase are compressed to form a pressure that is released through the breather valve. As the piston rises, a vacuum is formed and since the breather

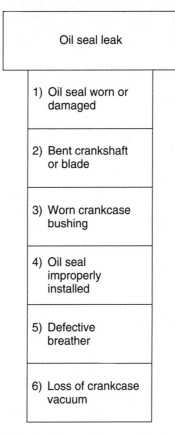

Fig. 10-20 The following are the steps from most common to least common solutions when an engine's oil seal leaks.

valve is a one-directional valve that lets air out of the crankcase but not back in, as shown in figure 10-21. A partial vacuum is formed.

As the combustion stroke drives the piston down, combustion pressures force a small amount of combustion gas past the piston rings into the crankcase, which reduces the vacuum. The added gas molecules are passed out the breather valve and are expelled into the atmosphere in the open system, as shown in figure 10-22, or are routed to the air cleaner side of the carburetor to be burned again in a closed system, as shown in figure 10-23.

Since the oil reservoir is also the crankcase in a four-stroke-cycle engine, the air that is expelled through the breather valve contains many oil droplets. A filter in the breather removes these droplets and returns them to the crankcase.

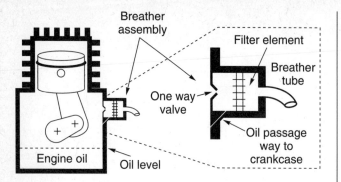

Fig. 10-21 As the piston rises, a vacuum is formed and since the breather valve is a one-directional valve that lets air out of the crankcase but not back in.

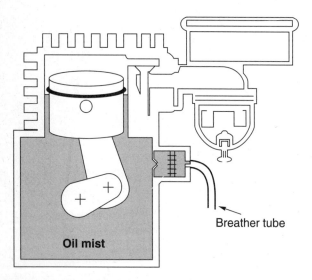

Fig. 10-22 The added gas molecules are passed out the breather valve and are expelled to the atmosphere in the open system.

A crankcase vacuum is necessary to control the amount of air exiting the crankcase and to halt the intake of possibly dirty air.

If the breather valve does not close, or is slow in closing, air is drawn back into the crankcase. Leaks in the crankcase from a oil filler gasket, or O-ring, a defective oil seal, or piston rings that are not sealing, can also admit air. When this happens, crankcase vacuum is reduced or lost. When the crankcase vacuum is reduced, additional oil-saturated air must be

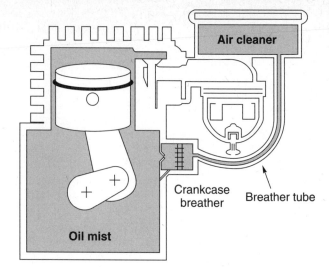

Fig. 10-23 Gas molecules routed to the air cleaner side of the carburetor to be burned again in a closed system.

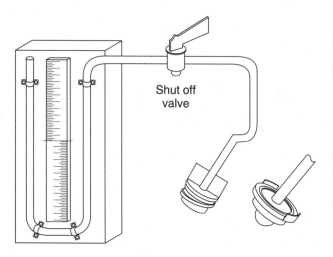

Fig. 10-24 The crankcase vacuum can be checked with a water manometer.

removed through the breather on each stroke. This excess movement of air through the breather may cause some oil droplets to pass through the filter, which can cause an oil loss that can splatter over the engine in an open system or be routed back to the carburetor to be burned and produce the symptom of a smoking engine.

Pressure build-up in the crankcase will push oil past the crankshaft oil seal and crankcase gaskets.

The crankcase vacuum can be checked with a water manometer, as shown in figure 10-24, or vacuum gage that is inserted at the oil filler hole. A vacuum of 9–15 inches of water, or 0.9–1.5 inches of mercury, should be present when the engine is running. The water manometer is the preferred tool because a pressure gage that measures in inches of mercury is not sensitive enough to detect a low vacuum. The manometer can be purchased or built.

To test the crankcase vacuum:

1. *Install one end of the hose from the manometer into the oil fill hole. The other end should be open to the atmosphere, but a shut-off clamp must be in place and closed while starting or stopping the engine. Without the shut-off valve, the water from the manometer may be sucked into the crankcase.*

2. *Start the engine and operate it at high speed (3200–3600 RPM).*

3. *Open the shut-off valve and note the water levels. Measure the distance in inches between the water levels on both sides of the tubes. The level on the engine side should be 5 to 10 inches above the level in the open side. If there is no vacuum, both sides will be the same. If there is a pressure, the engine side will be lower than the other side.*

The most common cause of low crankcase vacuum is an oil fill cap that is leaking or missing. Engines have been received by a repair shop with a cloth or cork stuffed into the oil fill hole. Another easy check for low crankcase vacuum is to check for a defective breather valve with the specifications in the engine's service manual.

PROBLEM-SYMPTOM SECTION

This section consist of a compilation of different problems and what some of the symptoms may be. For example: **a bent crankshaft** can cause an **engine to vibrate** or the **oil seal to leak.**

Problem	Symptom
Bent crankshaft or blade	Engine vibrates
	Oil seal leak
Blade loose	Engine knocks
Blade not aligned	Engine vibrates crankshaft
Blade out of balance	Engine vibrates
Breather drain holes plugged	Excessive oil out of breather
	Uses excessive oil
Carbon in combustion	Engine knocks chamber
	Engine overheats
Carburetor adjustment	Engine lacks power
	Engine misses under load
	Engine overheats
	Engine speed fluctuates
	Fuel flow problems
Carburetor float vent problem	Engine runs but stops
Cooling air flow obstructed	Engine overheats
	Engine speed fluctuates
Counter balance not aligned	Engine vibrates
Cylinder head leaking	Engine lacks power
Defective breather	Oil seal leak
Dirty air cleaner	Engine lacks power
Dirty cooling fins	Uses excessive oil
Engine mounts not secure	Engine vibrates
Engine timing (ignition)	Engine knocks
	Engine lacks power
	Engine misses under load
	Engine overheats
Excessive crankshaft endplay	Engine smokes
Excessive engine heat	Engine lacks power
	Engine runs but stops
Excessive engine loading	Engine overheats
Excessive engine speed	Excessive oil out of breather
	Uses excessive oil
Exhaust system obstructed	Engine lacks power
Faulty spark plug	Engine misses under load
Fuel tank contamination	Fuel flow problems
Fuel tank vent problem	Engine runs but stops
	Engine speed fluctuates
	Fuel flow problems
Governor linkage adjustment	Engine speed fluctuates

Governor linkage binding ....Engine speed fluctuates

Improper engine installation ....Engine overheats

Improper flywheel torque ....Engine knocks

Lack of lubrication ....Engine overheats

Loss of crankcase vacuum ....Engine smokes

....Excessive oil out-of-breather

....Oil seal leak

Oil level too high........................Engine smokes

....Excessive oil out-of-breather

Oil seal improperly installed....Oil seal leak

Oil seal worn or damaged ....Oil seal leak

....Uses excessive oil

Operation angle incorrect ....Excessive oil out-of-breather

Piston rings problems ....Compression problems

Use of multi-viscosity oil ....Uses excessive oil

Valve assembly defective ....Engine smokes

Valve guides worn.....................Uses excessive oil

Valve problems..........................Compression problems

Valves leaking ....Engine lacks power

Valves sticking ....Engine misses under load

....Engine runs but stops

Worn connecting rod ....Engine knocks

Worn crankshaft bushing..........Oil seal leak

Worn cylinder ....Engine smokes

....Uses excessive oil

Worn or broken piston rings.....Engine lacks power

....Engine smokes

....Uses excessive oil

SUMMARY

A professional technician is an individual who can identify the problem efficiently and solve it with minimal expense. The ability to troubleshoot a problem quickly and accurately is one of the most valuable talents, or tools, that the repairman can have.

The technician increases the probability of efficiently correcting any problem when he logically collects correct, relevant information and interprets it properly. Troubleshooting is a systematic approach of problem solving that requires a special plan of attack that is used from one engine to another.

The **Systematic Check** is an efficient, simple routine that is a logical approach to troubleshoot starting problems. The preliminary examination of the engine with the customer will lead to a guess about the cause of the malfunction. The next step involves a methodical, or well-ordered, testing to verify the assumption and arrive at the solution.

Questions

1. **An engine has failed because of lack of lubrication. Technician A says that the cause could be low oil or overspeeding. Technician B says that it could be dirt or the use of the wrong oil.**
   Who is correct?

   A. Only Technician A
   B. Only Technician B
   C. Both Technician A and B
   D. Neither Technician A or B

2. **Looking at the underside of a flywheel, a crack to the left of center is observed. Technician A says that this is due to a sudden stop which occured when hitting something. Technician B says that this is due to improper torque on the flywheel nut.**
   Who is correct?

   A. Only Technician A
   B. Only Technician B
   C. Both Technician A and B
   D. Neither Technician A or B

3. The carburetor can be checked for dirt ingestion. Technician A says that this can be done by removing the air cleaner and checking the choke and throttle shafts. Technician B says that this cannot be done because dirt passes through the carburetor and into the combustion chamber, leaving no deposits. Who is correct?

   A. Only Technician A
   B. Only Technician B
   C. Both Technician A and B
   D. Neither Technician A or B

4. In the problem solving sequence, Technician A says that the problem must be defined before it can be solved. Technician B says that the possible solution to the problem should be pursued after researching what is known about the problem. Who is correct?

   A. Only Technician A
   B. Only Technician B
   C. Both Technician A and B
   D. Neither Technician A or B

5. The systematic check is a routine that is a logical approach to the troubleshooting process. Technician A says that the systematic check should commence without the customer. Technician B says that it is important to have the customer present during the initial stages of the systematic check. Who is correct?

   A. Only Technician A
   B. Only Technician B
   C. Both Technician A and B
   D. Neither Technician A or B

6. The ignition should be checked with the spark tester connected to the spark plug and again when it is connected to the engine metal. Technician A states that this is necessary to check if the failure is caused by a defective spark tester. Technician B states that if there is no spark in either condition, the spark plug is defective. Who is correct?

   A. Only Technician A
   B. Only Technician B
   C. Both Technician A and B
   D. Neither Technician A or B

7. A defective carburetor can be identified if an engine will only start when WD-40 is sprayed into the carburetor with the air cleaner removed. Technician A states that it is because the WD-40 reduces the required oxygen to the combustion chamber. Technician B states that since the WD-40 is easily vaporized and combustible, the spray will provide a fuel charge that will be ignited if other engine components are working properly. Who is correct?

   A. Only Technician A
   B. Only Technician B
   C. Both Technician A and B
   D. Neither Technician A or B

8. The ignition system will not produce a spark. Technician A states the most probable cause is that the magnets have lost their magnetism. Technician B states that the possible cause is that the primary circuit is grounded to the engine frame. Who is correct?

A. Only Technician A
B. Only Technician B
C. Both Technician A and B
D. Neither Technician A or B

9. If the engine mounts are not secure, Technician A says that the engine will produce a knocking noise. Technician B says that the lawnmower will vibrate. Who is correct?

A. Only Technician A
B. Only Technician B
C. Both Technician A and B
D. Neither Technician A or B

10. If there is a fuel tank vent problem, Technician A states that the engine may run, but will stop, or the engine speed may fluctuate. Technician B states that it could be a fuel flow problem and the engine will run too rich. Who is correct?

A. Only Technician A
B. Only Technician B
C. Both Technician A and B
D. Neither Technician A or B

GLOSSARY

Aftermarket item An item that is not normally added to the engine by the manufacturer.

Air A mixture of gases found in the atmosphere. The major gases are nitrogen (78%) and oxygen (21%).

Air bleed Part of the carburetor that allows air to break up the fuel before entering the air stream in the venturi.

Air cleaner An appliance used to remove dust, dirt, and other airborne material from the air as it is drawn into the engine.

Air cooling The process of cooling an engine by moving cooler air over the hot fins of the cylinder.

Air horn The flanged surface located at the inlet of the carburetor's air channel on which the air cleaner base rests.

Air-fuel mixture The mixture of air and vaporized fuel provided to an engine through a carburetor.

Alcohol A fuel produced by the distillation of wood or grain products or derived from natural gas. The two major types of alcohol are ethanol (ethyl alcohol), made from grain, and methanol (methyl alcohol), made from wood.

Alloy A metal composed of two or more metals.

Alnico magnets Permanent magnetic fields made by a combination of aluminum, nickle, and cobalt.

Alternating current (AC) An electrical cur-rent which constantly reverses direction and polarity.

Alternator An electrical generator that produces alternating current which may be converted to DC current.

Aluminum A silvery metal that is lightweight, easi-ly machined, and resistant to corrosion.

Ambient air The outside air surrounding an engine.

Ammeter A gage used to measure the amount and direction of electric current flow.

Ampere (Amp) The electrical unit of measurement used to indicate the electric current flow rate in an electrical circuit.

Annealing The process of making a metal less hard by slowly heating and cooling it.

Antioxidant A fuel additive which reduces the for-mation of gums and varnishes and maintains the octane rating.

Arc An electrical flow across two conductive elec-trodes, separated by an air gap, that produces intense light and heat.

Atmospheric vent The opening used to allow atmospheric pressure to the fuel in the float area of a carburetor.

Atom The smallest particle of any element.

Atomization The process of breaking a liquid down into small droplets.

BDC Bottom Dead Center. The lowest piston posi-tion in the cylinder.

BTDC Before Top Dead Center. When the piston is on the upward stroke of the compression cycle.

Ball Bearing An antifriction bearing consisting of an inner and outer race separated by hardened steel spheres.

Bench grinder A grinding wheel, made of abrasive material, that is rotated by an electric motor. The grinder can be used to shape and sharpen tools.

Bi-metallic Material made of two different metal substances that expand at different rates when heated.

Blade brake A device used to stop the rotation of the lawnmower blade.

Blade tip speed The speed in feet per second of the tip of a lawnmower blade. It is calculated by knowing the speed of the engine and the total length of the blade.

Blow-by A condition caused when exhaust gases escape past the piston rings into the crankcase.

Bore Inside diameter of an engine cylinder.

Boss A raised part of a flat surface.

Boundary lubrication A partial lubrication is always present during engine start-up, initial operation of a remanufactured engine, and in places where the oil supply is limited and tem-peratures from the combustion are high.

Bowl vent Part of the carburetor that allows atmos-pheric air pressure to enter the carburetor system.

Brake horsepower (BHP) The actual horsepower delivered at the crankshaft of an engine.

Breaker points Contact points used to open and close the primary circuit of the ignition system.

Bronze A metal alloy composed mostly of copper and tin.

BTU The British Thermal Unit is a measurement for a quantity of heat.

Burr A sharp, rough area on a metal component.

Butterfly valve A pivoting metal plate used to control the flow of air into the carburetor.

C-clamp A moveable or portable device for holding pieces of material together while working on them.

Caliper An adjustable tool used to measure inside and outside diameters.

Cam A curved teardrop shape surface found on a camshaft that is used to move another component as the shaft rotates. It is also used on the crankshaft to open and close the contact points, and also opens and closes the valves.

Camshaft A shaft which consists of a set of cam lobes (one for each valve) that control the timing of the valve opening and closing.

Capacitor Two metal sheets separated by an insulator used to store an electrical charge.

Carbon dioxide A gas that is colorless, tasteless, odorless, and non-combustible.

Carbon monoxide A gas that is colorless, tasteless, odorless and highly poisonous. It is commonly produced by the incomplete combustion of gasoline.

Carburetor A fuel and air metering appliance used to change liquid gasoline into fuel vapor and mix it with the proper amount of air.

Carburetor repair kit A kit composed of repair parts and gaskets for overhauling a carburetor.

Centrifugal force A force created which tends to pull an object away from a center when rotated.

Chamfer A flat surface which is beveled or angled off from an adjacent surface.

Check valve A valve designed to permit a one-way flow of a fluid or gas.

Chisel A tool which is used to cut metal by driving it with a hammer.

Choke plate Part of the carburetor that partially blocks off air flow, creating low pressure throughout the carburetor to provide a rich-in-fuel mixture for cold starting.

Circlip A circular clip or snap ring that fits into a groove which is used to retain a shaft or component.

Circuit The path over which a current of electricity flows.

Closed circuit A complete electrical circuit.

Cobalite See **Stellite**.

Combustion The burning of the air-fuel mixture. Combustion is the first step in changing the energy in the fuel into a mechanical force that can be used to propel the piston.

Combustion chamber The part of the cylinder into which the fuel charge is compressed and then ignited. It is enclosed and formed by two units called the cylinder and cylinder head.

Compression The process of pressing into a smaller space; the process of condensing or reducing size or volume.

Compression ratio A comparison of two quantities, stated as a numerical relationship. It is based on the engine combustion chamber volume while the piston is at the bottom of its travel compared with the volume when the piston is at the top of its travel.

Compression release A device that reduces the compression pressures during the starting cycle of an engine.

Compression stroke The stroke in the four-stroke-cycle internal combustion engine in which the upward movement of the piston "squeezes" the air-fuel mixture into the combustion chamber. The greater the compression, the greater the power produced by the engine.

Condenser See **Capacitor**.

Conductor A material that has many free electrons, allowing the unrestricted flow of current.

Connecting rod A part which furnishes a means of converting the reciprocating motion (up and down) of the piston to a rotating movement of the crankshaft.

Contact points A switching appliance used to start and stop the flow of current.

Continuity A continuous path for electrical current to flow.

Conventional theory A theory that states that current flows from positive (+) to negative (−).

Corrosively Describes a substance which dissolves metals and other materials or burns the skin.

Counterbalance Extra masses (weights) geared to the crankshaft to reduce engine vibrations.

Crankcase The lower part of an engine block that houses the crankshaft.

Crankpin The offset journal of the crankshaft that

supports the connecting rod. The crankpin allows the crankshaft to change reciprocating motion into rotary motion.

Crankpin journal A round metal surface on the crankshaft to which the connecting rod is attached.

Crankshaft Converts the up and down movement of the piston, from the power stroke of the piston, into a rotational force called torque, which turns the pulley, blade, etc.

Cubic centimeter The metric system unit for volume. One cubic centimeter represents a space that is one centimeter high, one centimeter wide, and one centimeter deep.

Cubic inch An English system unit for volume. One cubic inch represents a space that is one inch high, one inch wide, and one inch deep.

Current The flow of electrons through a conductor.

Cylinder A long, round object with flat ends and which, in the internal combustion engine, is hollow and houses the piston.

Cylinder head A detachable component of the engine into which is machined the combustion chamber and socket for the spark plug.

Cylinder leakage test A test that will indicate the condition of the cylinder and valve train by filling combustion chamber with air through the spark plug opening. The leakage of air identifies the combustion chamber problems.

Detergent oil An oil which keeps contaminants and debris in suspension so that it may be removed when the oil is changed.

Detergents A fuel additive that reduces the buildup of varnish and carbon deposits within an engine.

Diagnosis The process used to find the cause of a problem by determining its nature and circumstances.

Dial caliper A measuring tool that provides a quick method of making a direct inside or outside measurement of an object.

Dial indicator A tool used to detect slight movement and display it on a circular gage face.

Diameter A straight line which runs through the center of a circle with both ends ending at the intersection of a circle's circumference.

Diaphragm A flexible rubber-like sheet used to cover an area, separating it into two different compartments. It is commonly used in fuel pumps where its back and forth movement pumps fuel.

Die Runs over the outside of a rod to make external or outside threads. A die of the correct size is placed in the handle and is turned.

Diesel engine An engine built to ignite diesel fuel solely by the heat of highly compressed air within the cylinder.

Diode A solid state electronic appliance that permits current flow in one direction only.

Direct current (DC) A flow of current in the same direction and polarity.

Displacement The volume of air moved by a piston which is traveling from the bottom to the top of its stroke.

Drill motor Used to power the drill bit.

Drill press Used to drill a hole in material where positioning and depth angle is crucial.

Drill, twist Tools for making holes and are constructed from a round bar with grooves cut in it.

Dynamometer A device used to measure the working ability of an engine. It actually "loads" the engine, causing it to work.

Eddy currents Currents induced in the core of a coil by the changing magnetic fields. These currents cause great losses of energy and emit heat.

Electricity A form of energy created by the movement of electrons from one atom to another.

Electrode Conductors at the center and side of the spark plug that provide an air gap for an electric arc to start combustion process in an engine.

Electromagnet A magnet formed by passing a current of electricity through wire that is wound around an iron core.

Electron An atomic particle with a negative charge.

Electron theory The theory that states that current flows from negative (−) to positive (+).

Element The simplest form of matter composed of atoms that are all of the same kind.

EMF ElectroMotive Force (voltage).

End gases The combustion gases formed at the end of an air-fuel mixture's combustion process.

EPA The abbreviation for the Environmental Protection Agency.

Evaporation The process that occurs when a substance is changed from a liquid to a vapor.

Exhaust The by-products generated in a combustion chamber as a result of burning the fuel.

Exhaust stroke The stroke of the four-stroke-cycle internal combustion engine in which the movement of the piston expels the exhaust gases from the combustion chamber.

External combustion engine An example is the steam engine, in which burning of the fuel in an <u>externally</u> fired boiler generates steam which is then transmitted to the engine cylinder where it is expanded against the piston.

Extractor A tool which is used to remove a bolt that has broken off.

Fastener A component used to attach one part to another, such as a nut and bolt, screw, rivet, etc.

Feeler gages Strips of hardened steel that are rolled to the proper thickness with extreme accuracy.

Field An area of magnetic force that surrounds a magnet.

File, double cut A tool that is used for fast removal of metal and easy clearing of chips.

File, single cut A tool that is used when a smooth finish is desired or when hard materials are to be finished.

Flange A projecting rim or collar used to give added strength to a metal fitting and often help to hold it in place.

Float A flotation device used in a carburetor float bowl with an inlet valve to regulate fuel flow and level.

Float bowl The fuel reservoir located within a carburetor for receiving fuel from a fuel supply and storing it at a constant level.

Float level The level of fuel maintained in the carburetor float bowl that is controlled by a float, inlet needle, and seat.

Fluid Any substance characterized by its ability to flow readily and which requires a vessel to contain it. Gases and liquids are fluids.

Flux A chemical used to clean a joint to be soldered of dirt and oxidation and keep oxygen away from the molten metal.

Flywheel A weight or mass attached to the crankshaft to maintain the inertia of its spinning during the three non-power producing strokes in a four-stroke-cycle engine.

Fogging The process when engine oil or fogging oil is squirted into the carburetor air horn with the air cleaner removed while the engine is operating in such quantities to stall the engine.

Force A cause that puts a resting object into motion or, if an object is already moving, will change the motion or momentum.

Four-stroke-cycle engine The five parts of combustion occur for every four strokes (two up-strokes and two down-strokes) of the piston.

Free electrons Electrons held loosely in the orbit of an atom.

Friction The resistance generated between two objects touching and moving by each other.

Fuel Any substance characterized by readily burning and furnishing heat.

Fuel pump An appliance used to move fuel from a fuel tank to an engine's carburetor.

Fuel system A system that performs many functions which include storing the fuel, delivering the fuel to the carburetor, and changing the liquid gasoline to a vapor.

Fuse An appliance used to protect electrical circuits from overloading. A link in the fuse melts and opens the circuit if the current is above the normal rating.

Gap The distance measured between a pair of contact points when they are fully open.

Gasohol A fuel mixture, usually composed of 90% gasoline and 10% alcohol.

Gasoline A volatile liquid hydrocarbon fuel used to power the internal combustion engine. Gasoline is one of the products produced when crude oil is refined.

Governor (mechanical) This type of governor system senses the engine speed by a mechanism that is geared to the crankshaft or camshaft.

Governor (pneumatic) This type of governor system senses the engine speed by the force of air blowing on a flap attached to the engine.

Governor adjustment The adjustment procedure necessary to remove any possible play between the components, which permits a quicker adjustment to different loads.

Governor hunting The most common problem encountered with the governor system is identified by engine pulsations or fluctuations varying between wide open and idle positions of the throttle creating a wide range of engine speeds.

Governor link A wire link between the throttle of the carburetor and the engine speed sensor.

Governor sag The term used to denote the temporary reduction of engine speed when more effort is required (increased load).

Governor spring A part of the governor system which, when stretched, will attempt to open the carburetor throttle.

Governor systems Systems which are included on almost all air-cooled engines. They regulate the carburetor's throttle opening to achieve the engine speed desired by the operator.

Ground An electrically conductive body that is a return path for an electrical circuit.

Hacksaw A tool which is used for cutting metal.

Hammer, Ball-peen A steel hammer that has a flat face for hammering and a ball part for rounding off rivets.

Hammer, Brass faced A soft faced hammer with the weight and impact force of steel.

Hammer, Rawhide A soft faced hammer like the rubber hammer that will not damage fragile parts and is intended to protect any machined surfaces.

Hammer, Rubber A soft faced hammer like the rawhide hammer that will not damage fragile parts and is intended to protect any machined surfaces.

Head gasket A gasket material placed between the cylinder head and the engine block to seal the combustion chamber.

Heat sink A heat-dissipating mounting for diodes and other components. A heat sink prevents those parts from damage caused by overheating.

Helicoil A threaded insert product used to restore damaged threads by drilling them out, retapping the new hole, and then screwing a springlike set of threads into the newly tapped hole.

Hone A rotating tool for machining the inside surface of a cylinder to a smooth surface and remove any glazing.

Horsepower The measurement of an engine's rate of doing work. One horsepower is defined as the unit of energy needed to lift 550 pounds a distance of one foot in one second.

Hydrocarbon An organic compound composed of only hydrogen and carbon.

Hydrodynamic lubrication A type of lubrication that occurs when two mating metal surfaces are separated by a layer or film of oil.

Idle mixture adjustment screw An adjustable screw mounted near the base of the carburetor for

metering the air-fuel mixture into the airway during low speed operation.

Idle speed The engine rotation speed set when an engine is running and the carburetor throttle plate is nearly closed.

Ignitability A term to describe a liquid with a flash point below 140°F or a solid that can spontaneously ignite.

Ignition system The system that boosts the voltage produced or supplied to generate a spark.

Ignition timing The relation of the piston's position in the combustion chamber when the spark plug fires.

Impeller A circular, rotating, fan-like appliance used to pump air or fluid.

Induced voltage Voltage generated in an electrical conductor as a result of the magnetic field moving past the conductor.

Induction The transfer of electricity by means of magnetism.

Inertia The tendency exerted by matter to remain in its state of motion. A moving body will tend to move and a stationary body will tend to remain stationary.

Inlet needle A carburetor part that opens as the float level drops to allow fuel to enter the bowl area.

Inner race A smooth hardened surface designed for bearing rollers of balls to ride on. Roller bearings typically have the bearing race and the rollers as one integral unit.

Insulator A material that will not conduct an electrical current.

Intake stroke The stroke of the four-stroke-cycle internal combustion engine in which the movement of the piston draws air and fuel into the combustion chamber.

Internal combustion engine The air-fuel mixture is burned and expanded within the combustion chamber, which is that portion of the cylinder above the piston at top dead center (TDC).

Jet A mechanical appliance used for controlling fuel flow in various parts of the carburetor.

Journal A bearing surface designed to support machine parts and allow them to slide or rotate, such as the rod journal or crankshaft.

"L" head engine An engine built with both valves located in the engine block on one side of the cylinder.

Lapping The process of fitting two surfaces together by placing an abrasive material between them and then rubbing or moving the surfaces against each other.

Lean mixture An air-fuel ratio that contains extra air compared to fuel.

Liquid cooling The use of a liquid to remove the heat of combustion from the cylinder.

LPG Abbreviation for Liquefied Petroleum Gas such as propane, butane, or natural gas.

Magnet A piece of alloy steel that is permanently magnetized. It will attract all ferrous material without the need for electricity.

Magnetic field Invisible lines of force (flux) surrounding a magnet or a conductor with current flowing through it.

Magnetic poles The points where magnetic lines of force enter and leave a magnet.

Magnetism An invisible force which attracts ferrous metals.

Magneto An electrical appliance used to generate low voltage that can be stepped up to high voltage sufficient to jump a spark plug gap. It contains rotating, permanent magnets mounted on the outer edge of the flywheel and a stationary ignition.

Main bearing A part which functions to support the crankshaft and to transmit oil to and from the main bearings.

Metal deactivator A fuel additive that neutralizes the effect of copper alloys, which can cause a gel that can plug gasoline filters.

Micrometer A precision tool which will accurately measure in thousandths of an inch.

Mixture adjustment (high speed) A screw found on the carburetor that is used to control the amount of fuel entering the air stream at high speed.

Mixture adjustment (idle speed) A screw found on the carburetor that is used to control the amount of fuel entering the air stream at idle speed.

Muffler A metal chamber that is constructed of materials which cool the exhaust gases and help quiet the combustion sounds.

Muratic acid (hydrochloric) An acid that will remove aluminum and rust from iron or steel components. (**Caution:** Use only in a well ventilated area.)

Needle bearing An antifriction bearing using hardened steel needle rollers between hardened races.

Negative terminal The terminal with an excessive amount of electrons which flow toward the positive terminal.

Neutron An atomic particle with a neutral charge.

Nitrogen Oxide Nitrogen gas found in the air is combined with oxygen molecules to form different compounds that are a major contributor to atmospheric smog.

NPN transistor A semiconductor device used to rectify or amplify an electric signal.

O-ring A ring made of neoprene that is used to provide a positive seal. It usually fits into a groove and is mated against a flat surface to provide a seal for oil, fuel, or air.

Ohm The standard unit for measuring resistance to electrical flow in an electrical circuit.

Ohmmeter An instrument used to measure the amount of electrical resistance in a length of wire or component in a circuit.

Oil seal (magneto) The oil seal is found in the cylinder block on the flywheel side of the engine. It is used to prevent the crankcase oil from leaking where the crankshaft extends out of the cylinder block.

Oil seal (PTO) This power take off (PTO) seal is found in the cylinder block where the crankshaft extends from the crankcase and connects to the equipment.

Open circuit A circuit with a break that prevents the flow of current.

Operating temperature The temperature reached by an engine when it operates most efficiently.

Out-of-round The wear condition observed in worn engine cylinders where the bore wears to an oval shape.

Overhaul The process by which a worn-out engine is restored to new operating condition.

Overhead Valve Engine (OHV) An engine configuration in which the valves are located in the cylinder head and the camshaft is mounted in the engine block.

Oversized An engine modified to larger-than-stock cylinder diameters.

Oxygen/Acetylene torch This type of torch produces intense heat that can be used for heating rusted parts, welding metal to metal, cutting through metal, and in methods for extracting bolts broken off at or below the surface level of the hole.

Permanent magnet A piece of steel or alloy that acts as a magnet without the need for an electric current to create a magnetic field.

Petroleum A complex, naturally-occurring mixture of solid, liquid, and gaseous hydrocarbons composed of compounds of hydrogen and carbon and small amounts of sulfur, nitrogen, and oxygen.

Piston A round metal plug which slides up and down inside a cylinder. The piston is connected to the crankshaft by the connecting rod.

Piston displacement The volume of space through which the piston travels from the very top of its stroke to the bottom limit, which can be determined by multiplying the stroke by the area of the circle made by the cylinder wall.

Piston groove One of several grooves cut around the circumference of the piston to receive a piston ring.

Piston land The portion of the piston's outer surface located between the rings.

Piston pin The steel pin which attaches the piston to the connecting rod. The piston pin must be machined smoothly to be able to pivot.

Piston ring compressor A tool used to compress the piston rings in order to insert a piston into a cylinder.

Piston ring expander A spring used with a piston ring to help expand it against the cylinder wall.

Piston ring gap The clearance measured between the piston ring ends when compressed in an engine cylinder.

Piston rings Rings that are installed on the piston to form a seal between the piston and the cylinder wall.

Piston rings (chrome) Rings that are used in a slightly worn cylinder because the ring metal is hard enough to wear down the cylinder wall so that a good seal soon develops between the piston and cylinder.

Piston skirt The outer surface of a piston located below the piston rings.

Plastigage A soft plastic material placed between two surfaces to measure the clearance between them. The width of the deformed material specifies the thickness.

Pliers, Adjustable A pliers that has many different channels to allow for different opening sizes.

Pliers, Combination A pliars that is used to hold parts, rather than the hand, to prevent injuries and to bend or twist thin materials.

Pliers, Diagonal cutting A pliars that contains extra hard-cutting jaws for cutting wire and cotter pins 1/8" or smaller.

Pliers, Needle nose A pliars that is used for gripping small objects, such as pins and clips, positioning small parts, and bending or forming wire.

Pliers, Snap ring A pliars that is used to spread snap rings just the right amount to remove or install them. There are two versions: one for inside retainers and one for outside retainers.

Pliers, Vise grip (locking) A pliars that is used like the combination pliers for holding, but the clamping action can be locked.

Point gap The distance between contact points when they are held open in the highest position by the cam or plunger.

Positive terminal The terminal with a lack of electrons to which they are drawn.

Power The time rate at which work is done. Power = work/time. The unit for power is the watt in the metric system and horsepower in the English system.

Power stroke The stroke of the four-stroke-cycle internal combustion engine in which the compressed air-fuel mixture is ignited in the combustion chamber. The power stroke is considered the "working" stroke of the engine.

Preignition Premature ignition caused by "hot spots" in the combustion chamber that ignite the air-fuel mixture before the spark plug "fires."

Primary winding Hundreds of turns of wire in an ignition coil to provide build-up and collapse of a magnet field, inducing a voltage in the secondary circuit.

Primer A system that provides a rich mixture by increasing the pressure in the top of the fuel bowl to force extra fuel into the venturi of the carburetor.

Prony brake A piece of equipment for determining engine power.

Proton An atomic particle with a positive charge.

Punch, Aligning A tool used to align holes during an assembly.

Punch, Center A tool used to mark parts before disassembly so they can be reassembled in the same relative position.

Punch, Pin Used to drive out shafts, pins, and bearings without the punch jamming in the hole.

Punch, Starting Used to loosen a frozen bolt or pin.

Radiator A heat exchanger which reduces coolant temperature in a liquid cooling system.

Reactivity Any material that reacts violently with water or other materials, releasing cyanide gas, hydrogen sulfide gas, or similar gases when exposed to acid solutions. This also includes materials that generate toxic mists, fumes, vapors, and flammable gases.

Ream To enlarge a hole with a reamer.

Rebuild The process of restoring an engine to the original operating condition by dismantling the parts and reassembling them to a new condition.

Rectifier An electric appliance used to permit current flow in one direction only.

Reed valve A one-way valve built in the shape of a flat strip. It is used on some two-stroke-cycle engines.

Reject gage A reject gauge is used to determine if hole size is acceptable. If the end of the gage can be inserted into the hole, the hole is too large and must be overhauled.

Resistance The opposing force offered by a circuit or component to the passage of electrical current. Resistance is measured in ohms.

Resonance A vibration of a large amplitude in a mechanical system caused by a periodic stimulus of the same period as the natural vibration period of the system.

Rich mixture An air-fuel mixture that contains too much fuel for the amount of air in it.

Ridge reamer A tool used to remove carbon deposits from the ridge at the top of an engine cylinder.

Rod cap The semicircular lower portion of a connecting rod that sandwiches the rod to the crankshaft.

Rod journal See **Crankpin journal**.

RPM Abbreviation for Revolutions Per Minute.

Rust inhibitor A fuel additive that reduces rusting and corrosion of steel parts within the engine.

SAE Abbreviation for Society of Automotive Engineers.

Safety Safety is a state of mind which safeguards the technician from possible danger.

Sandblast The process used to clean a surface by mixing sand with compressed air and directing the mixture to the surface.

SCR Silicon Controlled Rectifier.

Screwdriver, common A tool that is composed of a handle and a shank. The end of the shank is flattened to from a blade that fits squarely into the screw slot.

Screwdriver, offset A tool that is used on screws where there is limited space and the screw is hard to reach.

Screwdriver, philips A tool that has two slots at right angles. It is commonly used when it is desirable to reduce the chance of the screwdriver slipping from the head and marring the surrounding surface.

Screwdriver, TORX A specially designed slot that permits more precise torquing of the screw with a low incidence of slipping of the workpiece.

Seal An appliance used to prevent leakage.

Shim A spacer used between two parts to achieve proper clearance.

Short circuit A defect in an electrical circuit that allows current to return to the power source before passing through the load.

Shortblock A new factory-assembled cylinder, crankshaft, piston, piston rings, valves, springs, and oil seals.

Shroud The sheet metal panels used to conceal the engine compartment.

Sludge A thick, black, slimy material deposited in various parts of an engine.

Solder A metal alloy used to join two metal surfaces together. The most common is a 60–40 mixture of lead and tin that melts at just under 500° F.

Soldering gun A gun-shaped soldering device used to melt and apply solder with an electrically heated tip.

Spark An electrical discharge created between two separated points in an electric circuit. When electric voltage is great enough, the resistance of the open air gap is overcome and electricity flows in the form of a spark.

Spark advancement The adjustment of the ignition timing so that the spark occurs before TDC of the compression stroke.

Spark plug An appliance that provides a fixed air gap across which current jumps to provide a spark.

Spark plug seat The location where the spark plug touches the cylinder head and forms a leakproof seal.

Spigot A nozzle, spout, or tap that is inserted into a container.

Spring An elastic device built to yield when a pull is applied and then returns to its original shape when the pull is removed.

Steel A metal composed of iron and a small percentage of carbon.

Stellite A metal alloy composed of chrome, cobalt, and tungsten that is used for some exhaust valves because it is hard and has a high melting point.

Stroke The distance moved by a piston in a cylinder from TDC to BDC. Each stroke is one-half revolution of the crankshaft.

Synthetic base lubricant The base stock of a synthetic base lubricant is a synthesized hydrocarbon or organic ester.

Tachometer An instrument used to measure an engine's rotating speed in revolutions per minute.

Tap A tool used to cut internal threads. It is made from high-quality, hardened, ground, tool steel. The tap is similar to a screw, but flutes or grooves range the length of the thread that allow chips cut from the metal to escape.

Tappets See **Valve lifters**.

TDC Top Dead Center. The uppermost piston travel in the cylinder.

Tear-down To disassemble an engine to repair or rebuild it.

Thermal efficiency A measurement that compares the heat developed into useful work to the heating value of the fuel.

Third party A manufacturer who makes a part for use on another manufacture's machine.

Throttle plate Parts of the carburetor that controls the air flow through the venturi, thereby controlling the fuel flow to the engine.

Throw The offset portion or distance measured from a crankshaft centerline to a connecting rod shaft or crank pin centerline.

Torque A force characterized by exerting a turning or twisting effort.

Torque wrench A wrench used to tighten nuts or bolts to a specified torque value.

Transfer port An opening in the cylinder wall of a two-stroke-cycle engine which connects the cylinder to the crankcase.

Transistor A semi-conductor appliance used in electronic circuits. It is used to switch a specified current flow according to a small current flow through an auxiliary circuit.

Trigger coil A coil of wire that sends an electric pulse to a certain electronic circuit to start an operation when a magnetic field passes by it.

Troubleshooting A systematic approach of problem solving that requires a special plan of attack that is used from one engine to another.

Two-stroke-cycle engine The five parts of combustion occur for every two strokes (one up-stroke and one down-stroke) of the piston.

VOA meter An appliance or instrument capable of measuring volts, ohm, or ampere.

Valve adjustment The process by which the valve clearance between the valve lifter and the valve of the intake and exhaust valve is adjusted.

Valve face A beveled edge of a valve head used to contact a valve seat and provide a seal to the combustion chamber.

Valve face grinder A tool used to cut the intake and exhaust valve.

Valve guide A lubricated metal sleeve used to keep a valve centered when opening and closing. When a guide becomes worn, it allows excess oil to flow into the combustion chamber.

Valve lash The valve clearance or gap measured between the end of a valve stem and a valve lifter. If the valve lash is too wide, the valves open late and close early, decreasing engine performance.

Valve lash The gap between the valve stem and the valve lifter when the piston is at TDC of the compression stroke.

Valve lifters Parts which are driven up by projections or lobes on the camshaft.

Valve margin The distance measured between the

edge of a valve head and the outer surface edge of a valve face.

Valve seats Areas which are ground to an angle to mate the valve face so that a perfect seal is created when the valve contacts the seat.

Valve timing The relationship created between opening and closing of the intake and exhaust valves and up and down movement of the piston.

Vaporization The process when a liquid changes to the gaseous state.

Venturi The portion of a tube that is built to taper to a small diameter. A gas flowing through this narrowed down area will speed up and cause a partial vacuum to be formed.

Viscosity The value used to represent an oil's ability to flow. A low viscosity oil flows more easily than does a high viscosity oil.

Vise, bench Used to hold things in the workplace. The jaws should be covered with soft metal to prevent marring the surfaces.

Vise, drill A device used to hold round, rectangular, square, and odd shaped pieces for any operation that can be performed in a drill press.

Volatility The tendency of a fluid to form a vapor.

Volt A unit of measurement of electrical pressure.

Voltage A unit of electromagnetic force (pressure) which causes current to flow in a circuit.

Voltage drop A loss of voltage at the end of a conductor due to high resistance.

Water cooled An engine that dissipates the generated heat by circulating water around the hot areas of an engine.

Watt The unit of measurement for power in the metric system.

Welch plug A circular, soft , metal plug used to seal circular openings in a carburetor.

Wet test This is a test where a teaspoonful of engine oil is put into the cylinder through the spark plug hole and then the compression is tested. If the pressure increases from the earlier test, bad rings are indicated. If the pressure differs only slightly from the original reading, the valves are the cause of the low compression.

Work When a force is put into action, it does work, which is measured by multiplying the force acting and the distance moved. Work = distance x force. The unit for work is the Erg in the metric system and the Foot-Pound in the English system.

Wrench, adjustable The adjustable wrench has an adjustable jaw that allows the use of one tool to fit different sized bolts or nuts.

Wrench, Box end The jaws of the wrench fit completely around the bolt or nut to provide better contact.

Wrench, Combination The wrench has an open-end jaw on one end and a box-end on the other.

Wrench, Hex (Allen) The wrench that fits a set screw that are found on may pulleys having hexagonal recesses.

Wrench, Open end A tool that is used to fit snugly on two sides of a bolt or nut so the bolt or nut can be tightened or loosened easily.

Wrench, Socket An essential tool for working on an engine. The socket is firmly and snugly seated before force is applied to the handle.

Wrench, Torque A special wrench that indicates the amount of torque or twisting force applied to a nut or bolt.

Wrist pin A short steel shaft that is mounted inside the piston. It attaches the piston to the upper end of the connecting rod.

Zener diode These diodes are constructed from semiconductors which will allow current to flow above certain voltage levels. Below these levels, the diode behaves in a normal manner.

INDEX

Page numbers in *italics* denote figures